Builder's Guide to DRAINAGE & RETAINING WALLS

Max Schwartz

Builder's Book, Inc.
BOOKSTORE ■ PUBLISHER
8001 Canoga Avenue / Canoga Park, CA 91304
1-800-273-7375 / www.buildersbook.com

BUILDER'S GUIDE TO STUCCO, LATH & PLASTER

NOTICE TO THE READER / DISCLAIMER

This book is designed to provide general information about the planning, design and construction of drainage systems and retaining walls. Illustrations and specifications provided in this book are for illustrtion purposes only; not for construction. Before beginning any excavation or construction, the services and advice of a licensed engineer should be obtained, and local codes and ordinances should be followed.

The publisher has made every effort to provide complete and accurate information, but does not guarantee the accuracy or completeness of any information published herein. The publisher shall have neither liability nor responsibility to any person or entity for any errors, omissions, or damages of any kind (direct, indirect or consequential) arising out of use of or reliance on this information. This book is published with the understanding the publisher is not attempting to render professional services of any kind. If such services are required, the assistance of an appropriate professional should be sought.

ISBN-10: 1-889892-67-X
ISBN-13: 978-1-889892-67-2

For future updates, errata, amendments and other changes, contact Builder's Book, Inc., 1-800-273-7375.

Contents

INTRODUCTION

Drainage and Retaining Walls

I am writing this book under a strong incentive. At present, rainstorm after rainstorm is pelting most of California. Our central area is flooded as rivers rise above their berms. Countless landslides, mud slides, and washed-out roads are hitting much of California.

Much of the damage is due to substandard drainage design, construction and maintenance. This book is based upon the many standards and codes written toward prevention of these natural catastrophes.

The second part of this book is devoted to retaining walls, which are closely related to good drainage design and an integral part of hillside grading, roadwork, and flood control work. Master Builders have constructed retaining walls of concrete, masonry, timber, sheet piling or compacted earth. Their methods of work date back to Ancient Greece and the Biblical Times.

I hope that this Builder's Guide to Drainage and Retaining Walls will help in a small part to reduce the damages from future rainstorms.

CHAPTER 1

Rainfall and Runoff

Each trade or profession looks upon drainage from a different viewpoint.

The sheet metal contractor installs gutters, downspouts, or leaders around the perimeters of buildings. His main objective is to control the runoff from a roof and direct it away from the building walls. He must size the devices for anticipated rainfall for the area. He designs the drainage devices for hundreds of square feet of roof area and maximum rainfall over a five-minute period. Chapter 2 deals with Roof Drainage.

The lot-grading contractor must perform finish grading so that the runoff from the roofs and areas around the building are also controlled and directed to the street. Otherwise, the rainwater may pond on the property or run wildly and erodes the surface.

In addition to bare ground, the lot-grading contractor must prepare the sub grade for usable areas such as patios, driveways, and walkways that must be sloped properly to drain. This work may also include yard catch basins at low points that discharge into underground clay or concrete pipes, "V" or trapezoidal channels paved with concrete or asphalt to carry the rain water to the public street. You can base the design of all of these drainage devices on acres of watershed, maximum rainfall in one-hour, and a factor that determines the percentage of water the land will absorb. Chapters 3 and 4 provide detail information on Lot Drainage.

For tracts on hillsides, the grading contractor has a more difficult job. Where the building pad is above or below a steep slope, rough grading involves excavation of embankments to a firm strata, compacting fills, and cutting terraces to intercept the runoff before it cuts through the banks. The developer's civil engineers, in conjunction with the soils engineers, usually do the design of these devices.

When the runoff from a hillside tract is discharged onto a public or private street, the design work is usually done by the local government or private civil engineers who select the size and shape of roadways, side-ditches, curbs and gutters, street catch basins, culverts that are under-road drain pipes. They are usually dealing with maximum one-hour rainfall over the past 50 years, nature and slope of the landscape. Chapter 5, 6, and 7 describes the design and construction of Hillside, Street, and Slope Drainage.

Throughout the drainage work described above, the retaining wall is an important feature. Part two of this book deals with the subject of retaining walls.

Where natural slopes use up too much valuable land, a retaining wall is required. The wall holds the earth back that would otherwise slide over the leveled area. However, you must allow the ground water to flow through the wall by the way of weep holes or open joints. If you permit ground water to buildup behind the wall, the hydrostatic head could turn the wall over. Concrete and Masonry Contractors build the most common retaining walls.

On private property, concrete masonry walls are six to eight feet high. Higher walls are of reinforced concrete. Chapters 8 and 9 describe design and construction of concrete and masonry retaining walls.

Chapter 9 is about other types of retaining walls. Engineering contractors usually build the retaining walls alongside roads in mountainous areas. These walls must resist mud slides and landslides coming down from slopes hundreds of feet high. They often use the bin-type wall or cribbing. These are stronger, safer, and less expensive than the conventional cantilever concrete retaining wall.

Finally, Chapter 10 closes this book with the history of design and construction of retaining walls. The story goes back to Roman Era where Roman master builders built seawalls, fortress walls, and aqueducts that still stand.

There are many subjects that overlap between chapters and that can apply to more than one subject. This is also true with the figures.

DRAINAGE SYSTEM DEFINED

You can describe a drainage system as drainage piping within buildings, on public and private property. The system conveys waste, storm water, and other liquid wastes to a legal point of disposal, which may be the public street, a river, lake or the sea. It does not include the mains of public sewer system or a private or public sewage-treatment or disposal plant. The drainage system does not include any part of the venting system of a building, as that and the waste system are part of the plumbing system.

Drainage design is simply complying with gravity-flow of water, whether it is by a closed conduit like a pipe or an open channel. You base the rules of gravity flow on the *cross-sectional area of the water flow, slope,* and the *friction* between the water and surfaces of the conductor.

Different trades/professions may call a *slope* by different names.

- Civil Engineers who design roads and storm drains usually call the slope a *grade,* or *gradient,* measured in the amount of fall in 100-foot distance. In street or channel design, they call 100 feet a *Station.* A two-foot drop in a station is a 2 percent or 0.0200 grade.
- A plumber calls the incline of a pipe a *slope,* which he measures in fractions of an inch per foot, or a ¼-inch slope.
- A carpenter framing a roof uses the terms *slope* or *pitch,* which he measures in feet vertical to foot horizontal, or 4 in 12.

Some fields use an angular description such as a one or two degree grade.

By using the two basic formulas for gravity flow through open channels, you can select the most efficient size to handle the rainfall runoff. You call these the *Manning Formula* and the *Kutters Formula.*

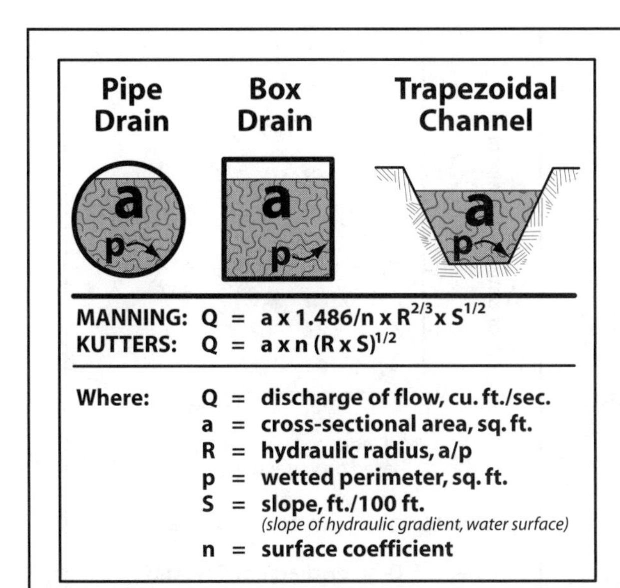

Figure 1-1. Capacity of Open Channels

Two formulas can be used to calculate gravity flow through three types of conduits. One is the Manning Formulas and the other is the Kutters formula. In either case, the flow (Q) is given in cubic feet per second.

HYDROSTATIC HEAD

In the case of a closed conductor, like a pipe, another important element is the head of water pushing the flow. You call this the *hydrostatic head.*

The art or science of designing a drainage system goes back to ancient times. One of the first was a Greek philosopher, Archimedes, who set up rules for the design of pipes and channels. He called this *Substructio.* Later, a Roman engineer, named Vitruvius, wrote about the design of water aqueducts, and recommended a slope of not less than 1:200, preferably 1:150. He used the principle of hydrostatic head in his design of stone conductors that carry water across valleys. To his credit, many aqueducts in Rome and Spain still carry water efficiently.

Fig. 1-2: Hydrology Formulas

Expresses the relationship in Drainage Pipe among Discharge, Wetted Perimeter, Area, Hydraulic Radius and Velocity

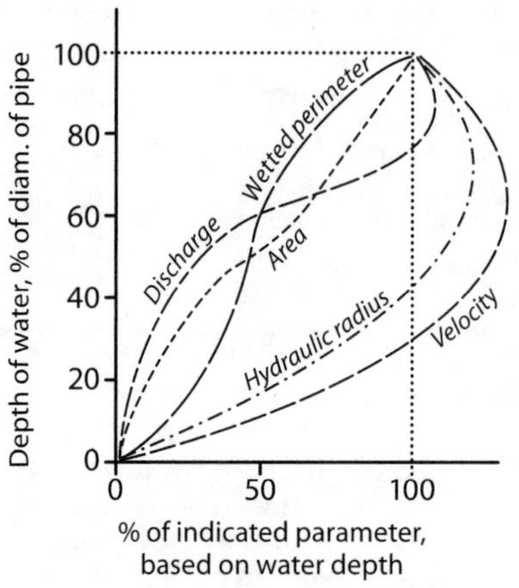

MANNING: $Q = a \times 1.486/n \times R^{2/3} \times S^{1/2}$

Where:

- **Q = discharge of flow, cu. ft./sec.**
- **a = cross-sectional area of flow, sq. ft.**
- **n = coefficient of roughness or surface coefficient**
- **R = hydraulic radius**
- **S = slope**

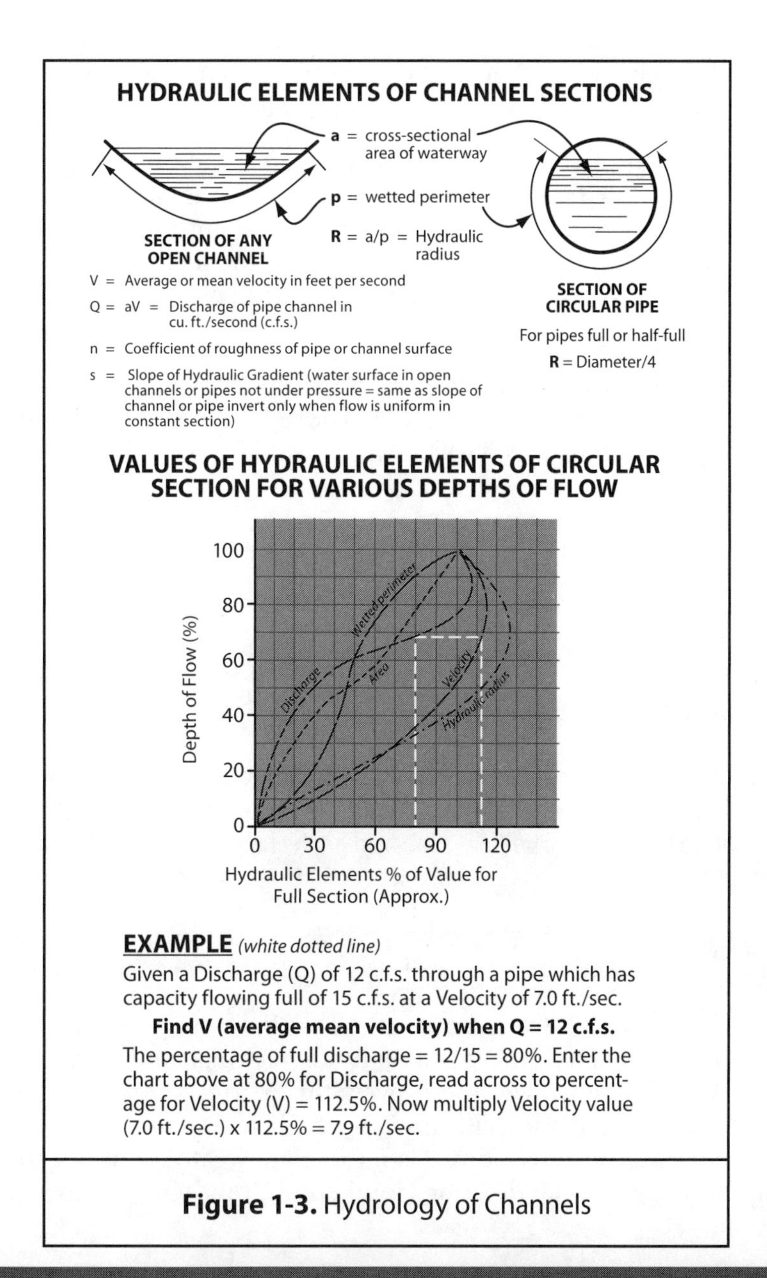

Figure 1-3. Hydrology of Channels

WEATHER RECORDS

You can obtain weather records from the federal, state, and county departments responsible for gathering and recording rainfall, run-off, temperature, and storm records. These records are measured at each weather monitoring station. You can term precipitation in inches per periods. Such periods may be per season, per storm, per 24-hour day, per hour, or per 10-minute interval. Records may extend as far back as 100 years, depending on the development of a particular area. State and county agencies involved in the design of structures affected by flooding generally maintain their own records of runoff and flooding conditions in their respective areas.

Runoff is measured as the volume of water per units of time flowing in creeks, streams, and rivers throughout the nation. You describe runoff in acre-feet per second. This refers to a volume of water one-foot deep covering one acre that flows past a point in one second. At critical points along most streams and rivers in the nation, there are gauging stations where the depth flow is measured and recorded at selected times. If you determine the cross section of the riverbed, the volume of water passing that point can be ascertained with the help of flow meters. You call this process *stream gauging*.

Many piers and bridge abutments across our rivers carry markings that indicate the depth of water. You can monitor the flow by reading the depth of water at various points in time. By these means, runoff over most of the country can be determined at any specific point. You can estimate the runoff from any watershed or tributary area by means of the history of runoff. Flood records are compiled in a similar manner. You may install gauges at critical points to measure water level during various flood conditions.

Newspaper publications of weather reports and storm forecasts furnished by the National Weather Service sometimes can provide useful information. Those involved with the design and construction of projects exposed to flooding and flash storms should be aware of the warnings given by the National Weather Service.

The following illustrates typical runoff tabulation by a county hydrology department:

RUNOFF RECORDING STATION NO. 5
ARROYO SECO AT CAMINO REAL NEAR SAN JUAN
LAT. 330°34'41", LONG. 117°30'25", approximately 300 feet downstream of freeway bridge, 1.8 miles north of San Juan.
Drainage Area: 45.6 square miles.
Gauge: Float-operated, dual-pen water stage recorder on the downstream side of the Camino Real bridge on the center wall of a double rectangular channel. Communication to float well by siphon pipe from silt box in each channel. Altitude of gauge is 153 feet (from channel plans).

WEATHER FORECASTS

A typical example of general weather records is the national and local weather forecast as published in a daily newspaper. Data may include a national, regional, and local weather map in which high- and low-pressure areas are shown; the movement of cold and warm fronts, stationary and occluded fronts; areas of rain and snowfall; low and high temperatures; wind direction; and the anticipated cloud condition at various cities. Also, there are weather reports and three- to five-day forecasts for geographical regions that describe cloud conditions, temperatures, rain and snow, and wind velocity.

Weather records can be important information in reconstruction related to flood damage. In one case, a property owner sued a railroad company for flooding his land as a result of an allegedly improperly designed culvert placed in fill in the roadbed. The court opined that evidence of rainfall surveys, including surveys of the United States Weather Bureau, which showed uniformly heavy rainfalls in general climatological areas in which this particular drainage basin was located, was proper evidence.

You can record maximum rainfall in inches per month and per year for the following durations:

- 5-minute duration
- 10-minute duration
- 15-minute duration
- 30-minute duration
- 1 hour
- 2 hours

- 24-hour period
- Calendar day
- Month
- Seasonal
- Storm
- Minimum recorded rainfall in inches per month, season, and year
- Longest period of consecutive days of rain
- Longest period of consecutive days without measurable rain

Based on the above data, it is possible to view a particular storm against the history of storms for that particular area.

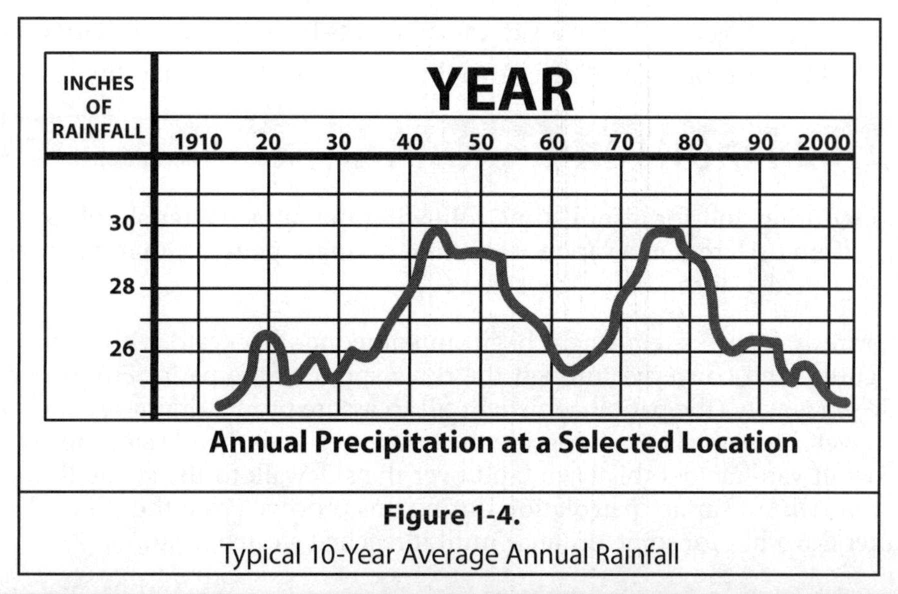

Figure 1-4.
Typical 10-Year Average Annual Rainfall

OVERVIEW OF DRAINAGE

Running water is the most important cause of erosion. Water from rainfall causes erosion, which probably does more to alter the surface features of the land than all the other agents combined. The average annual rainfall in many cities is approximately 35 in. of water. Of this, 20 to 30 percent becomes runoff water flowing from the land. In developed areas, this is as high as 90 to 95 percent, making much drainage more difficult. People prefer relatively dry areas and so we must drain them.

Drainage is a symbiotic adjunct to the grading process. Every grading plan must consider and solve the drainage problem specific to it. There are four methods by which rainfall is removed from where it falls:

- By surface runoff overland and downhill until, eventually, in many areas it reaches the ocean.
- By underground subsurface drainage, water infiltrates and moves through most soils both horizontally and vertically under the influence of gravity, although at a much slower rate than surface runoff.
- By evaporation from the leaves of plants, standing water, and from many different surfaces.
- By transpiration from trees and plants following photosynthesis.

These four methods of removing rainfall combine and become part of the hydrologic cycle. The cycle is complete when sufficient water evaporates, returns to the air, is transported, and dispersed over land, rises with topography and wind currents, cools, and returns to the earth as fresh rainfall.

Surface runoff is the primary method to remove storm water. Water is usually carried away in some sort of storm drainage system. A storm drainage system collects, conducts, and disposes of excess surface water caused by rainfall. Additionally, a storm drainage system can accomplish the following:

- Safeguard against, erosion by reducing the rate of flow and volume of water.
- Reduce flooding damage to property and increase usability through elimination of unwanted water.
- Eliminate unnecessary standing water that may lead to pollution and breeding of insects.
- Provide better growing conditions for trees and plants by reducing soil saturation.
- Improve load-bearing capacity of soils, thereby increasing the buildability of a site.

LAYOUT AND DESIGN OF DRAINAGE SYSTEM

Runoff occurs during and for a short time following the rains. Water should be collected and conducted away from use areas in a variety of natural drainage patterns, man-made open trenches, and closed pipe - called a storm drainage system.

A typical storm drainage system might begin in someone's backyard with water from the roof collected in a gutter, and conveyed through the downspout to the patio. From here, it may flow across the patio via gravity (the patio has a slight tilt to assure proper drainage) and onto the lawn. It would then travel across the lawn, alongside the house in a wide, gently sloping grass ditch called a swale to the front yard, across this lawn, and over the sidewalk to the street. Some of the water would be lost through subsurface percolation during this process. From the street, the water would flow in the gutter downhill for some distance until it reached a catch basin.

Once collected in the catch basin, the water would travel in a storm drain line (pipe) until it crosses a small natural stream, where the storm drain line would daylight (surface) and the water would spill into the stream. The water would flow through culverts (large pipes) where roadways crossed the stream, and eventually portions of the stream may have been rip-rapped (lined) with large rocks to prevent erosion.

As more and more storm drains feed into the stream, it may be lined with concrete to speed the flow of water, and may eventually be covered for safety and other engineering reasons. Eventually, this water will discharge into a larger river, a lake, or the ocean.

Obviously, there are other ways it could have been conducted from the house, including straight away in a storm drain to the lake or river or into the sanitary sewer line (combined with household sewerage), where it would be treated prior to disposal into the bay, river, or ocean. (This practice is disappearing as it causes overflow pollution during storms and owing to the expense of treating storm water.)

DRAINAGE PROCEDURE

It should be obvious that drainage is simply a matter of collecting, transporting, and disposing of water. In preparing a drainage plan, you must determine where all the runoff will be coming from, where you will eventually put it, and how to get it there. In addition, you must determine which areas you want to keep dry. All flat use areas should be properly drained so that they are usable; this includes paved surfaces, play fields, building entrances, parking, and roads. Additionally, all sloping areas should be designed not to dump runoff on adjoining flat use areas. A small swale between the slope and the flat area can carry runoff away.

Runoff can originate on your site or from adjoining property at a higher elevation. Your first task is to analyze the topography, including off-site adjoining lands, to determine the overall large-scale drainage pattern. What existing patterns of runoff affect the site? Where are the high points, ridges, valleys, streams, swales, etc.? This shows where water will be coming from, what quantities you must deal with, and how it will affect your site planning. For large projects, you may want to prepare an off-site drainage pattern plan to guide you. Remember, water always travels perpendicular to the contour and faster as contours are spaced closer.

Check in greater detail all on-site conditions to determine exact surface runoff pattern. Include high points, low points, ridges, valleys, streams, swales, points of concentration, etc. Note soil types, particularly gravelly or sandy soil, which percolate well, and clay or silt soils, which percolate very slowly. Generalized soil maps are available for most areas from the U.S. Soil Conservation Service. For specific sites, it may be necessary to dig several holes to sample the soil. There must, therefore, be a continuous minimum slope in the ground level to assure drainage.

These analysis steps are too easily overlooked or considered trivial and time consuming and ignored. However, exactly the opposite is true; it is not unusual for solutions to drainage problems to be discovered during this process. You must take the time and conscientiously locate all the factors mentioned. With practice, you will develop the skill to do it quickly.

Overlay the proposed development on a summary analysis sheet.

Note again where water is coming from (the highest points) and where it is going (the lowest point), and don't go against the pattern. Outline all areas that you want to keep dry, and determine fixed elevations, such as existing buildings, trees, roads, etc., that must not be changed and property lines that must not be graded or drained areas. This information is the primary data necessary to collect runoff; the next step will be to determine where we can dispose of it. Your task will be to devise drainage systems for removing excess rainwater using these and other techniques. Depending on the size and scope of the project, you will usually be designing that portion up to the public right-of-way and connecting to an existing public drainage system. However, you may eventually be called upon to develop a storm drainage system for a neighborhood or a larger portion of the public community.

Remember, a storm drainage system is designed to collect, transport, and dispose.

As usual, economy is the rule, and you will design the system producing the best results for the lowest cost. Generally, surface drainage across sloping paved and planted areas is cheaper than installing catch basins and underground drainpipes.

GOOD DRAINAGE PRACTICES

1. Gravity is the primary power for carrying away runoff.
2. Water always flows perpendicular to the contours.
3. It is ecologically better to slow down runoff water and let it be absorbed by the soil than to remove all of it through surface runoff. Duplicate natural runoff principles where possible.

4. Runoff water must never be purposefully directed from one property onto a lower neighboring property. It is acceptable for water that has flowed naturally from your property to the neighbors to continue, but you must never increase this flow artificially through grading.
5. Erosion is the biggest problem in drainage; slopes must be carefully calculated to ensure continuous flow, yet not so steep to erode. Plant all slopes immediately following grading.
6. Slow-moving water is likely to create a bog; water moving too fast will erode and form unwanted gullies. Somewhere in between is usually right.
7. Surface drainage is generally preferred to using underground pipes, as this eliminates the danger of pipes clogging, is less expensive, and allows more runoff to percolate into the ground.
8. Paved areas look better when graded almost level; avoid wildly sloping paved areas.
9. Large amounts of water (such as from a parking lot) should not cross a sidewalk to reach the street drain. Install a catch basin or trench drain before crossing the sidewalk.
10. Always design a secondary drainage route to handle runoff should the primary system become clogged or constricted. Runoff is disposed of either on or off site. Low-density projects with large vegetated open spaces can usually accommodate their own runoff needs by directing runoff onto the naturally vegetated open spaces or into a nearby stream. This is called on-site disposal. The main design task is to avoid concentrating runoff onto one location, which may cause erosion. Instead, devise a number of locations for disposal, and use particularly those areas with established vegetation or gravelly soils.

Higher-density developments usually have a public storm drainage system located in the street. Abutting projects can tie into this system to dispose of their runoff. This is called off-site disposal. This system consists of underground pipes buried 3 to 12 ft. deep, with a lateral pipe connecting to each property. On-site runoff is collected and placed in this lateral for disposal by the responsible public agency. It is normal in many high-rainfall areas to have roof and basement drains connected directly to this system. Runoff from landscape development can usually be included in this same lateral connection. Some on-site water can be directed across the sidewalk to the road gutter, but the amount must not make the walk impassable or cause ice conditions in winter.

CHAPTER 2

Roof Drainage

The roof is one of the most essential parts of a building as it protects occupants, contents, and interior of the structure from the elements. Once you determine the kind of roof you intend to use; you must give equal attention to the design of the roof drainage system. Runoff from a pitched roof is usually collected horizontal or slightly sloped gutters around the eaves. The gutters discharge the water into vertical conductor or pipe, called a downspout or leader. This may be sheet metal, rectangular, round or hexagonal. This discharges preferably on to a splash pan so as not to erode the soil surrounding the building.

A "flat" roof is usually slightly sloped and is surrounded by a parapet, which is a vertical extension of the outer walls. Water collected at the junction of the roof and the parapet flows to a drain, which is either through the parapet wall or through the roof to the interior pipe. The drain in the parapet discharges to hopper-like basin called a scupper, then into a downspout as with the pitched roof.

Factors you consider in the design of roof drainage systems are the area to be drained, size of gutters, downspouts, outlets, slope of roof, type of building, and appearance.

The capacity of a roof drainage system depends on the quantity of water to be handled. The quantity of water in turn depends on the roof area, slope, and rainfall intensity. To consider the roof area, remember that rain does not necessarily fall vertically and that maximum conditions exist only when rain falls perpendicular to a surface. Since the roof area would increase as its pitch increases, do not use the plan area of a pitched roof in the calculation of a drainage system.

Experience has taught us that use of the true area of a pitched roof often leads to over-sizing of gutters, downspouts, and drains. To determine the design area or a pitched roof, use Table 2-1.

Table 2-1 / Chart 1

Design Areas for Pitched Roofs

Pitch	*B
Level to 3 in./ft.	1.00
4 to 5 in./ft.	1.05
6 to 8 in./ft.	1.10
9 to 11 in./ft.	1.20
12 in./ft.	1.30

* To determine the design area, multiply the plan area by the factor in column B.

Divide these areas by the proper factor given in Chart 2, thus obtaining the required area in square inches for each downspout. From Chart 3 select the downspout.

Table 2-2 / Chart 2

Rainfall Data and Drainage Factors

AREA	A		B		C	
	STORMS WHICH SHOULD BE EXCEEDED ONLY ONCE IN 5 YEARS		STORMS WHICH SHOULD BE EXCEEDED ONLY ONCE IN 10 YEARS		MAXIMUM STORMS	
	Intensity in in./hr. lasting 5 minutes	Sq. ft. of calculated roof drained per sq. in. of downspout area	Intensity in in./hr. lasting 5 minutes	Sq. ft. of calculated roof drained per sq. in. of downspout area	Intensity in in./hr. lasting 5 minutes	Sq. ft. of calculated roof drained per sq. in. of downspout area
Alabama: Birmingham	7	175	7	175	9	130
Arizona: Phoenix	4	300	5	250	7	175
Arkansas: Little Rock	6	200	7	175	9	130
California: Los Angeles	3	400	4	300	6	200
Sacramento	3	400	3	400	5	250
San Diego	3	400	4	300	5	250
San Francisco	3	400	3	400	5	250
Colorado: Denver	5	250	6	200	11	110
Connecticut: Hartford	6	200	7	175	9	130
District of Columbia	7	175	7	175	10	120
Florida: Jacksonville	7	175	8	150	10	120
Miami	7	175	8	150	10	120
Tampa	8	150	9	130	13	95
Georgia: Atlanta	7	175	8	150	11	110
Illinois; Chicago	6	200	7	175	10	120
Indiana: Indianapolis	6	200	7	175	10	120
Iowa: Des Moines	6	200	7	175	10	120
Kansas: Wichita	6	200	7	175	10	120
Kentucky: Louisville	6	200	7	175	10	120
Louisiana: New Orleans	8	150	8	150	12	100
Maine: Portland	4	300	5	250	7	175
Maryland: Baltimore	7	175	8	150	11	110
Massachusetts: Boston	5	250	6	200	8	150
Michigan: Detroit	6	200	7	175	10	120
Minnesota: Minneapolis	6	200	7	175	10	120
Missouri: Kansas City	7	175	8	150	10	120
St. Louis	6	200	8	150	11	110
Montana: Helena	4	300	4	300	6	200
Nebraska: Omaha	6	200	7	175	12	100
Nevada: Reno	3	400	4	300	6	200
New Jersey: Trenton	6	200	7	175	9	130
New Mexico: Albuquerque	4	300	4	300	6	200
New York: Albany	6	200	7	175	9	130
Buffalo	5	250	6	200	10	120
New York City	6	200	8	150	9	130
North Carolina: Raleigh	7	175	8	150	10	120
North Dakota: Bismarck	6	200	7	175	10	120
Ohio: Cincinnati	6	200	7	175	10	120
Cleveland	6	200	7	175	10	120
Oklahoma: Oklahoma City	6	200	7	175	10	120
Oregon: Portland	3	400	3	400	5	250
Pennsylvania: Philadelphia	6	200	7	175	10	120
Pittsburgh	6	200	7	175	9	130
Rhode Island: Providence	5	250	5	250	7	175
South Carolina: Charleston	7	175	7	175	9	130
Tennessee: Memphis	6	200	7	175	10	120
Knoxville	5	250	6	200	9	130
Texas: Fort Worth	6	200	7	175	9	130
Houston	7	175	8	150	11	110
San Antonio	7	175	8	150	11	110
Utah: Salt Lake City	3	400	4	300	6	200
Virginia: Norfolk	6	200	7	175	9	130
Washington: Seattle	3	400	3	400	4	300
Spokane	3	400	3	400	6	200
West Virginia: Parkersburg	6	200	7	175	10	120
Wisconsin: Madison	6	200	6	200	9	130
Milwaukee	6	200	7	175	10	120
Wyoming: Cheyenne	5	250	6	200	8	150

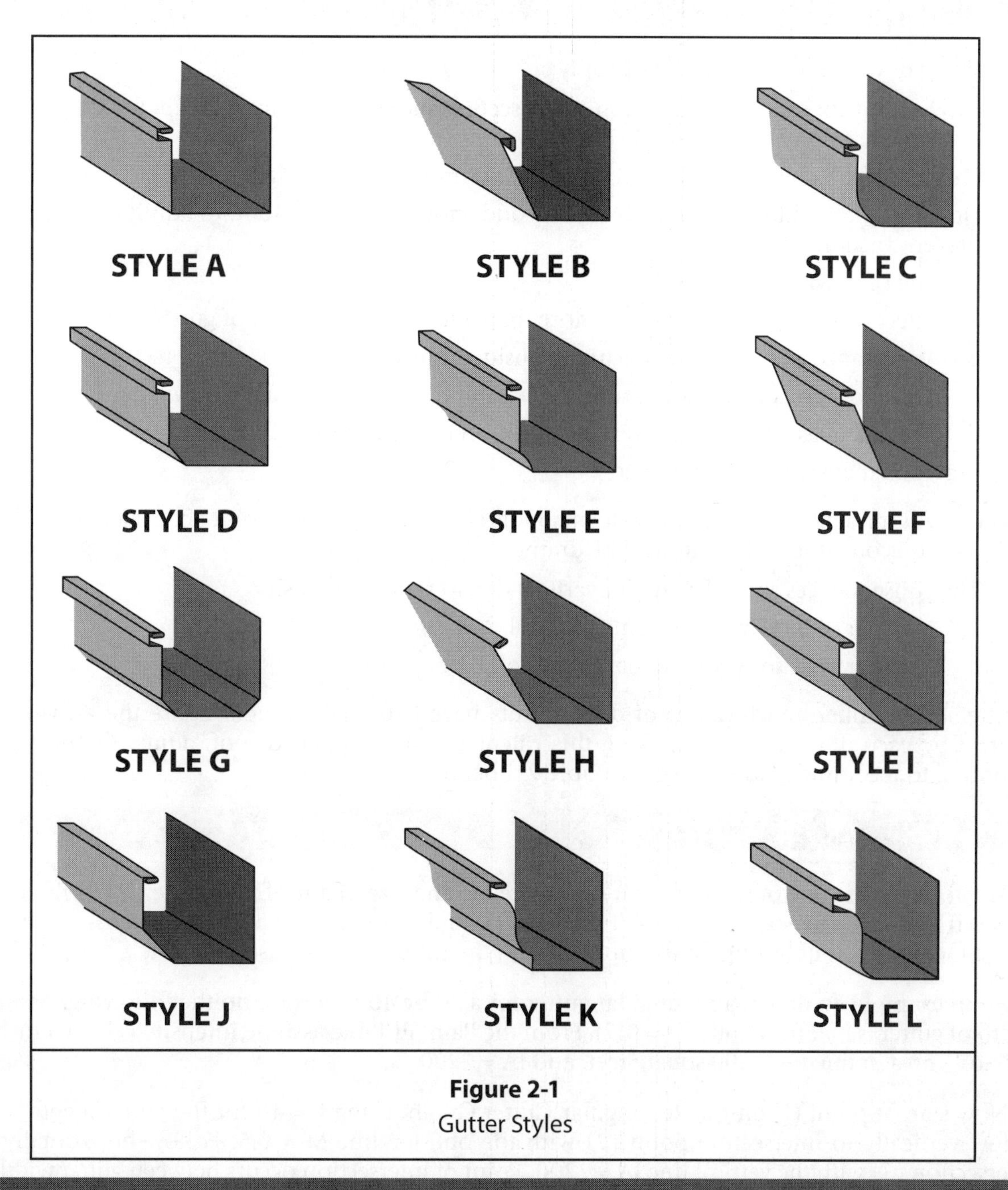

Figure 2-1
Gutter Styles

RAINFALL INTENSITY – DOWNSPOUT CAPACITY

You rate rainfall intensity in inches per hour. Table 2-2 above indicates the rainfall intensity that may be expected in 58 cities. (You can obtain rainfall intensity for other cities from the U.S. Weather Bureau.) This table includes the calculated roof area in square feet that can be drained per square inch of downspout. This factor is based on the assumption that 1 sq. in. of downspout will drain 1200 sq. ft. of roof when there is a maximum rainfall of 1 in. /hr. If the rainfall is 2 in/hr., then the capacity per square inch of downspout is reduced to one-half. If rainfall is 3 in./hr. downspout capacity is reduced to one-third.

DOWNSPOUT SIZING

In sizing downspouts, the following apply:

- Downspouts of less than 7.00 sq. in. cross section should not be used except for small areas such as porches and canopies.
- The size of the downspout should be constant throughout its length.
- Downspouts should be constructed with conductor heads every 40 ft. to admit air and prevent vacuum.
- Spacing of downspouts:
  a. Select a downspout to drain no more than a maximum of 50 ft. of gutter.
  b. Appearance and other architectural considerations.
  c. Avoid locations where water must flow around a corner to reach a downspout.
  d. In icing areas avoid downspout locations on north side of building.
- Provide expansion joints in gutter.

These will influence the location of downspouts since the expansion joint acts as a dam and the gutter sections on either side of it must be drained.

- The consequences should a gutter overflow due to rainfall intensity:
  a. A matter of inconvenience only (use col. A, Table 2-2).
  b. A matter of serious consequence (use col. B or C, Table 2-2).

After the number and location of downspouts have been determined, figure the areas to be drained by each downspout. In making this calculation for a pitched roof, adjust the plan area according to recommendations given in Table 2-3 below.

HOW TO SIZE A ROOF GUTTER

The following is one of the ways you can figure out the size of a roof rain gutter using Table 2-4 below. If you know the expected rainfall intensity (I) and tributary roof area (A), you use the intensity in inches per hour lasting for five minutes and the tributary area in square feet.

As an example, to design a rectangular gutter for a 20 by 40 foot roof, do the following. Assume width of gutter is half the depth (M = 0.5). From the Rainfall Table assume intensity is I = 9.5 inches per hour, area drained A = 800 square feet, and IA = 7200.

Now start at point (I) on the Rectangular Gutter Graph, using L = 40 for the gutter length, and follow vertically to intersection point (2) with the oblique line M = 0.5. Follow horizontally to Intersection (3) with the vertical line IA = 7200. Point of intersection occurs between gutter width 6 and 7 inches, and depth of 3.5 inches.

Another example assuming a semi-circular gutter (Table 2-4) is for a roof area A = 800 square feet. Intensity, I, is 10 inches per hour. The required gutter width is 8 inches.

For the design of a leader, or downspout, using a roof area of 3000 square feet, the Rainfall Table No. 2-2 and assuming one square inch of the leader serves 300 square feet of roof area, 15 square inches is required. From the leader dimensions Table No. 2-3, select either a five inch round, octagonal or square or 4 by 5 inch rectangular downspout.

Table 2-3 / Chart 3

Dimensions of Leaders (Downspouts)

TYPE	Area SQUARE INCHES	Nominal Size
Plain Round	7.07	3″
	12.57	4″
	19.63	5″
	28.27	6″
Corrugated Round	5.94	3″
	11.04	4″
	17.72	5″
	25.97	6″
Polygon Octagonal	6.36	3″
	11.30	4″
	17.65	5″
	25.40	6″
Square Corrugated	3.80	2″
	7.73	3″
	11.70	4″
	18.75	5″
Plain Rectangular	3.94	1¾″ x 2¼″
	6.00	2″ x 3″
	8.00	2″ x 4″
	12.00	3″ x 4″
	20.00	4″ x 5″
	24.00	4″ x 6″

Table 2-4 / Chart 4

Gutter Sizing

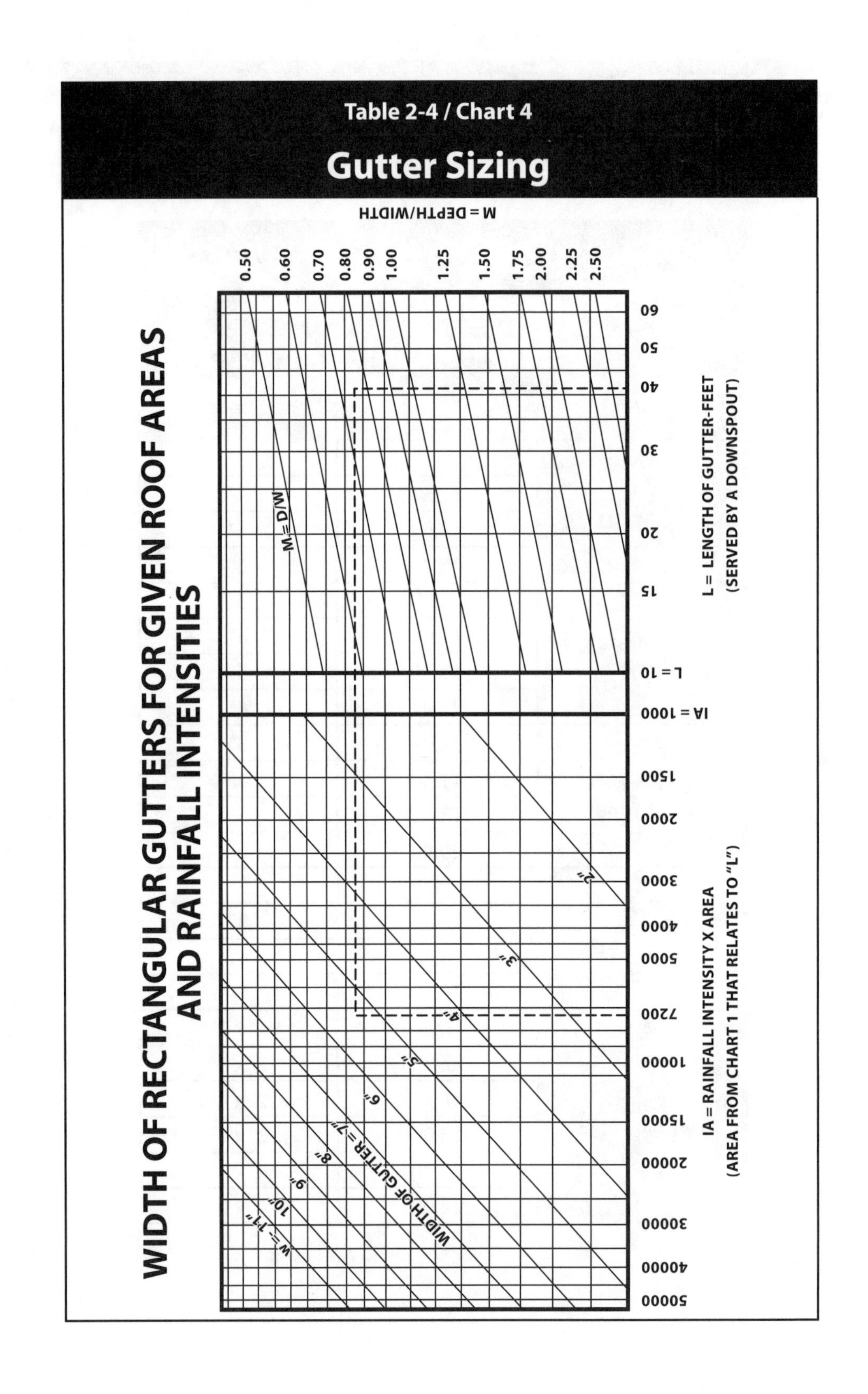

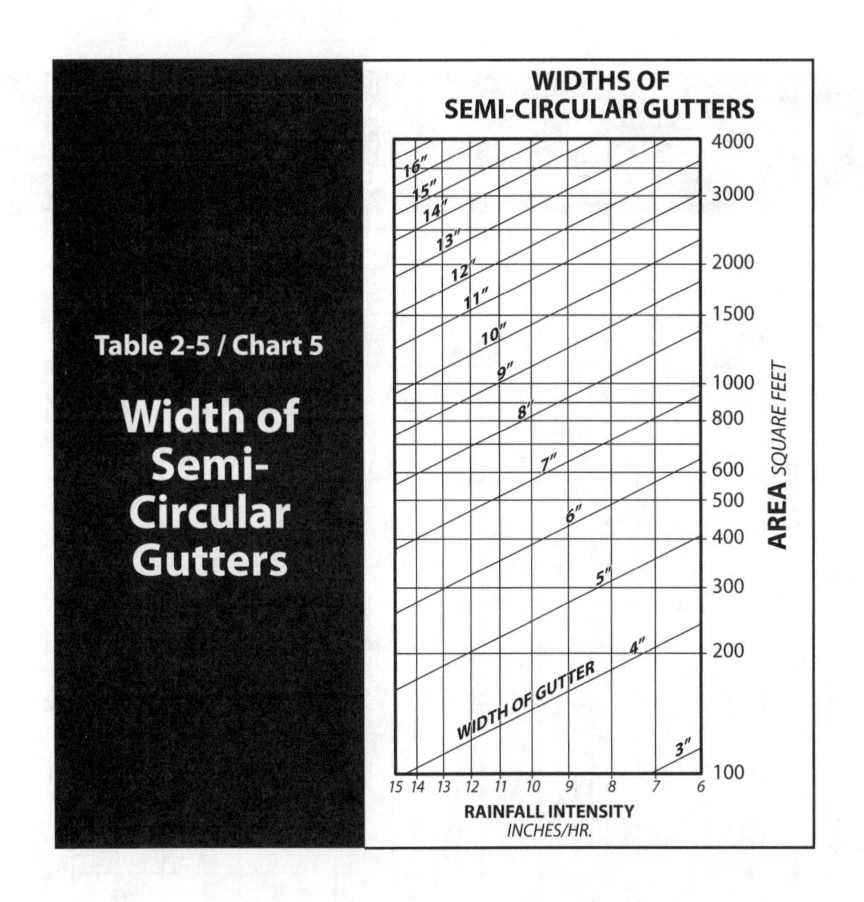
Table 2-5 / Chart 5
Width of Semi-Circular Gutters
WIDTHS OF SEMI-CIRCULAR GUTTERS
16″
15″
14″
13″
12″
11″
10″
9″
8″
7″
6″
5″
4″
3″
WIDTH OF GUTTER
4000
3000
2000
1500
1000
800
600
500
400
300
200
100
AREA SQUARE FEET
15
14
13
12
11
10
9
8
7
6
RAINFALL INTENSITY
INCHES/HR.

Table 2-6 / Chart 6

Roof Drainage Guide

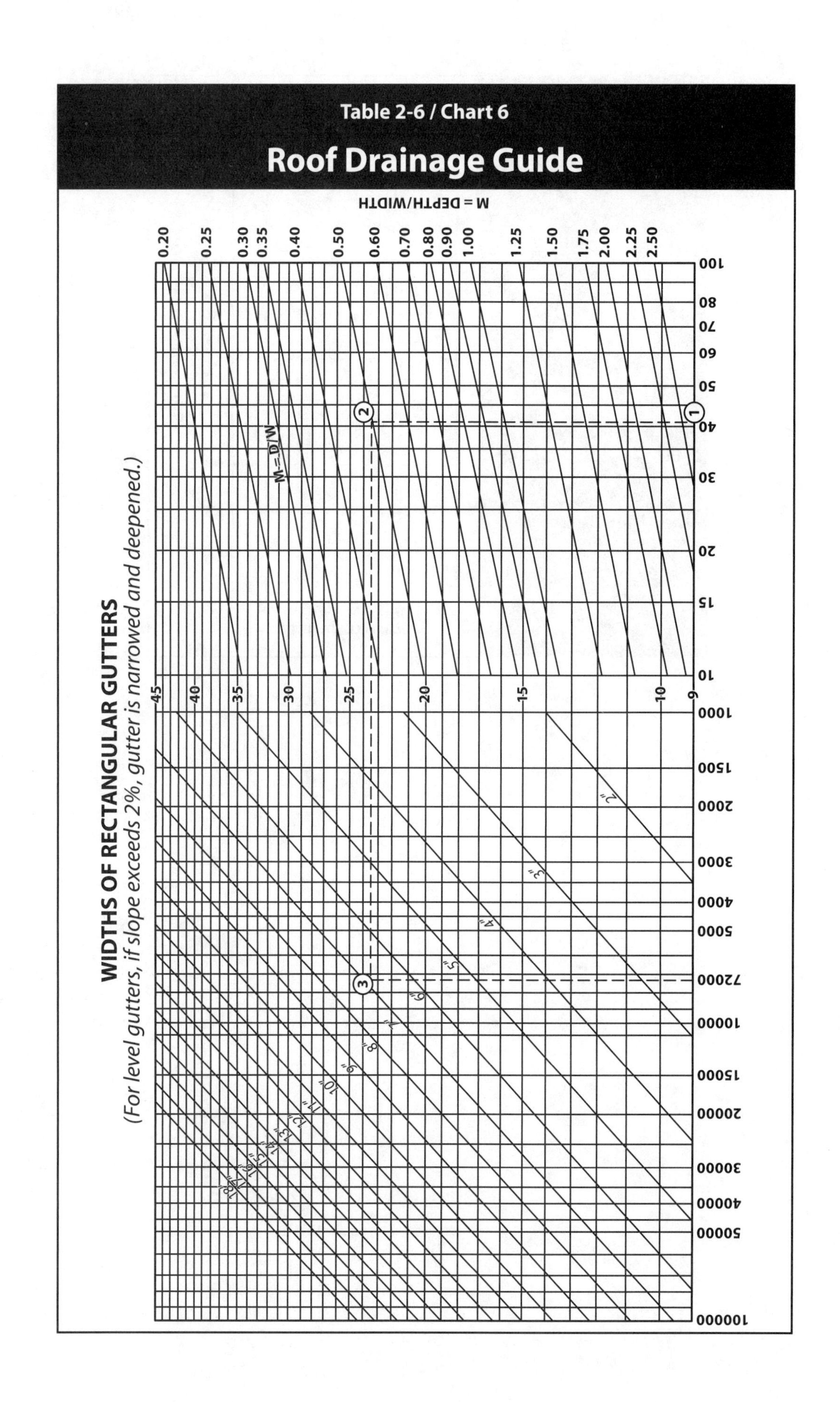

As a sample problem, select downspouts for a building, 100 x 85 ft. with a double pitched roof having a slope of 6″ in 1 ft. The slope is toward the 100-ft. side. Use maximum rainfall conditions to determine downspout size.

Drain the building with four downspouts located at each corner of the building. Install an expansion joint in each gutter between the downspouts. The plan area of this building is 8500 sq. ft. Assume the slope is 6 in./ft., use 1.10 factor (Table 2-1), making the design area 350 sq. ft. Thus, each of the four downspouts will serve a 2338 sq. ft. area. From column C, Table 2-2, you find that 1 sq. in. of downspout will drain 150 sq. ft. of roof area. Divide 2338 by 150 to determine that each downspout should have a minimum area of 15.59 sq. in.

From Table 2-3, choose a 5 in. Plain Round, a 5 in. Corrugated Round, a 5 in. Rectangular Corrugated, or a 5 in. Plain Rectangular downspout.

To size gutters, apply the following considerations:

- Spacing and size of outlet openings. (The gutter can never be any more effective than the downspout selected to drain it.)
- Slope of the roof. (The gutter must be of such a design that water from a steep pitched roof will not by its own velocity tend to spill over the front edge.)
- Style of gutters to be used. (All gutters are not effective for their full depth and width, see Plate 1 for design data.)
- Expansion joint location. (Water cannot flow past an expansion joint.)

The size of rectangular gutters depends upon these factors:

- Area to be drained. (A, Chart 4)
- Rainfall intensity per hour. (I, Chart 4)
- Length of gutter in ft. (I, Chart 4)
- Ratio of depth to width of gutter. (M, Chart 4)

The required sizes of gutters other than rectangular or round are determined by finding the .semicircle or rectangular area that most closely fits the irregular cross section.

The size of half-round gutters is directly related to the downspout size. Chart 5 is used for this purpose.

To size rectangular gutter for a building 120 x 30 with a flat roof with a raised roof edge on three sides. Locate a gutter on one of the 120 ft. sides. So that each section of gutter will not exceed 50 ft., three downspouts will be used with 2 gutter expansion joints. The area to be drained by each section of gutter is 1200 sq. ft., the rainfall intensity from Chart 2, Col A is 6 in., the length of each gutter section is 40 ft., and the ratio of gutter depth to width is 0.75.

On Chart 4 find the vertical line representing L = 40. Proceed vertically along this line to its intersection with the oblique line representing M = 0.75. Pass to B horizontally to the left to intersect the vertical line representing IA = 7200. The point of intersection occurs between the oblique line representing gutter widths of 5 and 6 in. The required width of gutter is, therefore, 6 in. and its depth need be only 41/2 in.

In designing rectangular gutter, the following considerations apply:

- The front edge of gutter should be lower than the back so that any overflow will spill over the front of the gutter. The elevation difference should be one-twelfth of the gutter width with I" minimum. Hanging gutters at sloped roofs should have front edge elevation coordinated with roof slope line; see Figure 2-1.

- A minimum ratio of the depth to width is 3 to 4.
- Avoid widths of less than 4 in. because they are hard to maintain.
- Install gutter expansion joints to compensate for temperature changes.
- Gutter is properly supported with hangers.
- Lap joints in gutters 1 in., riveted on 2 in. centers and soldered. (Materials that cannot be soldered, should be joined according to the manufacturer's recommendations.)
- In the selection of materials for gutters, take any corrosive environment under advisement.
- Hang gutter level or pitched toward the downspout. Where appearance is a consideration and building lines must be straight and true, level gutters are preferred.

The term "girth" means the width of material (in the flat) used to form a gutter. Recommended minimum gages for gutters are given in Table 2-6.

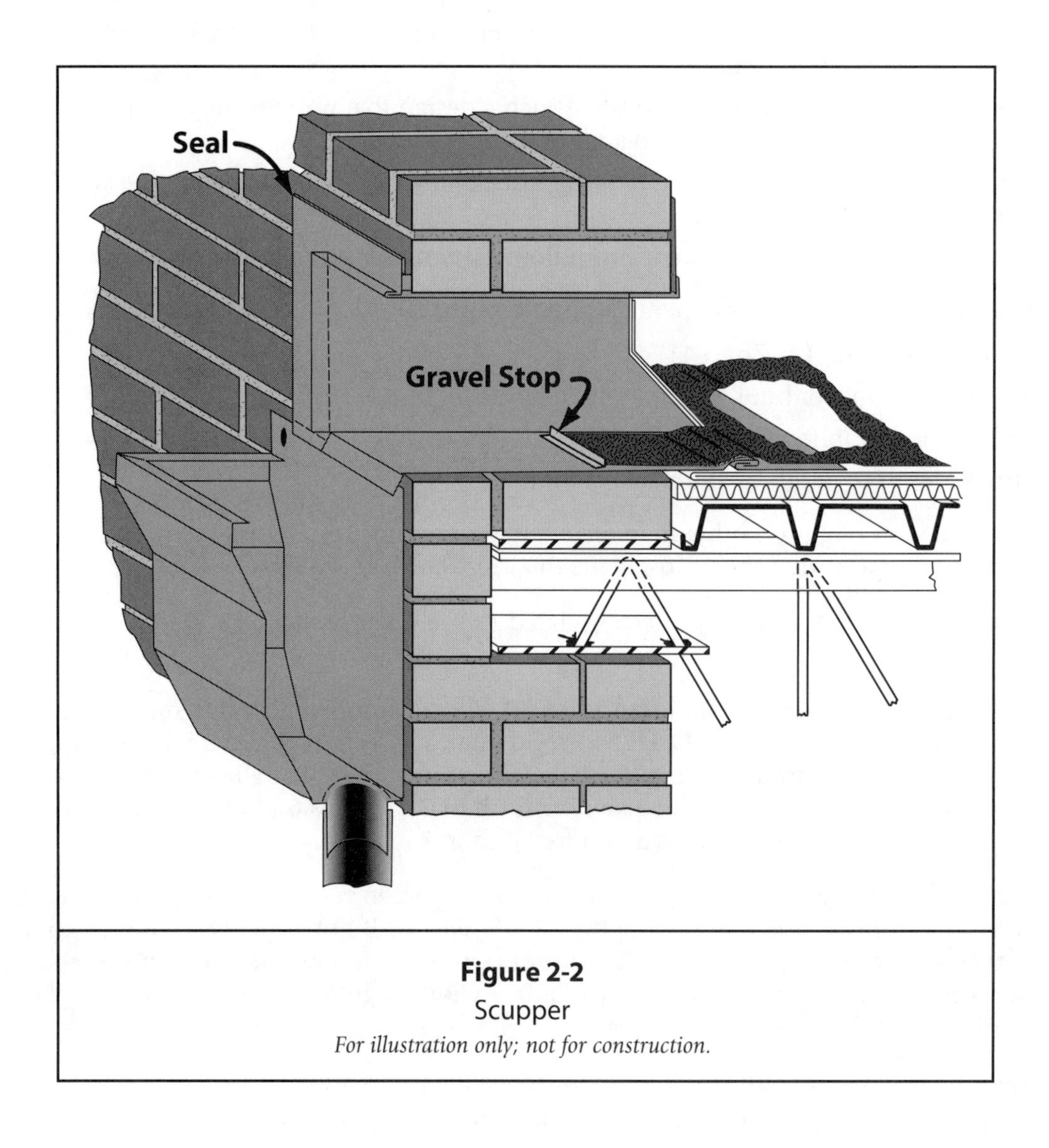

Figure 2-2
Scupper
For illustration only; not for construction.

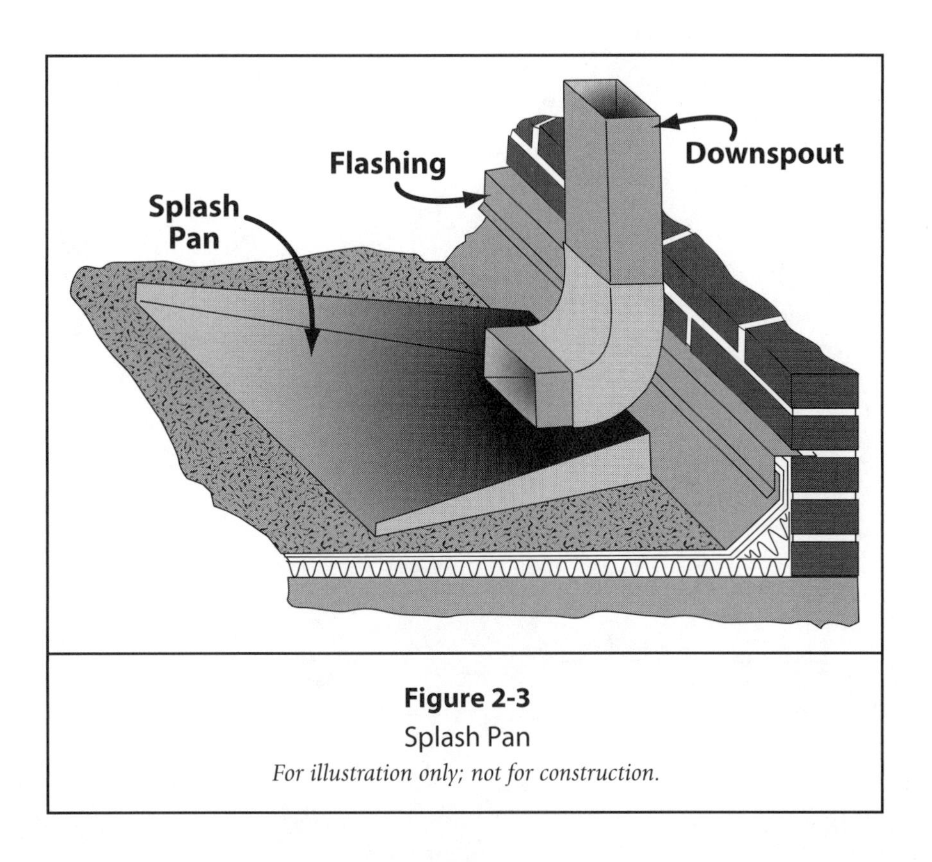

Figure 2-3

Splash Pan

For illustration only; not for construction.

CHAPTER 3

Basement Drainage

BASEMENT LEAKS

The most vulnerable area of a building for penetration of ground water is the basement. By nature, the basement is below grade and often below the ground water level. Most basement walls are made of reinforced concrete but concrete blocks are more common on the West Coast.

The floors of the basement areas are usually made of a four-inch-thick concrete slab. Ground water can penetrate the walls through osmosis, through the joints or under the footings. The water may rise through the floor slab again by capillary action or through the construction joints and slab.

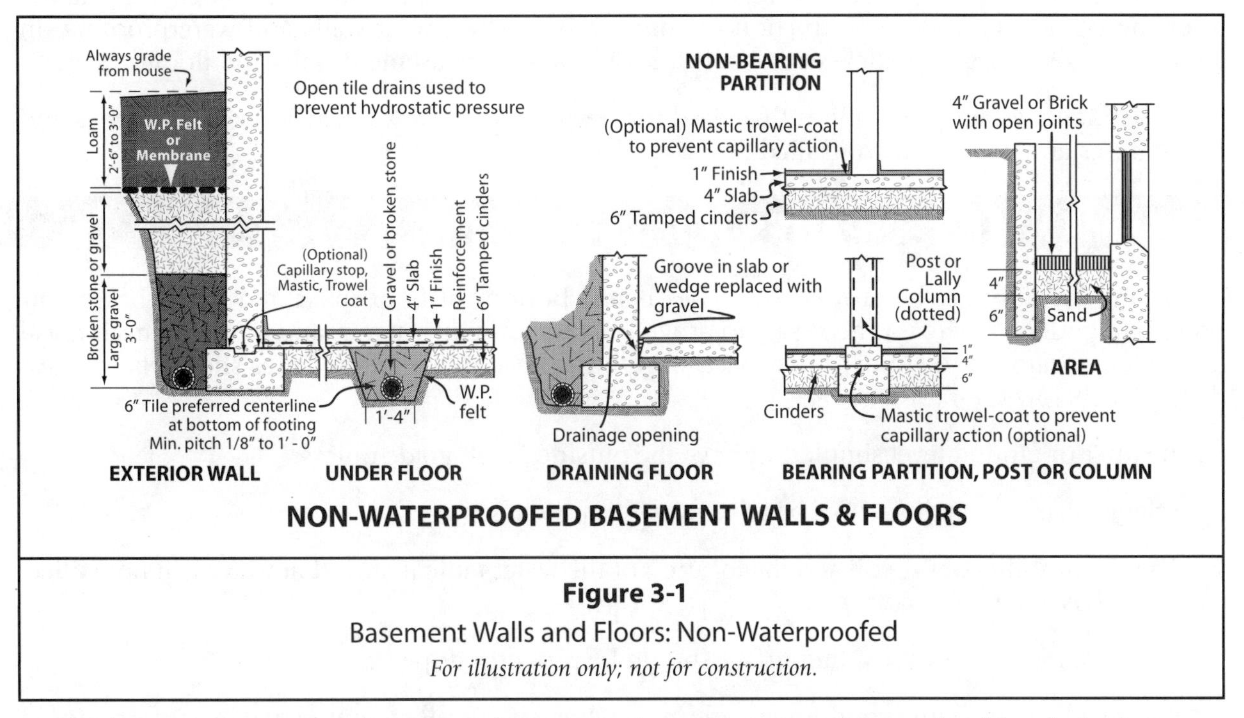

Figure 3-1

Basement Walls and Floors: Non-Waterproofed

For illustration only; not for construction.

The best way to prevent basement leaks is to lower the ground water level by drainpipes surrounding the building, set at a slope to drain to a watercourse or a sump. Pumps in the sump then pump the water to the street or other watercourse.

These pipes should be either perforated or set with open joints so that the water can enter the pipe. In addition, a bed of crushed rock should surround the drainpipe so that ground water has easy access to the pipe.

Ground water that passes under the drainpipe surrounding the foundation can be collected in drainpipes or gravel or crushed rock base under the slab, then to a sump pump.

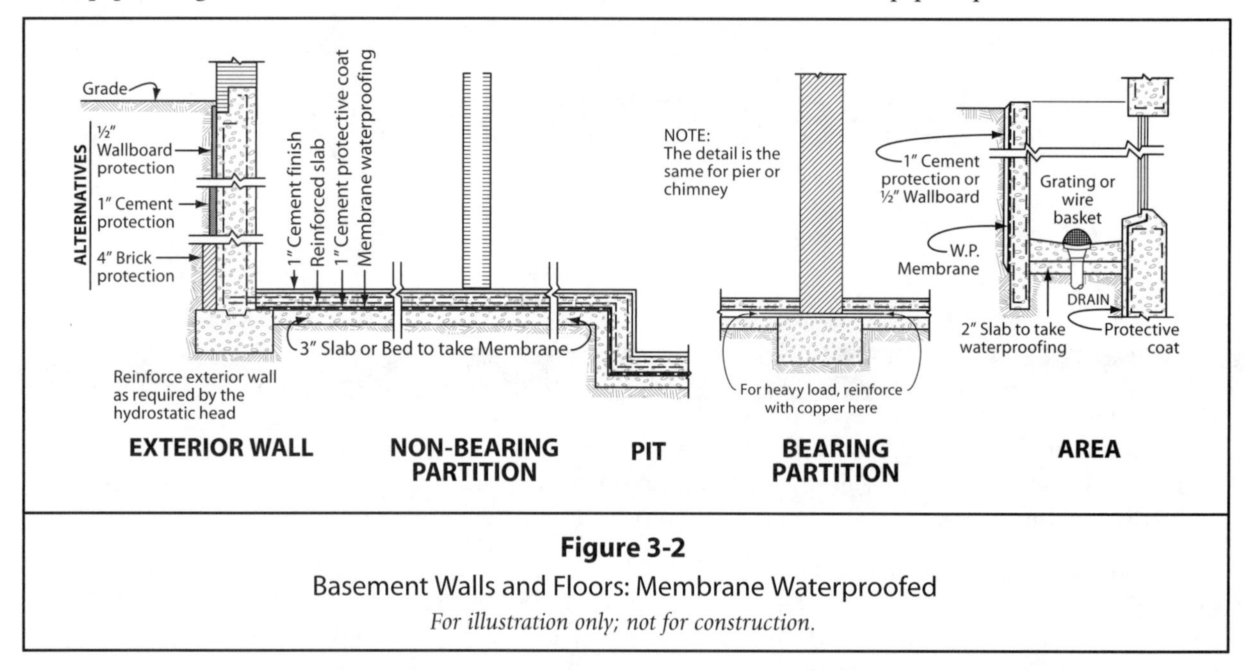

Figure 3-2

Basement Walls and Floors: Membrane Waterproofed

For illustration only; not for construction.

The exterior of the wall should be made waterproof to prevent penetration of water. See Chapter 9 for more information of construction of concrete block basement walls and waterproofing. In addition, Figure 3-2 shows suggested details for waterproofing basement walls and floors.

Proper drainage control requires that all of the devices carry rainwater away from structures and foundations and prevent uncontrolled flow over slopes.

BASEMENTLESS SPACES (CRAWL SPACE)

Ground level should be at least 18 inches below bottom of floor joists and 12 inches below bottom of girders. Where it is necessary to provide access for maintenance and repair of mechanical equipment located in the under-floor space, the ground level in the affected area should be not less than 2 feet below floor joists

The interior ground level should be above the outside finish grade unless:

a. Adequate gravity drainage to a positive outfall is provided, or

b. The permeability of the soil and the location of the water table is such that water will not collect in the basementless space.

When drainage is necessary, the surface should be properly sloped.

Where soil and moisture conditions warrant, or when specifically required herein, surface should be covered with a vapor barrier material.

Remove all debris, sod, stumps and other organic materials in crawl space, and provide a reasonably smooth surface.

DAMP-PROOFING AND WATERPROOFING

The objective is to construct a foundation that will prevent damage by infiltration of water or moisture. Basements and habitable spaces below grade should be designed and constructed to be adequately protected at all points against penetration of moisture. Normal precautions against water alternatively, dampness will require the measures contained herein.

In those locations where the foundation is subjected to a high water table or where surface or ground water drainage will present a problem, additional precautions may be required.

FOUNDATION DRAINS

Foundation or footing drains connected to a positive outfall should be provided around foundations enclosing basements or habitable spaces below grade. Install at or below area to be protected.

Clay, concrete or bituminized-fiber drain tile should comply with local governmental regulations.

Drainpipe should be installed having a minimum slope to outfall of 1/2 percent (approx. 1 inch in 20 feet). Protect top of joints in drain tile with strips of building paper. Cover drain tile with 6 to 8 inches of coarse gravel or crushed rock or blast furnace slag. Provide approximately 2 inches of aggregate material under tile.

Foundation drains may be omitted where well-drained soil exists or where ground or surface water will not present a problem.

FOUNDATION DAMP-PROOFING

Exterior foundation walls of masonry or of double-formed concrete enclosing basements or habitable spaces below grade should be damp proofed.

Asphalt and damp-proofing or waterproofing should comply with ASTM D-449, Type A.

MASONRY FOUNDATION WALLS

a. Apply at least one coat of Portland cement parging to wall from footing to finish grade. Minimum thickness, 3/8-inch.

b. Apply at least one coat of bituminous damp-proofing material over parging. Apply at rate recommended by the manufacturer.

CONCRETE FOUNDATION WALLS

a. Apply at least 1 coat of bituminous damp-proofing material to wall from footing to finish grade. Apply at rate recommended by the manufacturer.

Foundation damp proofing may be omitted in locations where well-drained soil exists or where ground or surface water will not present a problem.

Figure 3-3
Crawl-Space Drainage

NOTES:
1. *Provide perforated or other means to screen opening to drain.*
2. *Outfall to storm sewer or to natural drainage area.*

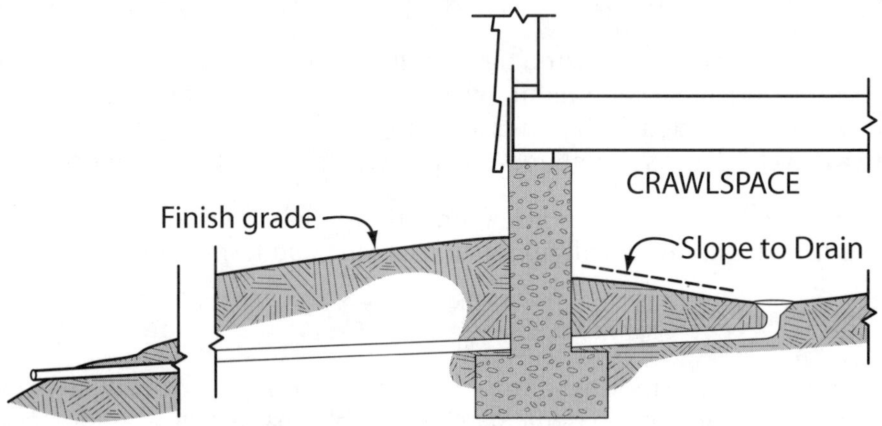

CRAWLSPACE
Finish grade
Slope to Drain

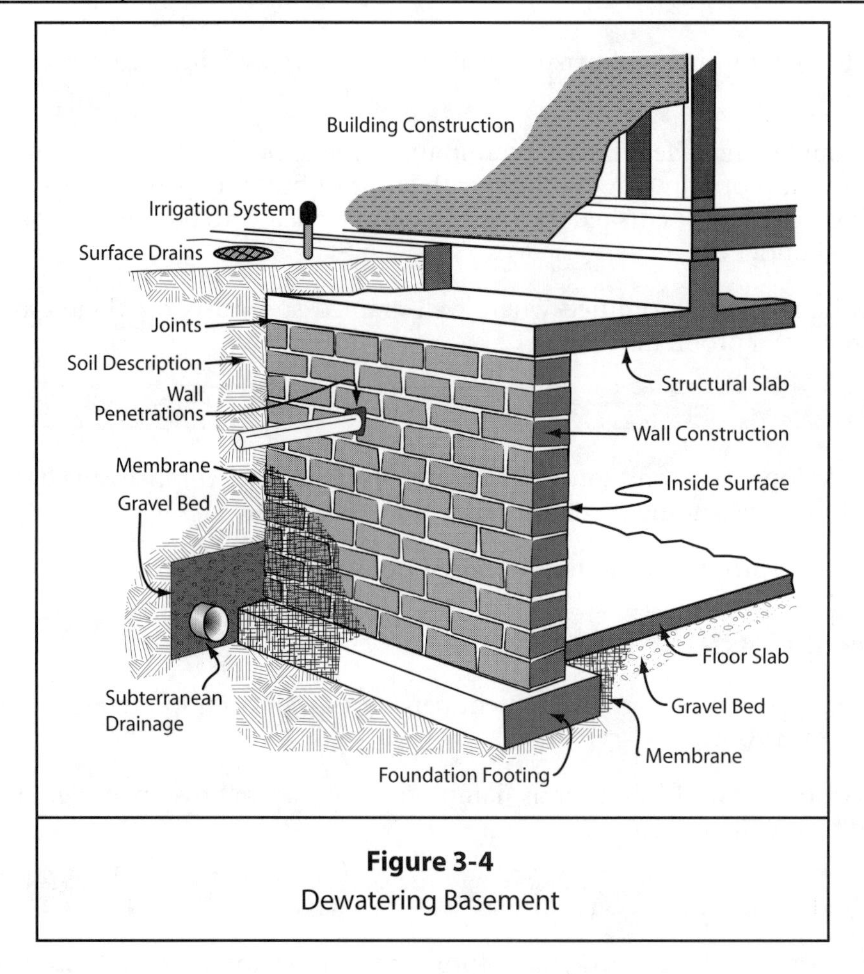

Figure 3-4
Dewatering Basement

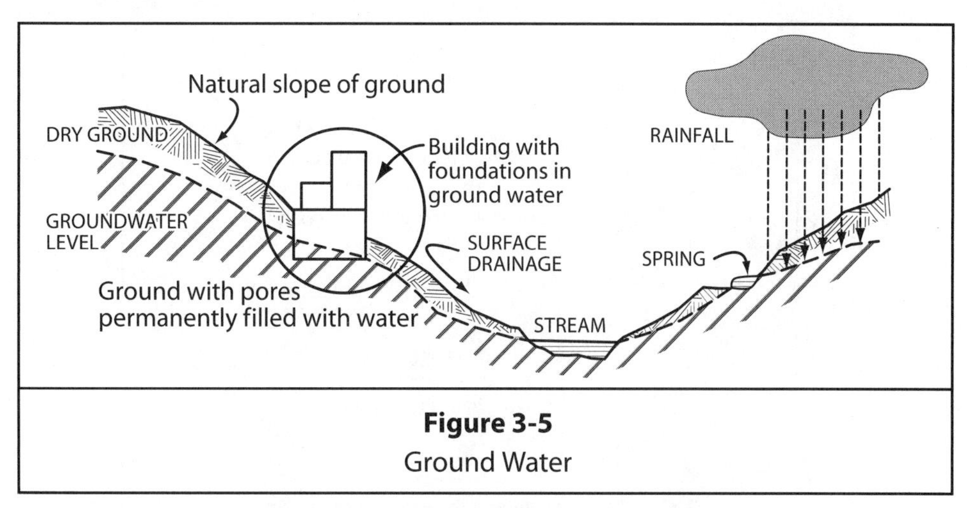

Figure 3-5
Ground Water

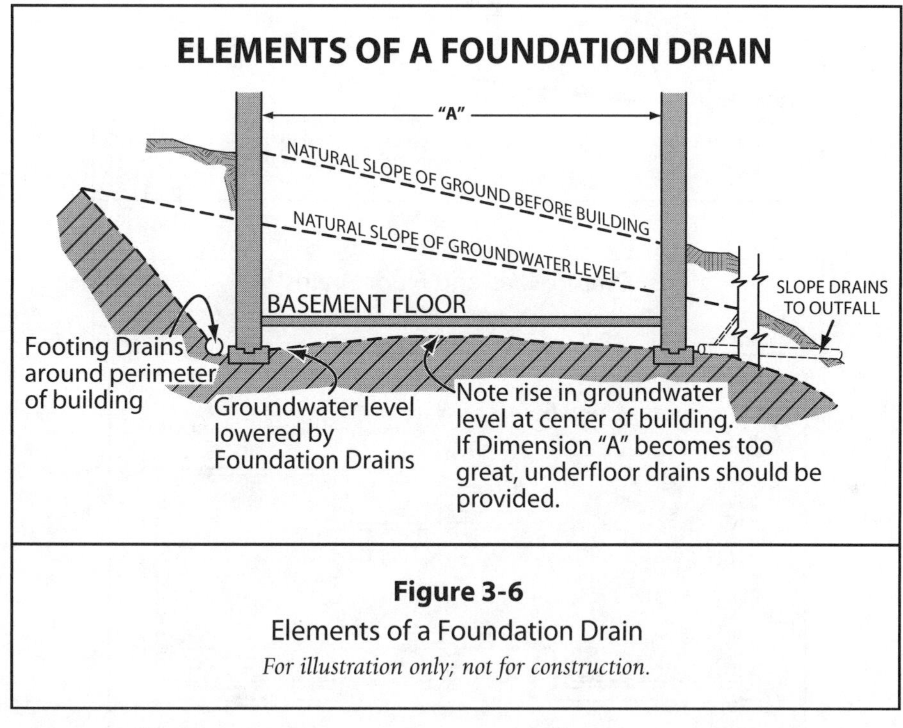

Figure 3-6
Elements of a Foundation Drain
For illustration only; not for construction.

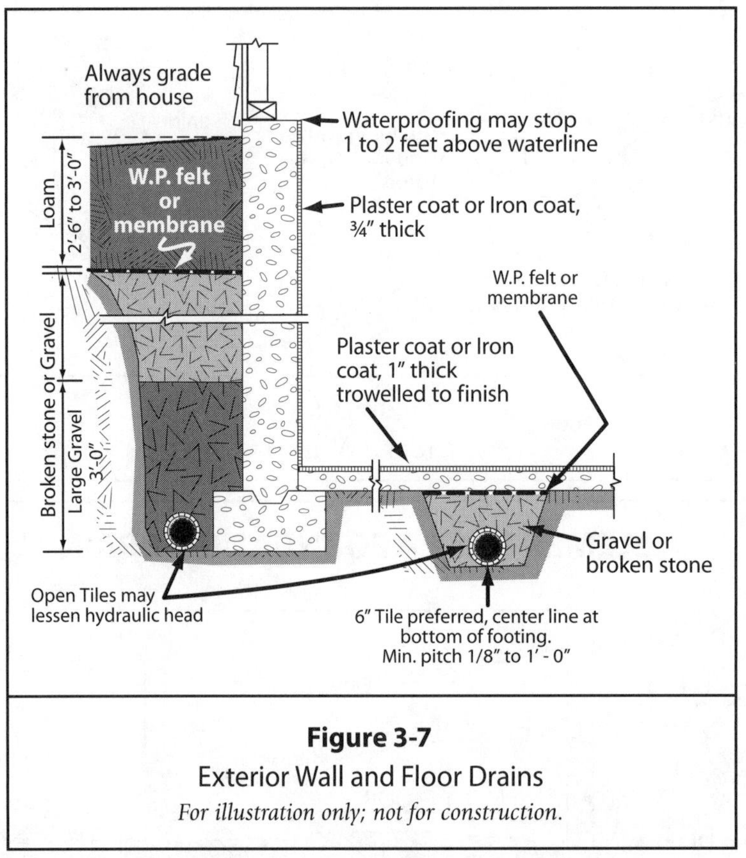

Figure 3-7

Exterior Wall and Floor Drains

For illustration only; not for construction.

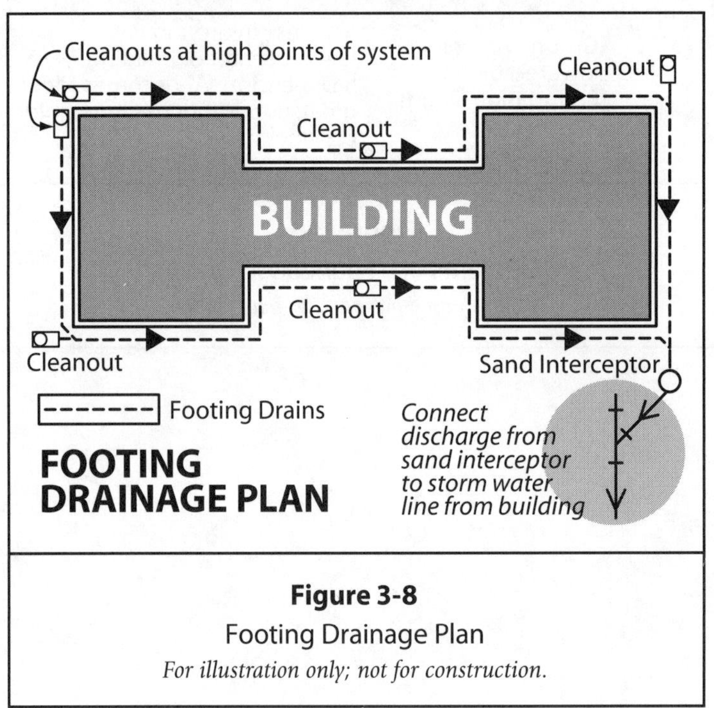

Figure 3-8

Footing Drainage Plan

For illustration only; not for construction.

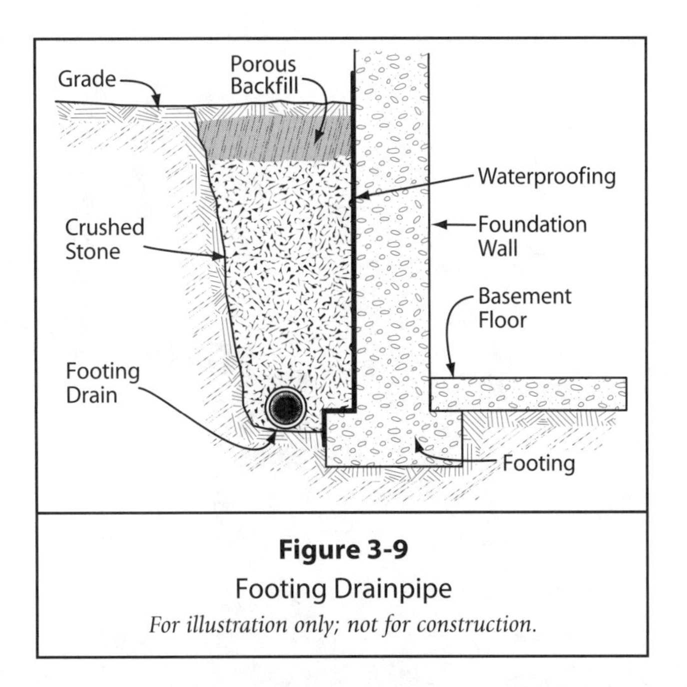

Figure 3-9

Footing Drainpipe

For illustration only; not for construction.

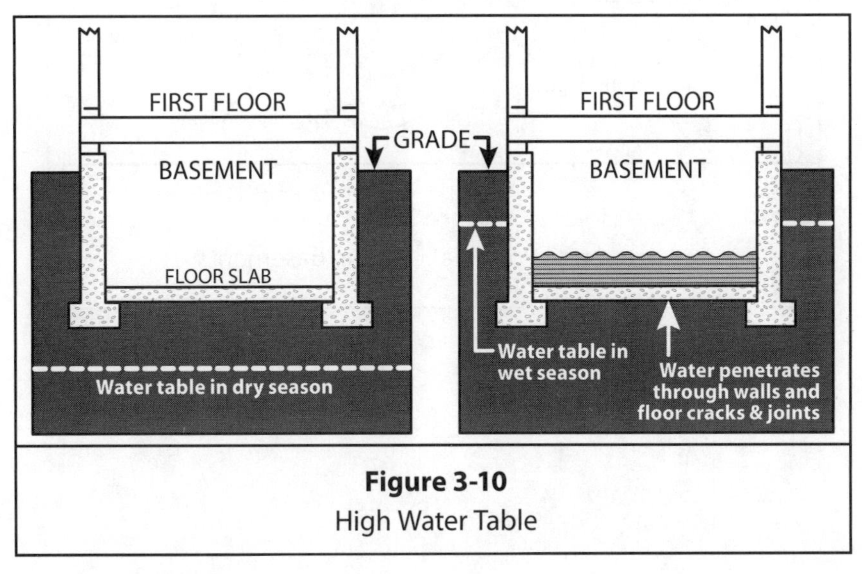

Figure 3-10

High Water Table

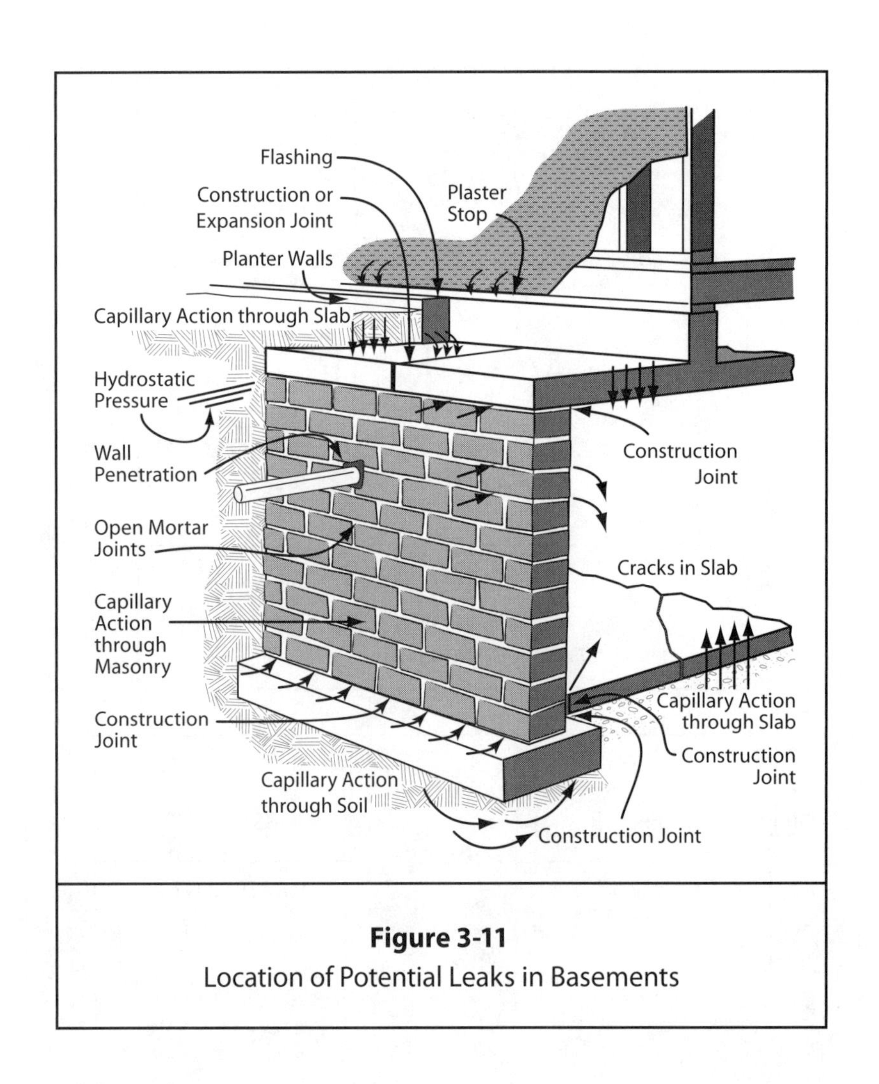

Figure 3-11
Location of Potential Leaks in Basements

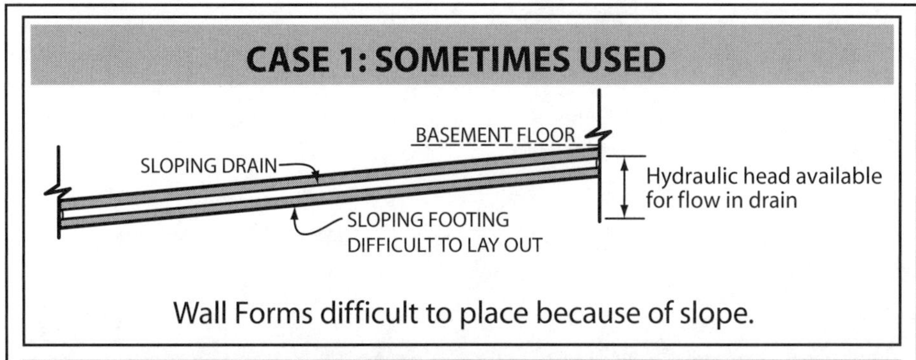

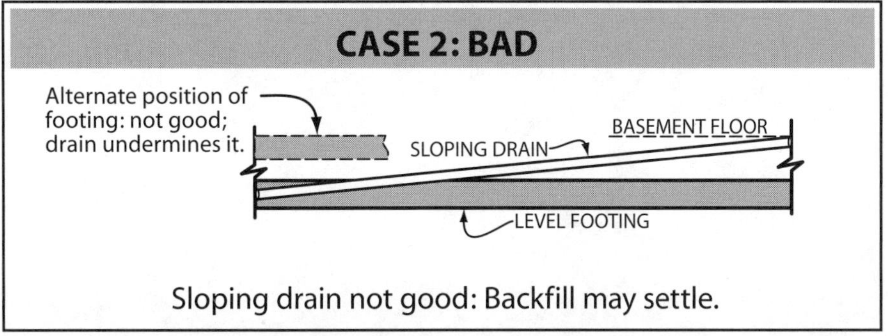

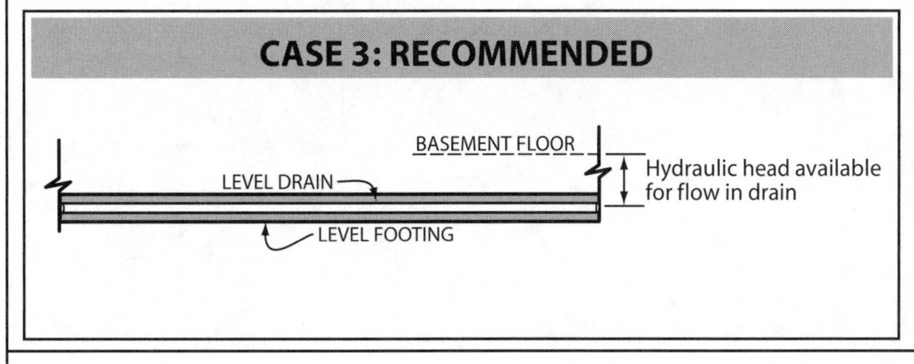

Figure 3-12
Relation of Footing Elevation to Drain

CHAPTER 4

Lot Drainage

Every residential lot should be finish graded so that rainwater flows away from the building and to the street or a watercourse. This requires knowledge of topographic surveying with transits, levels, measuring tapes.

FIELD SURVEYING

Before you start grading excavation or filling, you should know where to locate the building. Field surveying, sometime called construction surveying is a necessary part of grading and drainage work. It begins when the land surveyor, licensed by the state, has finished surveying the property lines. A construction surveyor does not have to be licensed, but should know how to operate and maintain survey equipment. In addition, he must understand trigonometry to accurately layout the drainage channels, pipelines, and retaining walls as called for on the plans. He should carefully check the dimensions shown on the plans and notify the job architect or engineer if a string of dimensions do not add up.

Plot plans, usually drawn to an engineering scale, may become a problem for laying out the job. This is because in a plot plan, a foot is divided into tenths of a foot, while building plans are dimensioned into feet and inches. So be prepared to have a chart ready to convert one scale to the other.

As a construction surveyor, you will probably do the following tasks:

- Set up transit lines, coordinate lines, and benchmarks that are necessary for laying out foundations, concrete structures, and roads.
- Control horizontal distances and elevations in formwork
- Perform a topographic surveying for grading, excavation and fill
- Prepare profiles and cross sections for road construction.
- Make a quantity and measurement survey for earthwork
- Prepare as-built plans

Before you start, study the plot plan to see how the land surveyor marked the property corners, and then, locate a datum elevation. A datum elevation is a point marked on a permanent feature, like a concrete curb or sidewalk that you can use as the arbitrary benchmark for the project. Usually you assign the elevation of 100.00 to this point, but always show its true elevation in reference to the public street elevation. For example, Elev. 100.00 = USGS Elev. 345.67. USGS means United States Geodetic Survey, usually used by public agencies.

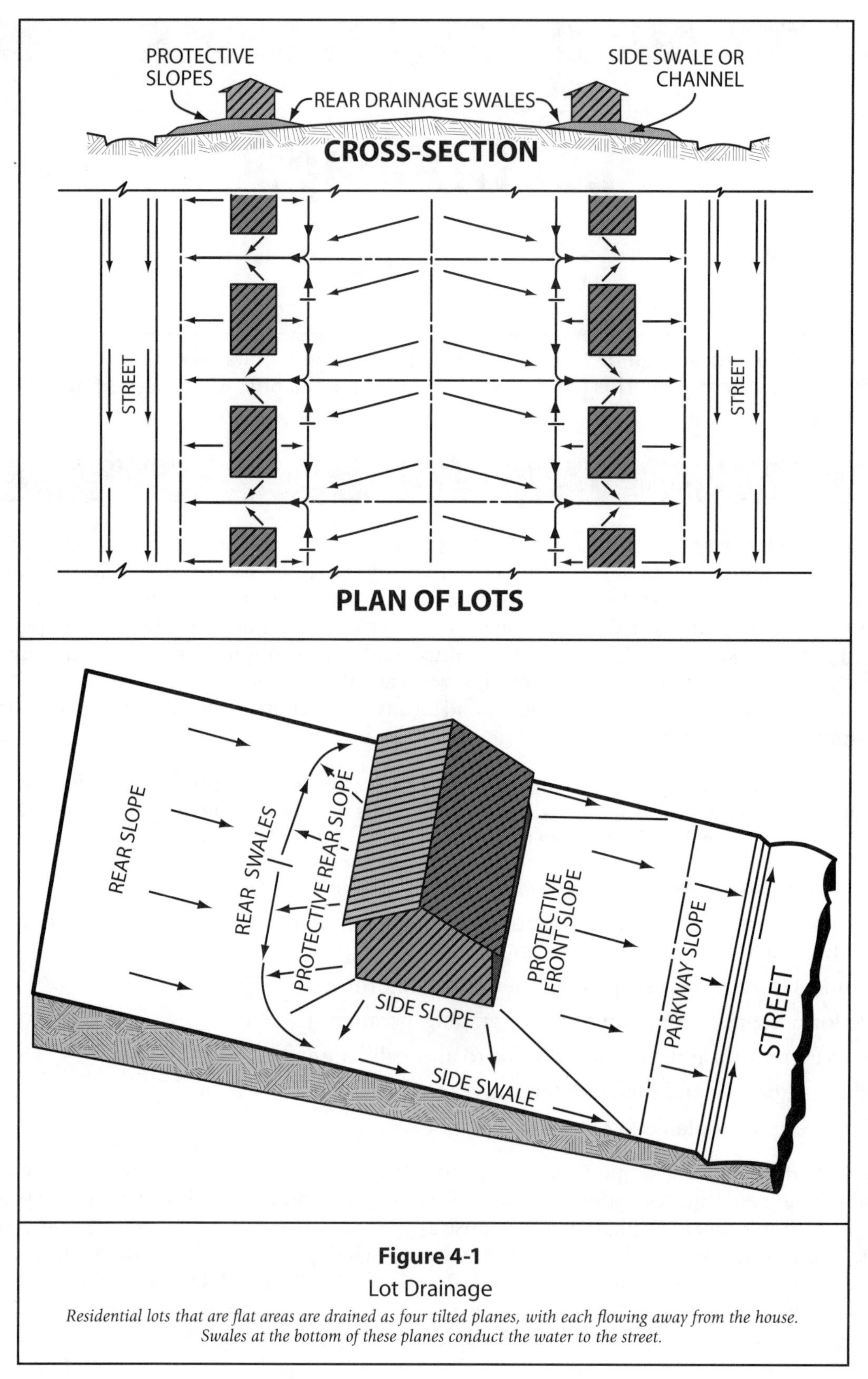

Figure 4-1

Lot Drainage

Residential lots that are flat areas are drained as four tilted planes, with each flowing away from the house. Swales at the bottom of these planes conduct the water to the street.

Set extra reference stakes around important points so a second survey will not be needed if the main control marks are disturbed during construction. Figure 4-2 illustrates reference stakes. Highlight these stakes with brightly colored cloth or plastic, called flagging. Also, make the wood hubs stand out with white paint, or by using shiny metal disks, called shiners. It's important to record the location of all stakes you set in a field logbook.

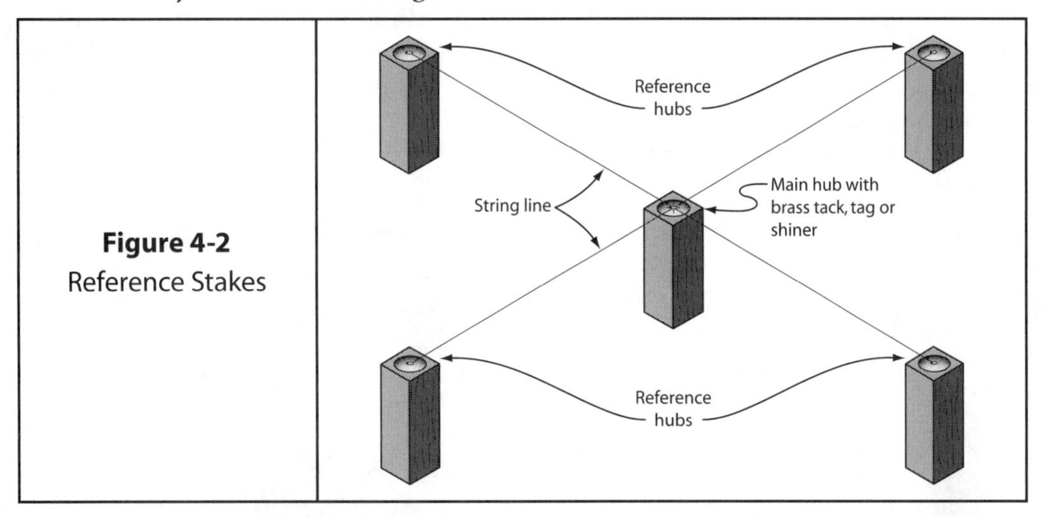

Figure 4-2
Reference Stakes

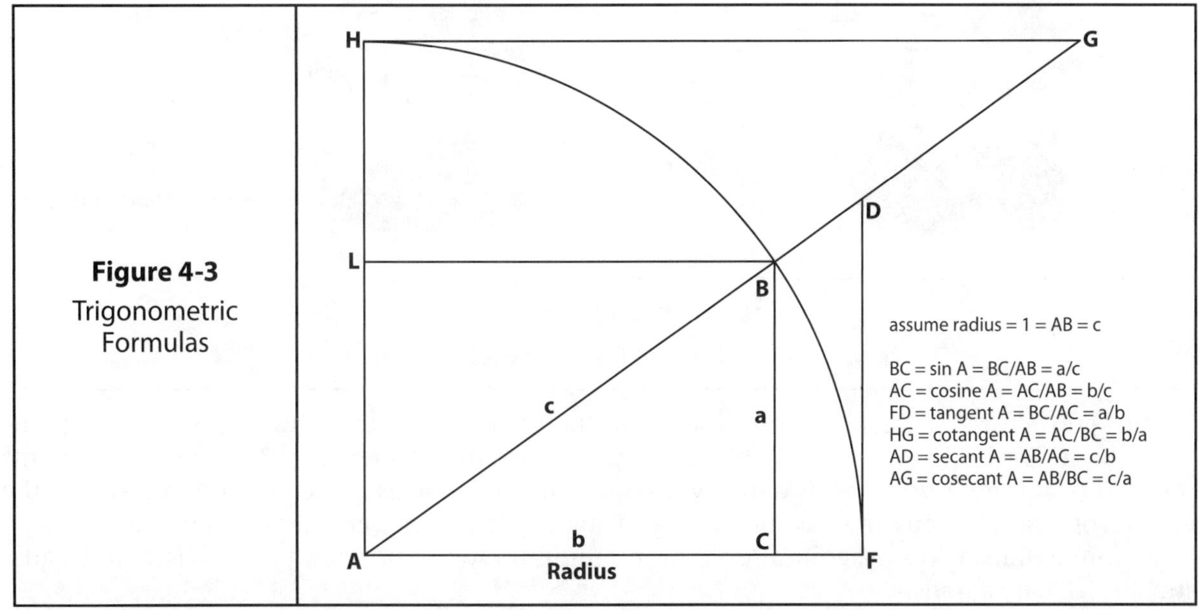

Figure 4-3
Trigonometric Formulas

In most of your field layout, you will have to use some trigonometry. The most important rules of trigonometry are shown in Figure 4-3 above. Use the following rules to remember the trigonometric relationships of a right triangle:

- Sine (sin) of an interior angle is equal to the length of the side opposite the angle divided by the length of the hypotenuse of the triangle.
- Cosine (cos) of an interior angle is equal to the length of the side adjacent to the angle divided be the length of the hypotenuse of the triangle.
- Tangent (tan) of an interior angle is equal to the length of the side opposite the angle divided by the length of the side adjacent to it.
- Cotangent (cot) of an interior angle is the reciprocal of the tangent of the angle, or the length of the side adjacent to the angle divided by the length of the side opposite to it.
- Secant (sec) of an interior angle is the reciprocal of the cosine of the angle, or the length of the hypotenuse of the triangle divided by the length of the side adjacent to the angle.

TRANSITS

Although most public works and large construction projects are laid out with electronic instruments, the traditional transit is still the basic tool for medium and small jobs. You use the transit for setting horizontal and vertical angles. If you're only setting elevations, a builder's level is more effective. Its telescope has greater magnifying power, and because it has fewer moving parts than the transit, it is easier to set up and keep steady.

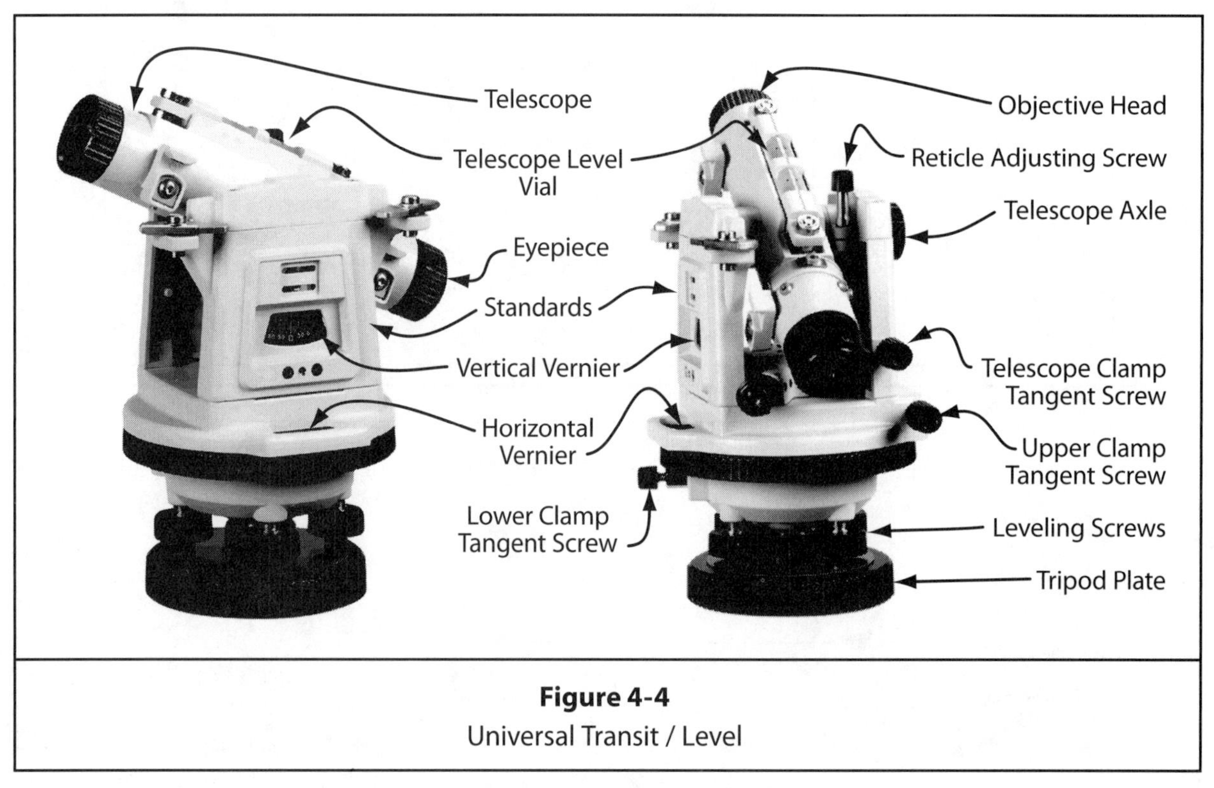

Figure 4-4
Universal Transit / Level

This figure shows the main parts of a standard transit. These are the telescope, standards, plate, and verniers. The horizontal and vertical circles are divided into degrees and 30-minute increments (one-half of a degree). You classify a surveyor's transit by the smallest angle that you can read on the vernier. For example, you can read one-minute of angle in either the vertical or horizontal planes on a one-minute transit. The magnification power, minimum focus distance, field of view, and stadia ratio also classify a transit.

The transit's telescope is suspended between standards and rotates. A leveling vial is attached to the telescope so you can use the instrument for leveling. The horizontal circle contains two movable plates that rotate or lock. The angles between the two circles are graduated and read with a vernier. You can move the horizontal circle by adjusting screws. Two leveling vials are mounted on top of the horizontal circle that you must level to set the instrument on a level plane.

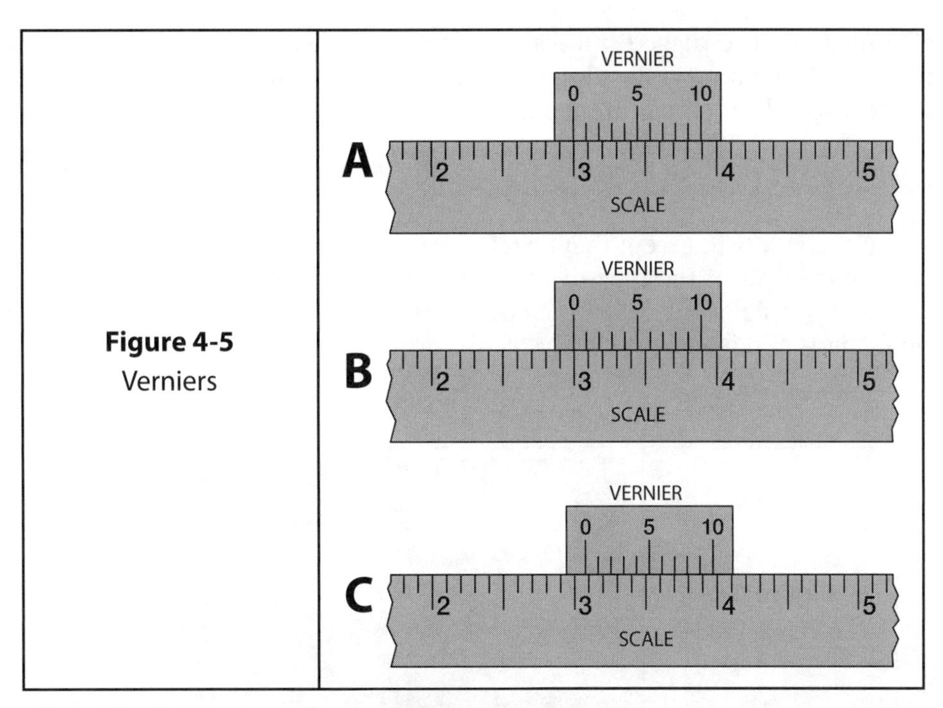

Figure 4-5
Verniers

The vertical standard also contains two movable plates. One is attached to the standard itself and the other is actually attached to the telescope. You can read the angles between the plates on a graduated vernier, as shown on Figure 4-5.

The vernier can be adjusted by the adjusting screw. The leveling base contains one or more leveling screws for adjusting the horizontal circle relative to the base and tripod.

Most transits come with a high visibility case and a carrying strap. The case contains a sunshade you put on the end of the telescope, a dust cap to protect the objective lens when it's not in use, and a rain cover. There's a brass plumb bob to center the instrument over a selected point, a magnifying glass to help you read the vernier, adjusting pins for tightening and correcting the leveling vials, a screwdriver for adjusting other components of the instrument. An instruction book is also included in the case. Here are suggested steps for setting up a transit:

- Mount the transit on a wood tripod.
- Hang the plumb bob, or plummet, from underneath the transit to a set point on the ground.
- Position the transit by setting the tripod firmly on the ground with the telescope approximately at eye level and the plumb directly over the intended point.
- Adjust the leveling screws so that the compass box is horizontal and the plate stays level no matter what direction you point the telescope.

You'll need to know about angles, bearings, and azimuths to use a transit. In addition, you should be able to add and subtract bearings to find angular distances, and to interpret the verniers on the horizontal and vertical planes to read the angles.

A bearing is the angle between any horizontal line and the north-south line, and is never more than ninety degrees. For example, a line of 30 degrees 20 minutes west of north is written as a bearing of N 30° 20′ W. A line 45 degrees 30 minutes east of south is written as a bearing of S 45° 30′ E. The bearing used in a property description is based on a previously recorded bearing, such as a street centerline or a section line.

A builder's transit is more rugged but less precise than a surveyor's transit. The telescope on a builder's transit usually has a magnification power of 10, which means an object 120 feet away looks like it's twelve feet away. The transit on the surveyor's transit may have twice the magnifying power.

LEVELS

A level is similar to a transit, except that its telescope is longer and has greater magnifying power. Figure 4-6 is a photograph of the type of level used in construction. It's used to set the height of concrete forms and floor slabs. A level is more practical than a transit in setting elevations because it has fewer moving parts and is easier to keep steady. Most levels don't measure vertical or horizontal angles.

Figure 4-6
Level

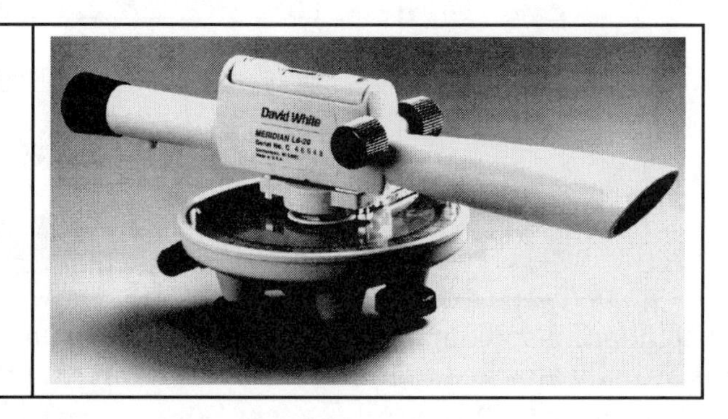

A basic level has a telescope with an attached level vial. The level is mounted on the ends of a straight bar. The bar rotates around its center on a vertical axis. Use the adjusting screws on the tripod to make the plate exactly horizontal. When you rotate the telescope on a level, you're sighting on a horizontal plane. Any object that is behind the horizontal cross hair in the eyepiece is at the same elevation as the telescope.

The builder's transit level serves as a transit and a level and is more rugged than either of the two. The smallest angle that can be read with it may be 5 to 15 minutes.

Figure 4-7
Electronic Level

The electronic level is also called the "Automatic-laser Level." It is an instrument that lets you do leveling work by yourself. It uses electronic and infrared technology to generate a reference plane around the instrument with a rotating emitter. It has a battery with a charger. Some laser levels have hinged top so that they can also generate a vertical reference plane. You use the bull's eye bubble to initially level the instrument. After that, it automatically keeps itself level. If you accidentally knock this instrument off level, it stops rotating and flashes red.

The automatic laser level (Figure 4-7) is an instrument that lets you do leveling work by yourself. It's an automatic level with electronics and infrared technology, mounted on a tripod. It has a battery with a charger and generates a reference plane around the instrument using a rotating emitter. Some laser levels have a hinged top so they can also generate a vertical reference plane. You use the bull's eye bubble to initially level the instrument, after that, it automatically keeps itself level. If you accidentally knock this instrument off level, it stops rotating and flashes a light.

Here's the procedure for using the automatic level:

- Measure the height of this type of instrument with a yardstick you set between the ground point and a mark on the housing.
- Use the detector or sensor mounted on a staff to take measurements. The detector has a battery- powered display and tone generator.
- Read the sensor from front or back. Several persons can use the automatic laser level at the same time.

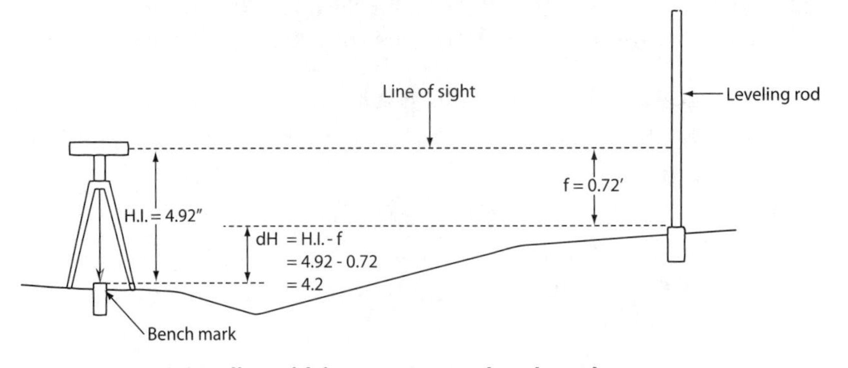

A. Leveling with instrument over bench mark

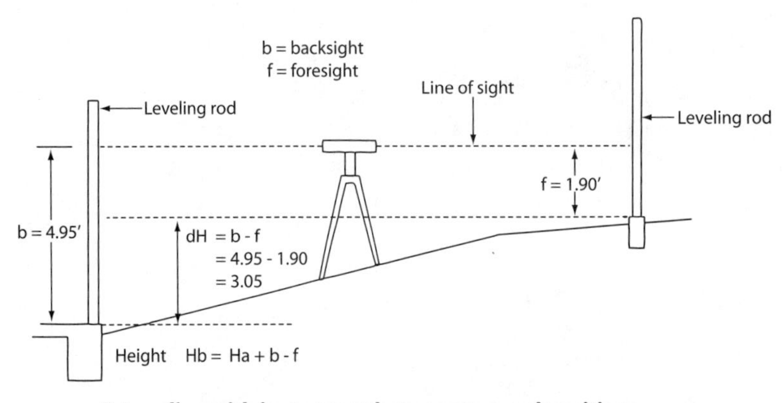

B. Leveling with instrument between two rod positions

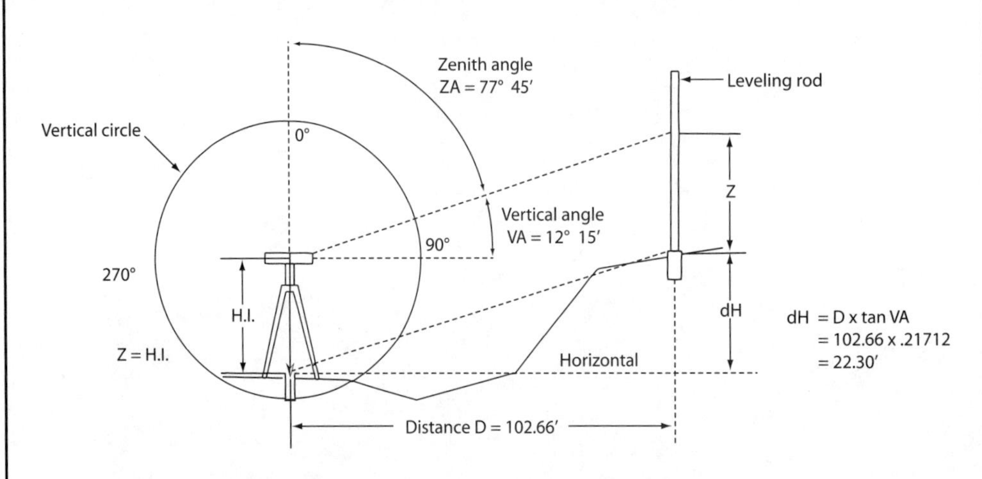

C. Leveling with instrument below rod botom

Figure 4-8. Leveling

Here is how to use an automatic level:

- *Measure the height of the instrument with a yardstick between the ground point and the mark on the housing.*
- *Use the detector or sensor mounted on a staff to take measurements. The detector has a battery-powered display and tone generator*
- *Read the sensor from the front or back. Several people can use an automatic laser level at the same time.*

Hand levels are handy to carry around but are not accurate enough for precise work. You can use them for rough grading, but follow up with a leveling instrument mounted on a tripod. See Figure 4-9.

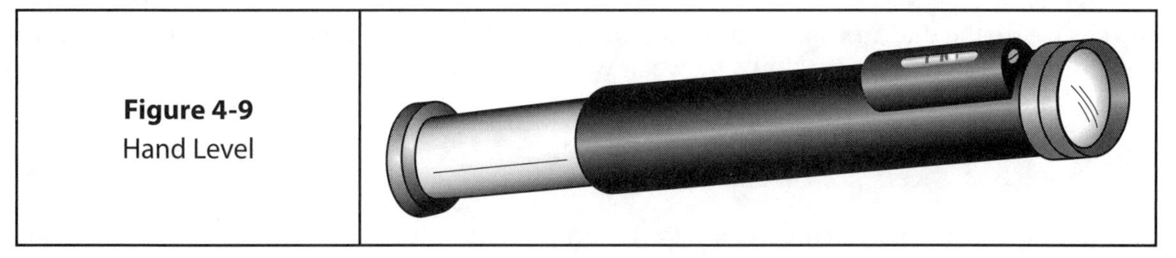

Figure 4-9
Hand Level

A portable leveling vial can be attached to a tight string or to the top of a batter board to make them horizontal. You can also use a four-foot long spirit level to keep the batter boards or forms level.

LEVELING ROD AND MEASURING POLE

There are many types of rods used for vertical measurement. The most common type is the Philadelphia rod. It's made of hardwood to resist shrinking and swelling. The face is made of white Mylar with markings and numbers that won't rub off. Every number indicating a foot is in red, tenths of a foot is marked in black Hundredths of a foot are marked in black bars along one edge of the rod. In the standard leveling rod the numbers increase from bottom up. In the direct elevation rod, the numbers increase from top to bottom.

Builder's rods are smaller but more ruggedly built than the Philadelphia rods. The Builder's rod is usually twelve feet long, and can be folded in two sections. It's marked in feet and inches to the nearest eighth of an inch. Engineer's rods are marked in feet and decimal parts of a foot. Rods are painted white with feet indicated in red, and fractions in red.

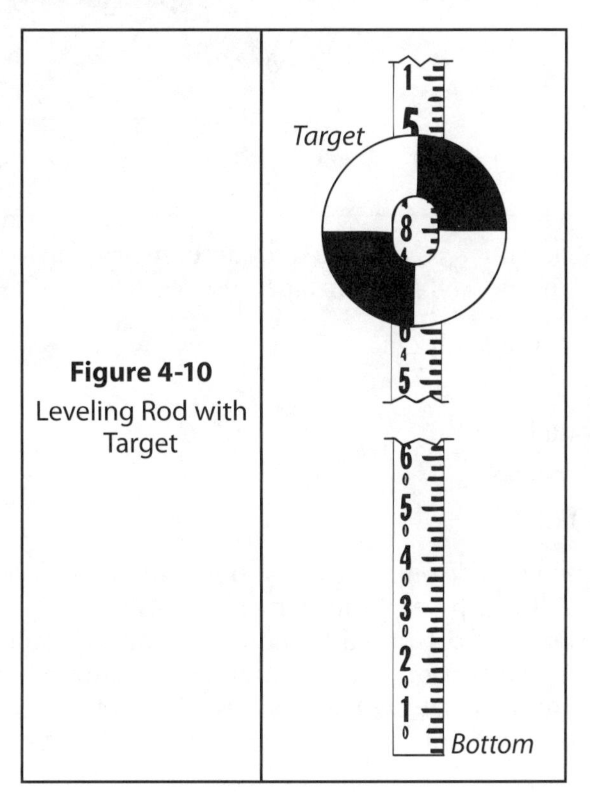

Figure 4-10
Leveling Rod with Target

Another version of the rod is the measuring pole. This device extends up to 26 feet in height and is self-reading so you can measure the height and depths using only a measuring pole.

Digital measuring poles are aluminum poles, which can be extended as much as 26 feet. You read distances on a digital dial attached to the lower section of the pole. The poles are useful in measuring the height of roofs, ceilings, or widths between walls. One person can use this instrument to make measurements in difficult locations where normally two people ad a ladder would be needed.

STEEL TAPES

Construction surveyors usually use steel tapes to measure distances. Tapes may be band chains, flat steel wire tapes, ordinary steel tapes, or metallic coated tapes. Most steel tapes are 100, 200, and 300 feet long. They're marked at one foot intervals. One end is marked in tenths and hundredths of a foot, reading from the right to the left. The 100-foot tapes are kept on reels when not in use. You can use leather thongs and holding clamps to keep the tape tight. Tension handles help reduce sag in the tape.

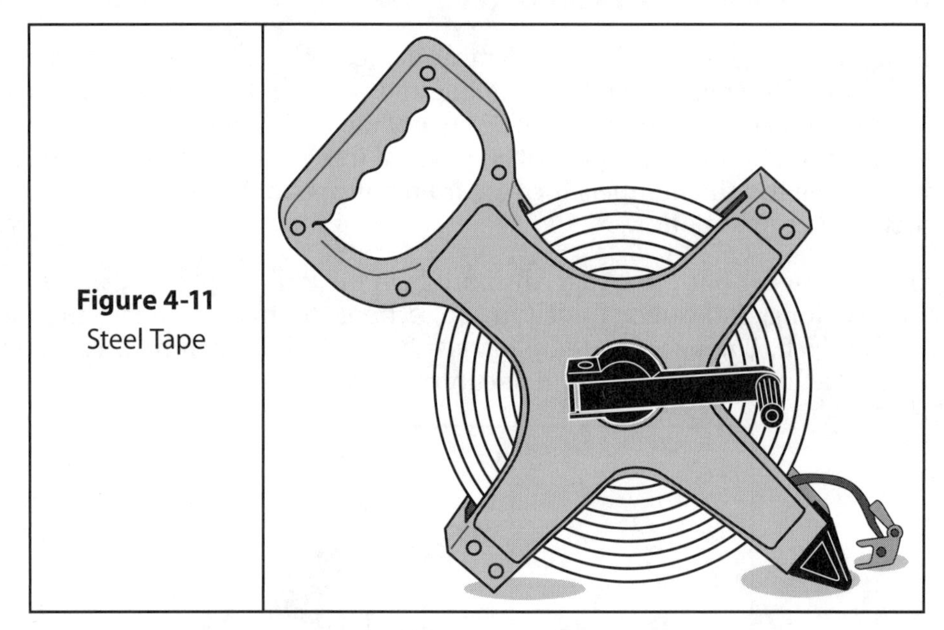

Figure 4-11
Steel Tape

Cloth and metallic tapes are not recommended for accurate layout work. Cloth tapes can stretch too much, especially when they're wet. Metallic tapes also stretch more than steel tapes.

Most measuring errors are the result of:

- Incorrect tape reading
- Tape not held horizontal
- Tape not straight
- Incorrect tension on tape

The most common reading error is omitting either one foot or 100 feet. You can use a hand level to keep the tape horizontal. To help keep a long measuring line straight, set temporary stakes in the ground along the measuring line. Normal tension on a 100 foot tape is 15 pounds. Most steel tapes are tested at the factory for tension in a 100-foot length required for accurate measurement. For example, the manufacturer may state that accuracy is within plus or minus 0.1 inch per 100 feet when the tape is supported throughout at 15-pound tension at 68 degrees F. See

TOPOGRAPHIC MAPS

A topographic map is a drawing of a portion of land in which the surface is illustrated by a series of contour lines. Each contour line represents a level of equal elevation. These contour lines are identified by numbers indicating the elevations they represent above some basic datum. The datum may be sea level or an arbitrarily assumed elevation.

A steep gradient causes the contour lines to group closer together, while a flattening makes them spread out. Shading often indicates a sudden embankment, and hachure marks mean a depression. The slope or grade between two points is determined by the vertical distance divided by the horizontal distance stated as a ratio or a percentage. The ridge, or crest line, divides two drainage areas. The region from which the rainfall will drain to a particular channel or area is referred to as the watershed of that area. The natural watercourse can be a gully, valley, or ditch.

Every change of the land surface made by a grading contractor requires a survey of the existing ground surface to establish the topography prior to the grading operation. Finish grades are then determined above or below existing grades and these are indicated on a grading plan. This represents the final topography. If the earthwork is under the control of a governmental agency, these topographic maps are retained as records by the local grading department, the department of building and safety, or the road department.

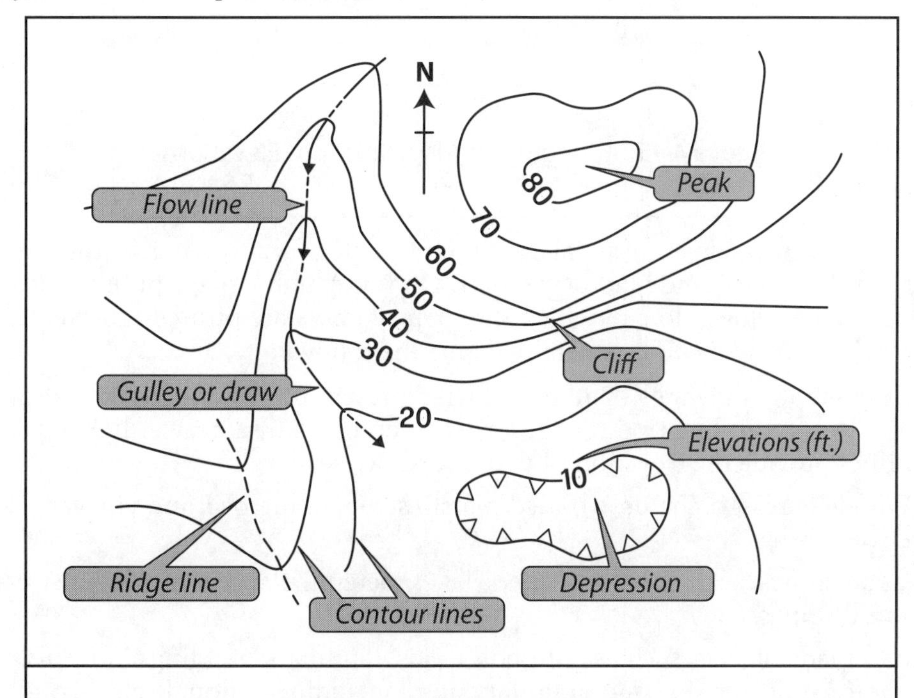

Figure 4-12. Topographic Map

A contour line connects all points of equal elevation above or below an assumed reference point. These lines form a topographic map that shows hills, valley, ridges, hogback, etc. By use of solid or dashed contour lines, existing and proposed topography are shown. Drainage is always perpendicular or at right angles to the contours. Hachures (shading) can also be used to show the direction of drainage.

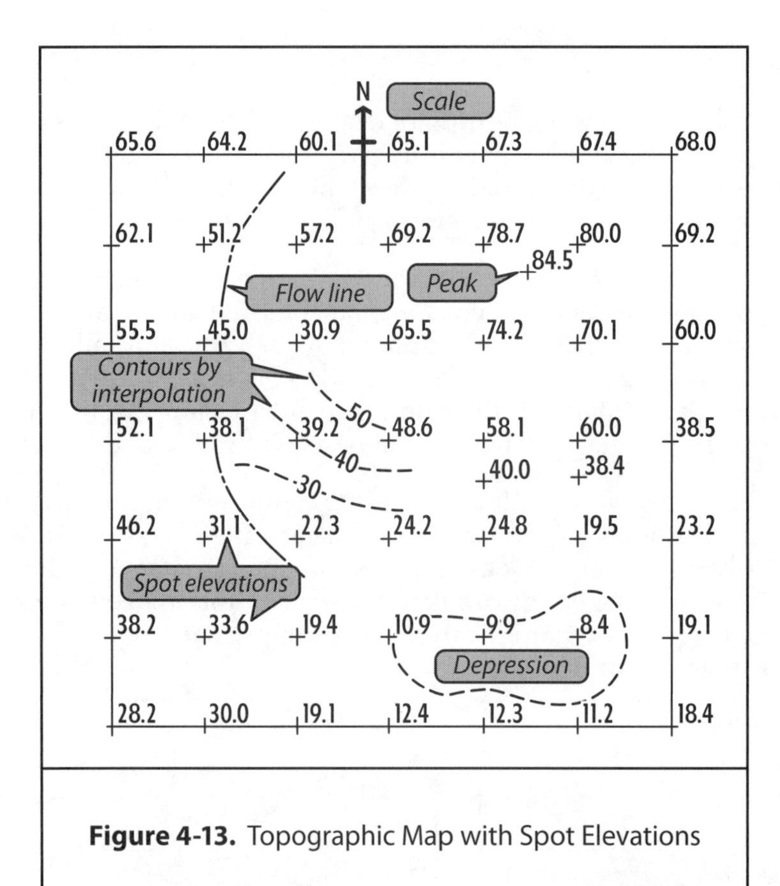

Figure 4-13. Topographic Map with Spot Elevations

Topographic maps are made using aerial photography and land surveying, by the highway department, flood control agencies, and land developers. On topographic maps published by the United States Geological Survey, the colors in which the map symbols are printed represent different features for easier reading. These colors generally depict the following:

- Brown: Topography shown by contours which represent points of equal elevation. Every fourth or fifth contour line (index contour) is accentuated by a heavier line. Figures along contour lines provide the elevation of that line above sea level.
- Green: Wooded areas, orchards, wooded marsh, scrub, mangrove, and vineyards as opposed to clearings.
- Blue: Water features, such as oceans, rivers, lakes, perennial and intermittent streams, washes, and springs.
- Black: Man-made objects, such as railroads, roads, landmark buildings (schools, churches, etc.), transmission and boundary lines, including national, state, county, parish, municipal, township, range, and section lines; important survey information, such as benchmarks (BM), vertical control stations indicating spot elevations, and horizontal control stations, and monuments used in the triangulation system.

CHARACTERISTICS OF CONTOUR LINES

- All points on a contour line have the same elevation. A contour line connects points of equal elevation.
- Every contour closes on itself somewhere, either within or beyond the limits of the map. In the latter case, the contours will run to the edge of the page.

- A contour that closes on itself within the limits of the drawing is either a summit or a depression. Depressions are usually indicated by a spot elevation at the lowest point, or by placing short hachure marks on the low side of the contour.
- Contour lines never cross other contours. The only exceptions are unusual landforms, such as a hanging cliff, natural bridge, or pierced or arched rock.
- Contours that are equally spaced indicate a uniform sloping surface.
- On a convex slope, contours are shown spaced at increasing intervals going up a hill; the higher contours are spaced farther apart than the lower contours.
- On a concave slope, the contours are shown spaced at increasing intervals with the lower contour lines spaced further apart than the higher ones.
- Valleys are indicated by contours pointing uphill. In crossing a valley, contour lines run up the valley on one side, turn and cross the stream, and run back the other side.
- Contours that are close together indicate a steep slope.
- Contours that are spaced far apart indicate a relatively level site or slight grade.
- Contours never split in two; however, you will occasionally see two side-by-side contours numbered the same. This indicates either a high or low area. It will be high if the numbers for both contours fall between contours, or low, if the numbers do not.
- Contours are usually labeled on the high side or on the line.

GRADING SPECIFICATIONS

The following are general recommendations for Grading Plans:

- Specifications have precedence over drawings.
- Keep the stamped set of plans on the job site at all times.
- Do not start any work whatsoever in or about a grading project without first notifying the grading inspector.
- Do not do grading work, import, or export, between the hours of 6:00 p.m. and 7:00 a.m. on any day and do not do work on Sunday at any time.
- Keep the construction area sufficiently dampened to control dust caused by grading and construction. Provide reasonable control of dust caused by wind.
- If a grading job extends over a period of time exceeding six months, the Governing Agency may require planting of those portions of the job where all other grading requirement have been met in order to prevent dust and erosion.
- Keep highway equipment in good operating condition and muffled as required by law.
- Control noise resulting from repair or heavy equipment after normal working hours by locating such activities as far as practicable from adjacent inhabited areas and so that such activities do not constitute a public nuisance or disturb the peace.
- Notify the Traffic Bureau of the local Police Department prior to the start of hauling.
- Detail a haul route on the plans satisfactory to the Grading Department.
- Complete the export or import of the fill material within the maximum time limit of from the start of hauling.
- Keep the fill material in each truckload low enough to prevent spillage and sufficiently wet down to prevent dust. .

- When hauling any earth, sand, gravel, rock, stone, debris, paper, or any other substance over any public street, alley or other public place, do not allow such materials to blow or spillover and upon the public street, alley or other public place or adjacent private property.
- When excavating, compacting, hauling or moving earth, sand, gravel, rock, stone, debris, or any other similar substance, cause, do not allow or permit any mud, earth, sand, gravel, rock, stone, debris or other substance to drop, be deposited, or fall from the body, tires, or wheels of any vehicle so used upon any public street or alley without immediately and permanently removing the same there from.
- Secure permission from the Department of Public Works if the trucks are loaded in the street.
- Obtain approval of the location of dumping excess soil from the grading inspector prior to starting excavation.
- Proceed brushing and scarifying of slopes only as far as periodically cleared by the grading inspector.
- Brush slopes prior to placing fill.
- Loose material should not exceed 3" in depth on a filled slope.
- Remove all debris and foreign material from the site.
- Remove and compact all loose fill.
- Locate buildings clear of the toe of all slopes, which exceed a slope of two horizontal to one vertical according to the following:
  a. A minimum of three feet provided the slope does not exceed six feet in vertical height.
  b. One-half the vertical height with a maximum of 15 feet for slopes exceeding six feet.

EXCEPTION: Attached or detached open carports or detached garages constructed a minimum of three feet clear of the toe of conforming cut or fill slopes. In the case of attached enclosed garages, all portions of the building shall comply with Subsection (b) above.

- Do not make tops of cut slopes nearer to a property line than one foot, plus one-fifth the height of the cut, but need not exceed a horizontal distance of ten feet.
- Driveway grades shall not exceed 20 per cent.
- Driveways exceeding 15% should have centerline profile on plans that clearly show radius at top and bottom sufficient for vehicular access.
- Pave driveway grades exceeding 10 per cent.
- If at any stage of work on an excavation or fill the Department determines by inspecting that further work as authorized by an existing permit is likely to endanger any property or public way, the Department may require as a condition to allow the work to continue that plans for such work be amended to include adequate safety precautions.
- Maintain sanitary facilities on the site from beginning to completion of grading operations.
- Include as part of the plans all recommendations and conditions of the Soils Report by dated; the Geological Report by dated and the Board of Building and Safety Commissioners approval letter file.
- The engineering geologist, soils engineer and civil engineer shall comply with governmental requirements and provide the Department with a grading certification upon completion of the job.

- Tract civil engineers should submit an as-graded plan at a scale of 1" = 40′ with the required grading certifications to the Department upon job completion.
- Provide supervisory control during the grading operation to insure compliance with approved plans and with the Municipal Code. When necessary, and avail yourself of geological and foundation engineering services to implement your supervisory control.
- Lay subdrains under all fills placed in natural watercourses.
- Place subdrains along the watercourse flow line and along the flow line of any branches tributary thereto.
- Install additional subdrains to collect any active or potential springs or seeps that may be covered by the fill.
- Install subdrains after the watercourse has been excavated to firm material in preparation for receiving the fill.
- Show on the plans individual design and recommendations of the soils engineer and geologist to the satisfaction of the Department.

EXCAVATION - GUIDE

- Do not cut a slope to exceed a vertical height of 100 feet unless horizontal benches with a minimum width of 30 feet are installed at each 100 feet of vertical height.
- Slope: Do not make an excavation with a cut face steeper in slope than two horizontal to one vertical.

EXCEPTION: The Grading Department, in case an appeal, may permit the excavation to be made with a cut face steeper in slope than two horizontal to one vertical if you show through investigation, subsurface exploration, analyses, and report by both a soils engineer and an engineering geologist, to the Department's satisfaction, that the underlying bedrock and the materials to be exposed on the slope have strength characteristics sufficient to produce a stable slope with a minimum factor of safety of not less than 1.5 for static loads.

- Cut existing or proposed slopes to be no steeper than the bedding planes in any formation where the cut slope will lie on the dip side of the strike line.
- Whenever grading at the top of any natural or man-made slope exposes soil or bedrock material that will allow the infiltration of water in a manner that would adversely affect the stability of the slope, the exposed area should be capped with a relatively, impervious compacted soil blanket seal having a minimum thickness of two feet. The soils engineer should certify in writing that the blanket seal is adequate to reduce water infiltration to permissible levels.
- If the material of the slope is of such composition and character as to be unstable under the anticipated maximum moisture content, reduce the slope angle to a stable value. This requirement should be confirmed by the soils engineer's written certification following laboratory testing.
- Intervening paved benches on cut slopes should have a minimum width of eight feet and shall be spaced at intervals of 25 feet measured vertically.
- Do not make tops of cut slopes nearer to a property line than one foot, plus one-fifth the height of the cut, but need not exceed a horizontal distance of ten feet.

FILLS - GUIDE

The following recommendations are base on the requirements of the Grading department of the City of Los Angeles Department of Building and safety. Check with your local authority before beginning work.

- Do not a fill slope that exceeds a vertical height of 100 feet unless horizontal benches with a minimum width of 30 feet are installed at each 100 feet of vertical height.
- Do not make a fill that creates an exposed surface steeper in slope that two horizontal to one vertical. The fill slopes abutting and above public property should be so placed that no portion of the fill lies above a plane through a public property line extending upward at a slope of two horizontal to one vertical.

EXCEPTION: The Grading Department, in case an appeal, may permit a fill to be made which creates an exposed surface steeper in slope than two horizontal to one vertical if you show through investigation, subsurface exploration, analyses, and report by both a soils engineer and an engineering geologist to the Department's satisfaction, that the fill to be used and the underlying bedrock or soil supporting the fill have strength characteristics sufficient to produce a stable slope with a minimum factor of safety not less than 1.5 for static loads. The soil engineer should verify by field-testing and observation and should certify attainment of the required strength characteristics in the fill material as specified in the approved report.

- Intervening paved terraces on fill slopes should have a minimum width of eight feet and should be spaced at intervals of 25 feet measured vertically.
- All man-made fills should be compacted to a minimum of 90 per cent relative compaction as determined by ASTM method D 1557.
- For slopes to be constructed with an exposed slope surface steeper than two horizontal to one vertical compaction at the exposed surface of such slope should be obtained either by over-filling and cutting back the slope surface until the compacted inner core is exposed, or by compacting the outer horizontal 10 feet of the slope to at least 92 per cent of relative compaction.

EXCEPTIONS:

Fills in non-hillside areas that do not exceed 12 inches in depth need not be compacted, but such fills should not change the existing drainage pattern.

The Department may approve uncompacted fills in self-contained areas where the fills are not to be used to support buildings or structures and no hazard will be created.

- Fill material placed in areas within cemeteries used or to be used for interment sites should be compacted to a minimum of 80 percent, unless such fill is placed on a slope steeper than three horizontal to one vertical, or placed on slopes adjacent to public properties or private properties in separate ownership. or is to be used to support buildings or structures, in which cases it should be compacted to a minimum of 90 percent.
- Fill slopes should be prepared for planting in one of the three following ways:
    a. The slope surface of fills may be prepared for planting by casting top soil over the slope surface. The top soil layer should not exceed three inches in depth.
    b. The slope surface may be scarified to a depth not to exceed three inches.
    c. Loose material not to exceed three inches in depth may be left on the slope.
    d. Toes of Fill Slopes. Toes of fill slopes should not be made nearer to a project boundary line than one-half the height of the fill, but need not exceed a horizontal distance of 20 feet.

e. Inspection and Control. Every man-made fill should be tested for relative compaction by a soil testing agency approved by the Department.

f. Certificate of compliance setting forth densities so determined should be filed with the Department before approval of any fill is given.

g. All man-made fills, whether compacted or not, which were placed prior to October 17, 1952, should be tested for relative compaction by an approved soil testing agency before any approval to build should be issued.

EXCEPTION:

Where proposed buildings are no higher than one story high and contain an area less than 1,000 square feet, the Department may waive this requirement, provided inspection shows the fill satisfactory for the proposed use.

h. Where a combined cut and fill slope exceeds 25 feet in height, the required drainage bench should be placed at the top of the cut slope. The effect of surcharge of the fill upon the cut bedrock should be considered by the soils engineer and engineering geologist, and specific recommendations should be made relative to the setback between the cut and fill.

i. Areas on which fill is to be placed should be investigated by the soils engineer or by the soils engineer and geologist to determine it is adequate to support the fill.

EXCEPTION: The Department may waive this investigation where it determines by inspection that the underlying material is adequate to support the proposed fill.

SEPTIC TANKS AND SEEPAGE FIELDS

HOUSE SEWER

The house sewer should have watertight joints. It should be on a grade of not less than 1/8 inch per foot. Ells or bends of 90 degrees is not acceptable. Cleanouts should be installed in accordance with the National Plumbing Code.

SEPTIC TANK

- Design should provide adequate volume for settling, for sludge and scum storage, and access for cleaning. The structural design and materials used should be in accordance with generally accepted good engineering practice providing a sound, durable tank, which will safely sustain all dead and live loads and liquid and earth pressure involved in each case.
- You should provide the minimum safe distances shown in Table 4-1.

Table 4-1
Minimum Safe Distances in Feet

FROM	TO			
	Septic Tank	Absorption Field	Seepage Pit	Absorption bed
Well	50	100	100	100
Property Line	10	5	10	10
Foundation Wall	5	5	20	5
Water Lines	10	10	10	10
Seepage Pit	6	6	*	–
*3 diameters.				

- Liquid capacity should be based on the number of bedrooms proposed, or that can be reasonably anticipated in the dwelling and should be at least that shown in Table 4-2.

Table 4-2
Minimum Capacities for Septic Tanks

Number of Bedrooms	Min. Liquid Capacity (gallons)
2 or less	750
3	900
4	1,000
Each addl. bedroom, add	250
NOTE: The capacities shown here provide for the plumbing fixtures and appliances commonly used in a single-family residence (automatic clothes washer, electric garbage disposal and dishwasher).	

- The liquid depth of the tank or a compartment thereof should be not less than 30 inches. A liquid depth greater than six feet should not be considered in determining tank capacity.
- No tank or compartment thereof should have an inside horizontal dimension less than 24 inches.
- Inlet connection should be submerged or baffled to divert incoming sewage toward bottom of tank.
- Outlet connections of the tank and of each compartment thereof should be submerged or baffled to obtain effective retention of scum and sludge.
- Scum storage volume (space between the liquid surface and the top of inlet and outlet devices) should be not less than 15% of the required liquid capacity.
- Outlet baffles and baffles between compartments, including pipe fittings used as baffles, should extend below the liquid surface a distance equal to approximately 40% of the liquid depth of the tank.

  They also should extend upward to within 1 inch of the underside of the cover. When a partition wall is used to form a multi-compartment tank, an opening in the partition wall may be used provided the minimum dimension is 4 inches, the cross-sectional area is not less than that of a 6-inch diameter pipe, and the mid-point of the opening is below the liquid surface a distance approximately equal to 40% of the liquid depth.

- When multi-compartment tanks are used, the volume of the first compartment should be equal to or greater than that of any other compartment.
- The tank should be constructed so that gases generated in the tank, absorption field, seepage pit or absorption bed can easily flow back to the main building stack.
- The inlet invert should be at least one inch above the outlet invert.

For examples of Septic Tank Construction, see Figure 4-14 below.

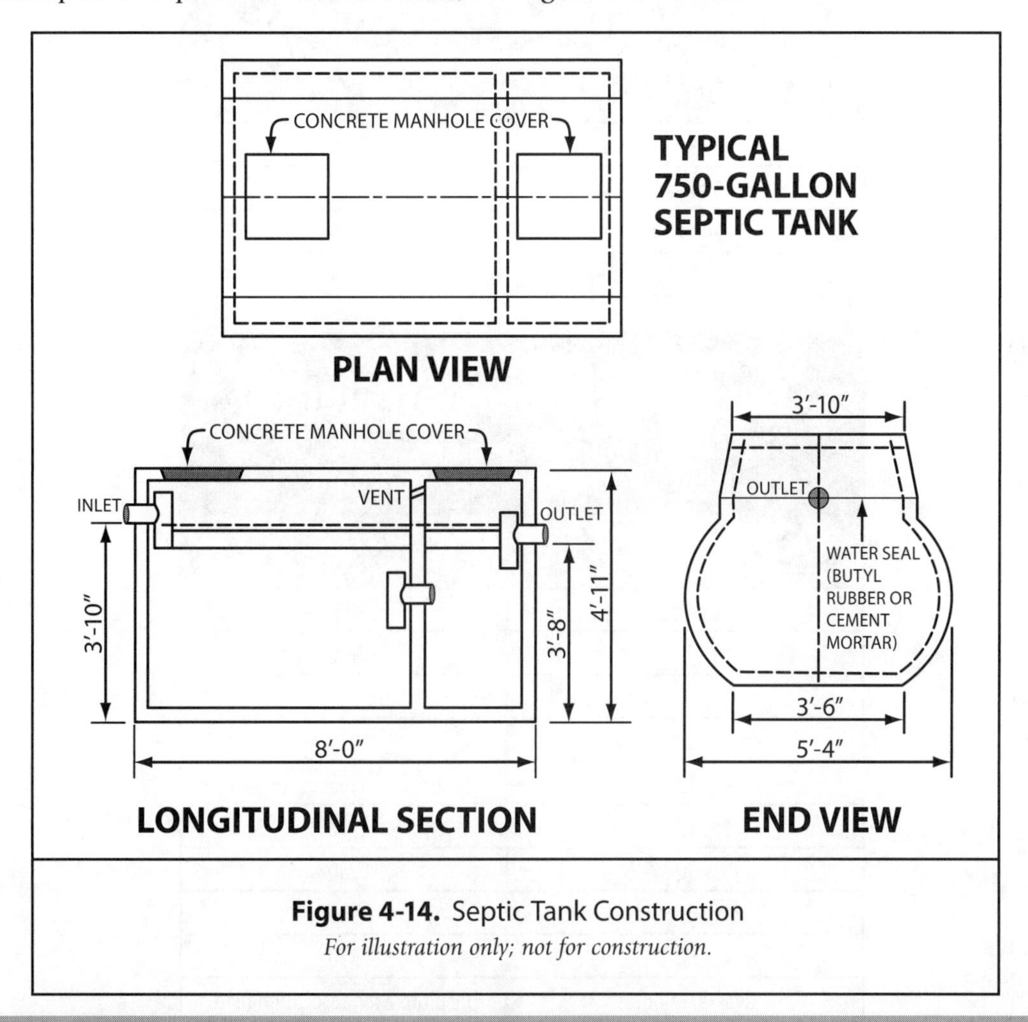

Figure 4-14. Septic Tank Construction
For illustration only; not for construction.

SUBSURFACE ABSORPTION FIELD (LEACHING FIELD)

- Location of the absorption field should be in an unobstructed area and should comply with the minimum distances given in Table 4-1.

NOTE: When existing wells are involved or exceptionally coarse soil formations are encountered the 100-foot distance from any water supply should be increased to protect any water supply source if subsoil and site conditions indicate an increase is necessary.

- One of the following methods should determine minimum absorption area (total bottom area of trenches) of the absorption field:

a. Results of percolation tests conducted in accordance with Percolation Tests. The trench bottom area required should be determined from Table 4-4.

b. Recommendation of the State Department of Health or the Health Authority having jurisdiction based on experience data and percolation tests.

- Tile lines should be spaced in accordance with Table 4-3.

Table 4-3
Size and Minimum Spacing for Disposal Trenches

Width of Trench at bottom	Min. Spacing of trenches, wall-to-wall (feet)
12 to 18	6.0
18 to 24	6.5
24 to 30	7.0
30 to 36	7.5

Table 4-4
Surface Absorption Fields
Minimum Required Trench Bottom Area per Bedroom

Average time in minutes for water to fall 1 inch	Min. required area (sq. ft.)
2 or less	85
3	100
4	115
5	125
10	165
15	190
30	250
45	300
60	330
Over 60 minutes	Unsuitable for absorption field.

NOTE: The capacities shown here provide for the plumbing fixtures and appliances commonly used in a single-family residence (automatic clothes washer, electric garbage disposal and dishwasher).

- That portion of an absorption trench below the top of the distribution pipe should be in natural or acceptably stabilized earth.
- All trenches in an absorption field should comply with Tables 4-3 and 4-5.

FIELDS IN FLAT AREAS

In locations where the slope of the ground over the absorption field area is relatively flat (6 inches fall or less in any direction within field area) the trenches should be connected to produce a continuous system and the trench bottoms should be level.

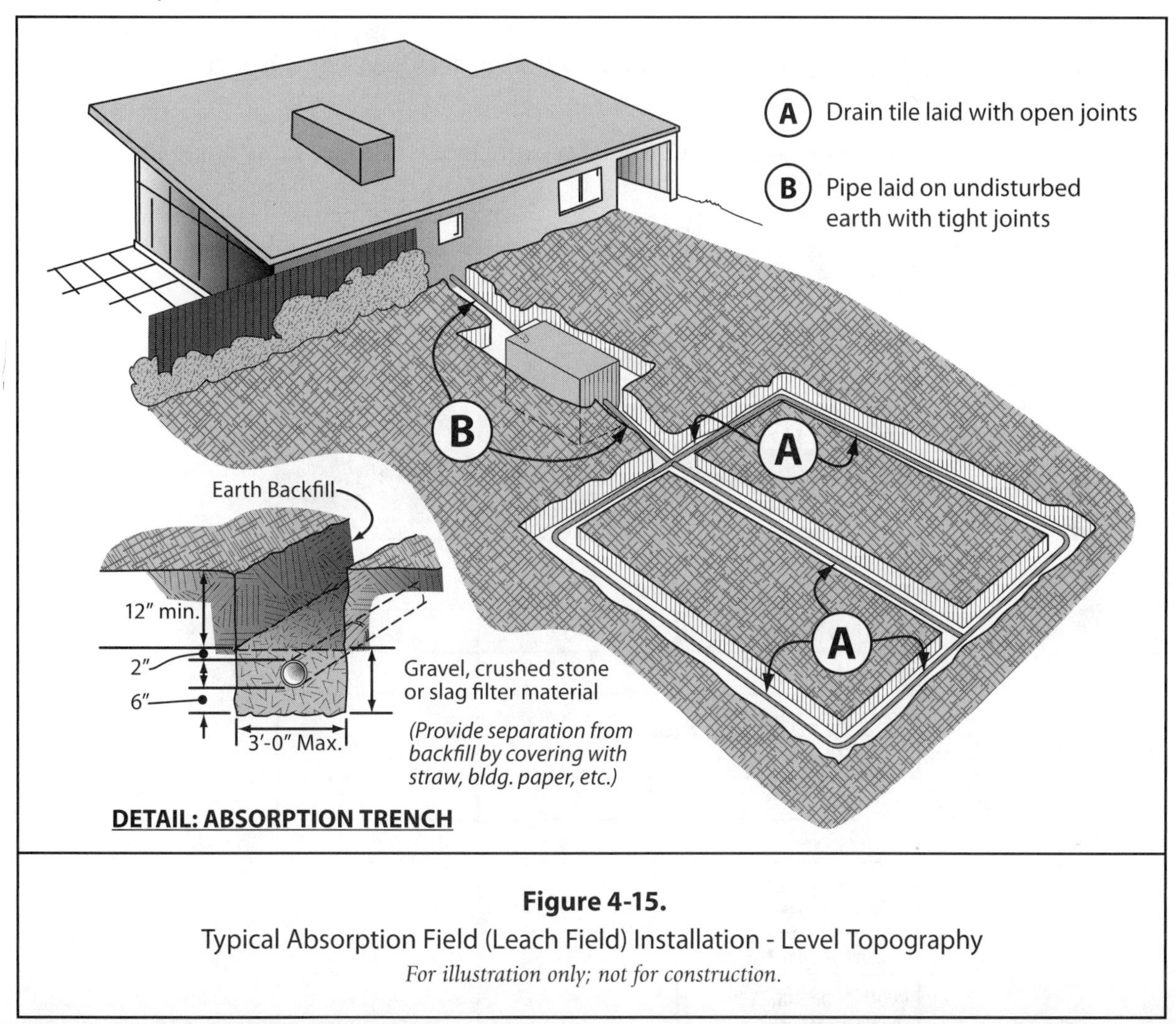

Figure 4-15.
Typical Absorption Field (Leach Field) Installation - Level Topography
For illustration only; not for construction.

FIELDS IN SLOPING GROUND

a. In locations where the ground over the absorption field ail slopes (fall greater than 6 inches in any direction within field area) a system of serial distribution trenches following the contours of the land may be used. The trenches could be installed at different elevations, but the bottom of each individual trench should be level.

b. Trenches should be connected with a watertight overflow line in such a manner that a trench will be filled with sewage to the depth of the gravel before the sewage flows to the next lower trench.

c. The overflow line should be a 4-inch watertight sewer with direct connections to the distribution tiles in adjacent trenches. Distribution tile lines should have a level grade.

d. There should be undisturbed earth between trenches. At the point where an overflow pipe leaves an absorption trench, the trench for this pipe should be dug no deeper than the top of the soil-gravel interface. Overflow line should rest on undisturbed earth and backfill should be carefully tamped. The inlet to a trench should be placed as far as practical from the outlet (overflow) from the same trench.

- Effluent from the septic tank should be conducted to the absorption field through a watertight line with a grade of at least ¼ inch per foot. Tees, wyes or other distributing devices may be used.

NOTE: If a distribution box is used, it should be of sufficient size to accommodate the necessary field lateral lines. The invert of all outlets should be level and the inlet invert should be at least 1 inch above the outlet inverts. Outlet inverts should be from four to 6 inches above the floor permitting water retention to act in lieu of a baffle for the purpose of securing equal distribution.

Table 4-5

Subsurface Absorption Field Construction Details

Items	Max.	Min.
Number of lateral trenches	–	2
Length of trenches	100′	–
Width of trenches	36″	12″
Depth of tile lines	6 in./foot	*
Depth of coarse material		
Under pipe		6″
Over pipe		2″
Under pipe located w/in 10′ of trees		12″
Size of coarse material	2½″	½″
Depth of backfill over coarse material		12″
Seepage Pit	6	*
*Level (preferred).		

- Pipe used for the line between the septic tank and the absorption field, all lines within 10 feet of dwellings and under paved areas should comply with the National Plumbing Code. Pipe used under driveways or other areas subject to heavy loads should be installed to withstand the imposed loads and should be watertight. Such sections should not be considered in determining the effective absorption area.
- Pipe used in the absorption field should comply with the National Plumbing Code. The openings between joints should be no greater than ¼ inch.
- All open joints should be protected to effectively eliminate soil infiltration.

- Filter material should be of crushed stone, gravel, slag, or similar material of equivalent strength and durability. It may vary from 1½ inch to 2½ inches and should be free from fines, dust, sand or clay. The filter material should completely encase the tile or perforated pipe.
- An effective barrier such as building paper or straw should be placed over the filter material to prevent infiltration of the backfill.
- Heavy equipment should not be driven over the trenches during backfilling or after completion of the absorption field.

SEEPAGE PITS

Use of seepage pits with septic tanks is acceptable only when soil conditions or topography indicate a need for this method of disposal. In addition, their use must be acceptable to the health authority having jurisdiction. Seepage pits are not acceptable in limestone areas or in localities where wells are used.

- When more than one seepage pit is used, the installation may be operated in series or in parallel. If operated in series each pit should be equipped with an inlet device (tee or ell). An outlet device (tee or ell) should be installed to prevent floating scum flowing into the second pit. If the installation is operated in parallel a tee, wye, or distribution box should be used.
- Location of seepage pits should be not less than the stated minimum distance in Table 4-1.
- Effective absorption area of a seepage pit should be calculated as the side area, below the inlet, exclusive of any hardpan, rock, or clay formations.
- Required seepage area should be determined in accordance with Table 4-6.

Table 4-6
Seepage Pits
(Min. Area Required)

CHARACTER OF SOIL	Sq. Ft. Wall Area Required per Bedroom
Coarse sand or gravel	50
Fine sand	75
Sandy loam or sandy clay	125
Clay with considerable sand or gravel	200
Clay with small amount of gravel	400
Heavy tight clay, hard-pan, rock or other impervious formations	Unsuitable
NOTE: The capacities shown here provide for the plumbing fixtures and appliances commonly used in a single-family residence (automatic clothes washer, electric garbage disposal and dishwasher).	

- A minimum depth of 4 feet of porous formation for each installation should be provided in one or more pits.
- Pits less than 20 feet deep should have a diameter of at least 4 feet
- No pit excavation should extend into the water table. Where ground water is encountered, the bottom of the pit should be backfilled, with clean coarse sand, at least 2 feet above the water table.

- Pipe with tight joints should be used in connecting the septic tank to the pit.
- All seepage pits should be either lined or filled with coarse stone.
- The lining may be brick, stone, block, or similar materials at least 4 inches thick, laid in cement mortar above the inlet and with tight-butted joints below the inlet.
- The annular space between the lining and the earth wall should be filled with crushed rock or gravel.
- The structural design and materials used should be in accordance with generally accepted good structural engineering practice providing a sound, durable structure, which will safely sustain all the dead loads, live loads, liquid and earth pressures involved in each case.
- A reinforced concrete top, at least 4 inches thick, bearing at least 12 inches on top of Soil outside of pit, unless supported by the pit wall, and with a watertight concrete pump hole and cover should be provided in all cases.
- The top should be not less than 18 inches below the finished grade.
- For examples of seepage pit construction, see Figure 4-16 below.

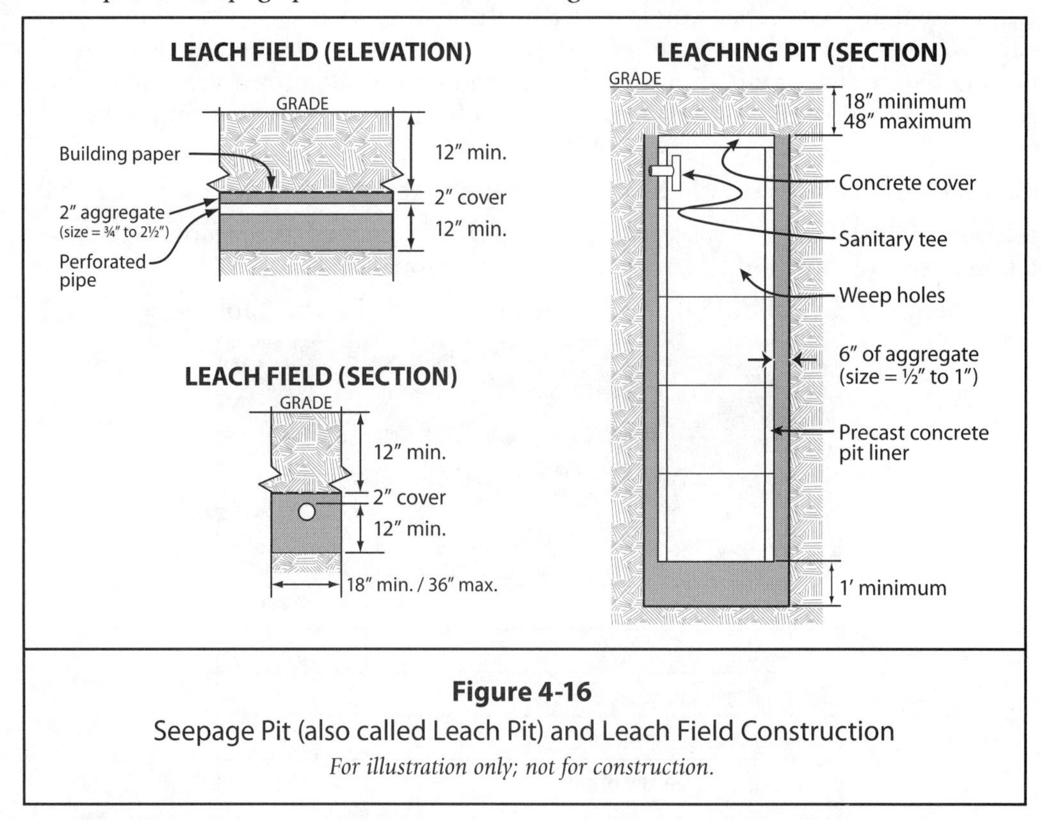

Figure 4-16

Seepage Pit (also called Leach Pit) and Leach Field Construction

For illustration only; not for construction.

ABSORPTION BEDS

Use of an absorption bed with a septic tank is acceptable only when necessary because of soil conditions or topography.

- Location of absorption beds should comply with the minimum distances shown in Table 4-1.
- Effective absorption area of an absorption bed should be calculated as bottom area.
- Total bottom absorption area should be determined from Table 4-7 using results of percolation tests conducted in accordance with Percolation Tests.

Table 4-7
Absorption Beds
Minimum Required Bottom Area per Bedroom

Average time in minutes for water to fall 1 inch	Min. required area (sq. ft.)
2 or less	170
3	200
4	230
5	250
10	330
15	380
30	500
Over 30 minutes	Unsuitable for absorption bed.
NOTE: The capacities shown here provide for the plumbing fixtures and appliances commonly used in a single-family residence (automatic clothes washer, electric garbage disposal and dishwasher).	

- The bottom of an absorption bed should terminate in a porous formation at least 4 feet in thickness.
- No bed excavation should extend into the water table. Where ground water is encountered, the bottom of the bed should be raised with filter material.
- Absorption beds constructed in unstable filled ground are not acceptable.
- The construction of the absorption bed should comply with Table 4-8.

Table 4-8
Absorption Bed Construction Details

Items	Max.	Min.
Distance between distribution lines	3′	–
Distance between distribution lines and wall	1½′	–
Depth to bottom of bed		1½′
Depth of filter material		
Under pipe		6″
Over pipe		2″
In bed w/in 10′ of trees		12″
Depth of backfill over filter material	24″	12″

SOIL CLASSIFICATION SYSTEM

The Unified Soil Classification System has been used by the Corps of Engineers and Bureau of Reclamation. It also identifies soils according to their textural and plasticity characteristics, dividing soils into course-grained soils, fine-grained soils, and organic soils.

The U.S. Department of Agriculture has combined several soil classification systems into a simple triad that places sand, silt, and clay on each side, and describes various soils as composed of percentages of the three.

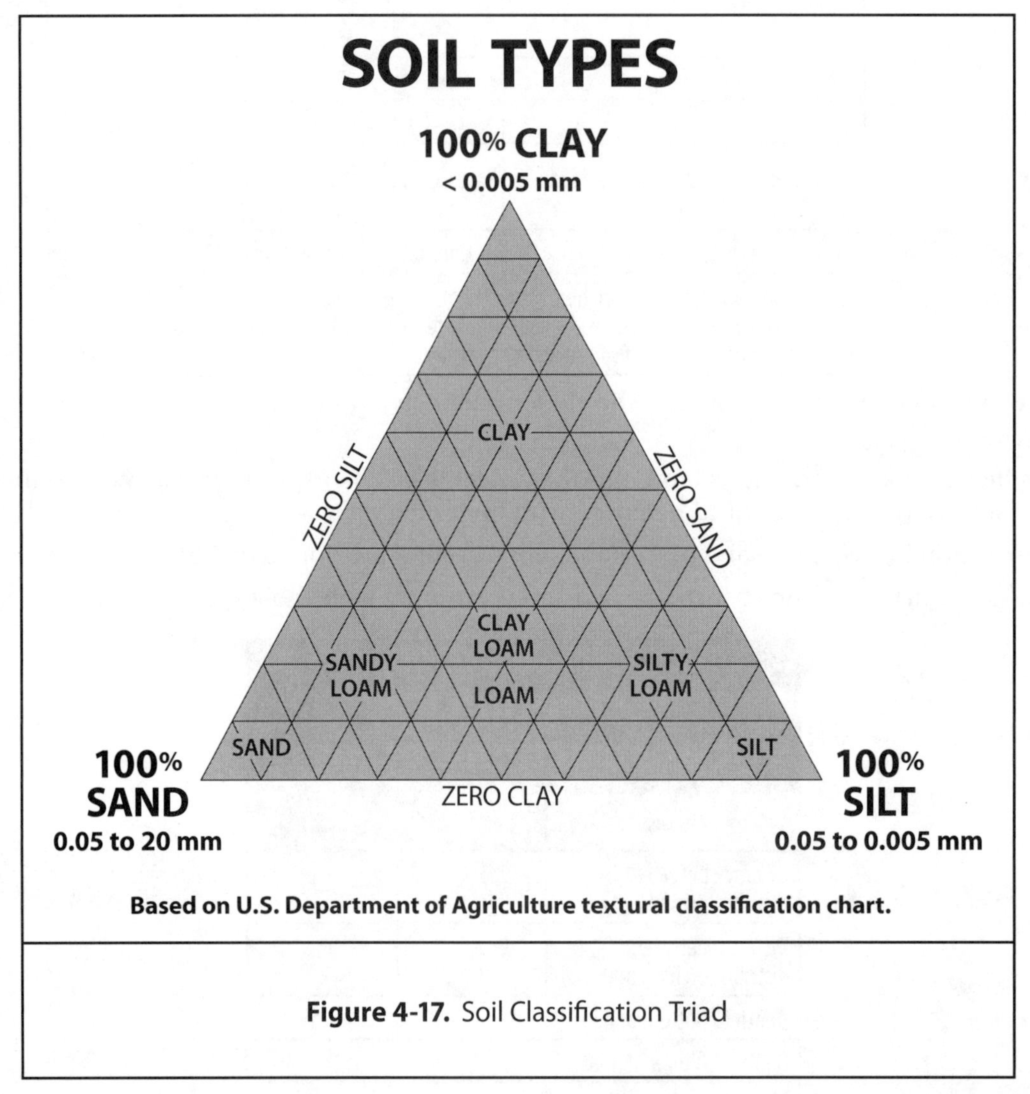

Figure 4-17. Soil Classification Triad

Knowledge of soil types and their characteristics interpreted for land-use decisions forms an important input to grading. In many metropolitan areas, the U.S. Department of Agriculture has reinterpreted agricultural soils information for use in making urban land-use decisions.

Their soil scientists analyzed what land-related activities occurred during urbanization, such as long-term problems (landslides, flooding, erosion, water pollution, etc.), what had to be constructed (foundations, septic drain fields, utility systems, roads, drainage, etc.) and broad land use categories (industrial, residential, recreation, woodland, ecological, etc.). They then looked at soil characteristics that might affect these activities, such as drainage, depth to bedrock, soil type, etc., and mapped soils according to the following limitations.

PROBLEM SOILS

According to the U.S. Department of Agriculture, the following lists problem soils:

- Gravelly or stony.
- Less than optimum bearing capacity.
- Drainage class unfavorable; depth to seasonal water table too shallow; drainage may be needed; flooding or ponding hazard.
- Slope or topography unfavorable; imposes use or construction hazard; slippage potential.
- Depth to bedrock or impervious material is limiting.
- Frost action potential is moderate or high.
- Shrink-swell potential is moderate or high.
- Permeability (percolation rate) is restricted.
- Accessibility by equipment is restricted.
- Excavation difficult owing to consolidated materials, or clay or silty clay soil texture.
- Piping hazard.
- Thickness of suitable material is limited.
- Erosion hazard is moderate, severe, or very severe.
- Shear strength is moderate or low; slope stability is fair or poor.
- Compaction characteristics are fair or poor.
- Permeability rate is too rapid.
- Low water-holding capacity.
- Organic soil.
- Possible pollution of groundwater even though site is at least well drained.
- Soil textures permit ready sloughing of sidewalls.
- Compressibility is medium, high, or very high.
- Moist consistency is loose, firm, very firm, or extremely firm.
- Excessive fines or limited supply.

ESTIMATED SOIL PROPERTIES

Depth to seasonal high water table refers to the highest level at which the groundwater stands for a significant period of time.

Depth from surface indicates the thickness of significant layers of a typical profile. The thickness of the horizons differs somewhat among mapping units of the small soil series.

Percentage passing sieve refers to the percentage of dry soil material that will pass sieves of the indicated sizes.

Permeability refers to the rate at which water moves downward through undisturbed soil. It depends largely on the texture, structure, porosity, and density of the soil.

Available water capacity represents the maximum amount of water that plants can obtain from the soil.

Reaction refers to the acidity or alkalinity of the soil, expressed in terms of pH. A pH of 7.0 is neutral. Values less than 7.0 indicate acidity, and values more than 7.0 indicate alkalinity.

Shrink-swell potential is an indication of the volume change that can be expected with a change in moisture content. It depends largely on the amount and type of clay in the soil. In general, soils classified as CH or A- 7 have a high shrink-swell potential; soils classified as SP, GW, or A-I have a low shrink-swell potential.

Corrosivity refers to the deterioration of concrete or untreated steel pipelines as a result of exposure to oxygen and moisture and to chemical and electrolytic reactions.

On a large grading project or one involving difficult terrain, soil, types should be mapped and understood to minimize problems during the grading. The principal soil characteristics that affect grading are as follows:

- Angle of repose
- Permeability
- Erosion hazard
- Slippage potential

ANGLE OF REPOSE

Each soil has an equilibrium angle at which it win stand naturally. If this angle is exceeded, the soil will slump or slip off. This angle varies with the soil grain size and shape, the character of the material, and the proportion of water present. Any cut or fill cannot exceed this natural angle of repose if stability is expected. Generally, sandy and gravelly soils have a fairly steep angle of repose, whereas clay, silt, or loams do not. Constantly wet or completely dry soils have a shallower angle of repose than wen-drained but moist soils.

Typical angles of repose are as follows:

- Firm earth in place, 1:1
- Loose earth or humus soil, 1.5:1
- Firm clay, 1.5:1
- Wet clay, 3:1
- Dry sand, 1.75: 1
- Wet sand, 1.25:1

These ratios vary with rainfall, moisture content, subsurface geology, plant cover, etc.

PERMEABILITY

This is particularly important when designing the drainage plan. Runoff can be disposed of in a closed system, or it can infiltrate into permeable soils as if it were a sponge. Disposing of runoff through permeable soils is good practice as it reduces cost and raises the water table. Generally, sandy and gravelly soils are more permeable than clay and loamy soils. However, many sandy soils are underlain with impermeable clay layers that reduce infiltration and may flood following severe rain. Additionally,

Paving permeable soils prevents water from percolating into the soil. This is unecological in principle, and should be done only where the situation is extreme. Build on impermeable soil and drain to permeable soils.

Planting exposed banks with thick, matty, and deep-rooted plants helps stabilize the soil. Planting can reduce minor slippage, but is of little effect against a major slide.

Terrace with cross ditches to prevent excessive water from crossing the bank. Drain ditches should be placed above the bank, with terraces at intermediate intervals to reduce the quantity of water.

Subsurface drains may be drilled into the bank to relieve groundwater. This method is expensive, and it is difficult to predict the results.

Soil permeability is the important factor in locating any septic system. Obviously, gravel or sandy soils will ensure proper operation of a septic system; however, topography and underlying impermeable soils are almost more important. An impermeable subsurface or steep topography may cause effluent to flow laterally and surface at a lower elevation raising health hazards.

In pervious soils, grading below the water table usually requires dewatering or drainage of the site during construction. Drainage is accomplished by constructing a sump, well point, or deep well at the lowest level and pumping out the water.

EROSION HAZARD

All grading causes erosion. All soils erode, although some more than others. There are two principal concerns: first, topsoil, which is the lifeblood of our plant community, should not be allowed to erode. The process of building this top layer of soil is time consuming and cannot be easily duplicated by man. Generally, the top layer of organic soil should be stripped carefully and stockpiled for redistribution and use following grading. The second concern is to be careful with soils subject to erosion.

Erosion occurs following the steepest terrain. It begins as rivulets and eventually becomes gullies, with sedimentation somewhere below. Erosion is part of consuming and cannot be easily duplicated by man. Generally, the top layer of organic soil should be stripped carefully and stockpiled for redistribution and use following grading. The second concern is to be careful with soils subject to erosion. Erosion occurs following the steepest terrain. It begins as rivulets and eventually becomes gullies, with sedimentation somewhere below. Erosion is part of the natural river-forming process and is quite normal as long as the time table is slow and long; however, man tends to hasten the process to the point of causing problems.

GRADING FOR EROSION CONTROL

On large projects where erosion is likely, it is becoming acceptable practice to prepare the grading plan, anticipate where erosion will occur, and devise a pre-grading plan to stop erosion. Pre-grading utilizes small check dams or holding basins below the area of major grading. The pre-grading work must be carefully carried out using small equipment (possibly a backhoe or by hand) and should be immediately seeded or planted so it will stabilize. Pre-grading adds to the overall grading costs, and benefits are not immediately realized, so there is some resistance to its use. However, grading permits are becoming more difficult to obtain, and the addition of an erosion-control plan may help, perhaps proving an immediate benefit from a client's point of view.

Grading during the dry season followed immediately by planting may reduce erosion damage. Scheduling sensitive projects should ensure that difficult grading operations are performed during the dry season. Small, robber-tired equipment or hand grading is economical in sensitive grading situations to minimize disturbance.

SLIPPAGE POTENTIAL

Slippage occurs when shear stress between layers of soil exceeds the shear strength of those soils. This can occur by increasing the stress or by decreasing the shear strength. The increased stress must be over a large area, and the slope must be relatively high and quite steep before the slip is likely to occur.

Increased soil stress is caused by removing the bottom (toe) of the bank or by increasing weight at the top of the bank. An excavation at the bottom of the bank for a building or a swimming pool is likely to cause increased stress. Steepening the slope beyond its normal angle of repose also causes stress. The addition of a building at the top of a slope with more drainage water increases the weight or pressure on the soil and is likely to cause slippage.

Reduction of shear strength is caused by a change in the soil property. Clay can be softened by the addition of water. As the surface dries, it can crack, allowing water to penetrate into the soil. Finally, water entering the soil can create friction between layers of noncohesive soil, which may result in a buoyancy that causes the upper layers of soil to float free of the soils below. Noncohesive soils such as sands and gravels are held together purely by friction between particles, which is easily dislodged.

Prevention of slippage requires study of the geology and soil types, with careful design to solve any uncovered problems. Possible solutions include the following:

- Avoiding all impossible sites.
- Reducing the proposed gradient by extending the bank. This uses more land and exposes more soil to erosion, but sometimes can reduce the slippage potential.
- Under draining the bank to allow groundwater to flow out, rather than be trapped within the bank. Under draining is expensive and not always successful, but may help.
- Building on pile or pole foundation to avoid disturbing the ground.
- Removing an amount of soil equivalent to the weight of the proposed structure. This balances the pre and post condition, by adding no weight.
- Installing concrete retaining walls with deep footings can produce satisfactory results, but is expensive.

STORM DRAINAGE

In designing culverts or storm sewers within his plant site, you are interested in three aspects of a storm: from the weather bureau records you must determine (t) the maximum rate of rainfall per hour of a maximum storm, (2) the duration in hours of the maximum storm, and (3) the frequency of the maximum storm. The frequency of a storm of a certain intensity is the probability that such a storm will occur within a certain number of years. As an example, a 5-year storm means one which will occur once every 5 years. You can expect a severe storm of a given magnitude to occur once in 5, 10, 25, or 50 years. Obviously, for economic reasons you may not be justified in constructing drainage devices for an intensity of rainfall that may occur only once in the life of the project.

For the design of a drainage system, which may include a channel, ditch, or pipe, the amount of expected flow must first be determined.

DESIGN CRITERIA FOR DRAINAGE

You have available well-tested standards for design of project drainage devices. These are the same standards used by the local city and county engineering departments dealing with storm drains and road design.

The following suggestions supplement the checklist to be followed in preparation of a drainage design.

1. A topographic plan of the plant site should be made with contour lines at 5-ft. intervals except on high banks shown. This map should show both existing and new conditions.
2. The scale of the topographic map should not be smaller than 1 in. to 20 ft.; 1/8 in. to 12 in. would be better if the plant is not too large.
3. Finished elevations for hard surfaces should be to the nearest 0.01 ft.; they should include floor elevations and the tops of manholes, curbs, concrete gutters, and all tank foundations.
4. Finished elevations for earth surfaces should be to the nearest 0.1 ft. These elevations should be in grid form, and all sudden changes of grade, ridge lines, flow lines, top and bottom of banks, and spot elevations at change of direction of these items should be shown.
5. All walks and paved areas must be pitched away from all buildings and equipment foundations.
6. Minimum slope for all paved surfaces should be no less than 1 percent grade.
7. All flow lines below a rate of 0.50 percent should be constructed in concrete gutters.
8. Minimum slope for flow lines should be at 0.75 percent minimum grade.
9. All building roof drains should be directed to storm drains if they are available.
10. Storm drainpipes with less than 1-ft. cover should be encased in concrete or be constructed of cast-iron pipe, except in the vehicular traffic areas where all pipes with less than 1-ft. cover should be encased in concrete.
11. All drainpipes should have cleanouts or Y's every 150 ft. and at each change of direction of line to facilitate maintenance.
12. Manholes should be spaced at least 400 ft. apart on long straight lines.
13. The crown of all roads should be sloped from the centerline 1.0 to 1.5 percent to allow drainage off the roadway.
14. Gutters adjacent to roadways which collect drainage from the roadway should have a minimum slope of 0.3 percent.
15. Parkways draining to gutter should have a slope of 2.0 to 3.0 percent.
16. When designing drain pipes, the maximum velocity of flow in pipelines should be 12 ft./sec.
17. The maximum slope for paved or planted banks is normally 2 ft. horizontal to 1 ft. vertical.
18. Planting areas should not drain across traffic areas. Drainage should be diverted along edge-of paving or pickup in drainage system.

19. Graded planes on the plant site should join public sidewalks where possible.

20. Planes should not intersect to concentrate drainage across public or plant sidewalks.

21. Where concentration in unavoidable, it should be diverted into a sidewalk culvert and spilled directly into the street gutter.

22. Roof downspouts spilling into unpaved areas must have splash blocks to minimize erosion.

23. On drainage plans, all storm-drain systems should indicate the following information:
    a. Size and type of drainpipe.
    b. Rate of slope, invert elevation at each change in rate of slope and/or direction, elevation and depth of catch basins, manholes, and junction chambers, and size of culverts.
    c. Profile of major storm-drain lines.

CHAPTER 5

Street Drainage

MINIMUM STREET GRADES

Drainage and disposal are important factors that control street design. The local storm drain department normally handles this phase of work but it is essential for the contractor to have some basic knowledge of the problems, responsibilities, liabilities, of the city, the property owner, and the builder.

One of the controlling factors of immediate concern to the builder is the city's policy of not permitting a street or alley to be improved or partially improved without providing a means for adequate drainage.

For example, you should not build a street improvement so that the water from the abutting property and the proposed improvement drains onto an unimproved street, or a partially improved street without drainage control, or other private property. However, if it can be established that the amount of water involved is negligible and will not damage, erode, or otherwise worsen the existing drainage conditions of the existing street section or of private property, construction may be permitted.

If a considerable amount of water is involved which drains onto private property, and a drainage easement or a waiver is obtained from the property owners, the proposed street improvement may be permitted.

Other controlling factors as dictated by drainage are based on studies that have to be made by the storm drain designer, such as a joint analysis of rainfall frequency and duration, the longitudinal street grade, pavement cross-slopes, curb and inlet types, and spacing of inlets or discharge points. This information is needed to determine whether or not the width of water on the pavement during selected storm frequencies is great enough to unduly interfere with traffic flow.

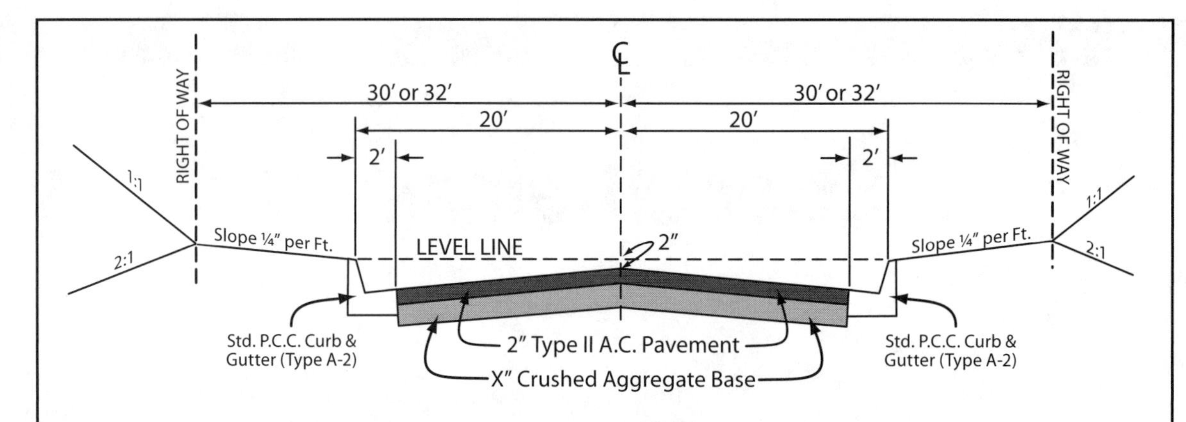

LOCAL STREET - LEVEL SECTION

NOTE: When parkway width is 6′ or less, use Type A-4 Curb & Gutter

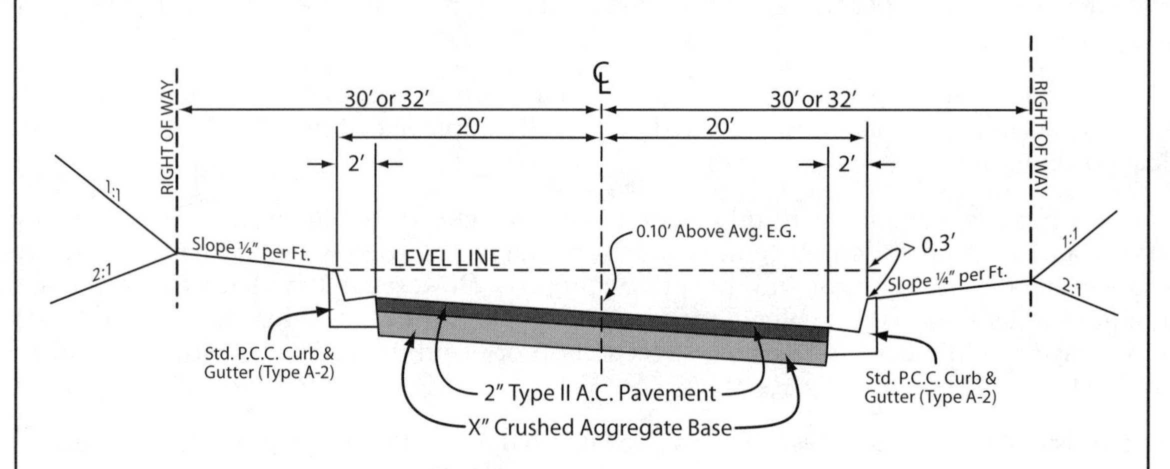

LOCAL STREET - LEVEL SECTION

Figure 5-1

Typical Street Slopes: Local Streets

For illustration only; not for construction.

You can drain roads in one of three ways:

1. *A crown is formed in the center of the road and the water is drained to both sides toward gutters.*
2. *Road is pitched toward its center while it slopes down its length. This type of road is often used for alleys.*
3. *The road is sloped toward one side toward a gutter. This is common on highway curves where the road must be super-elevated to counter the speed of the vehicles.*

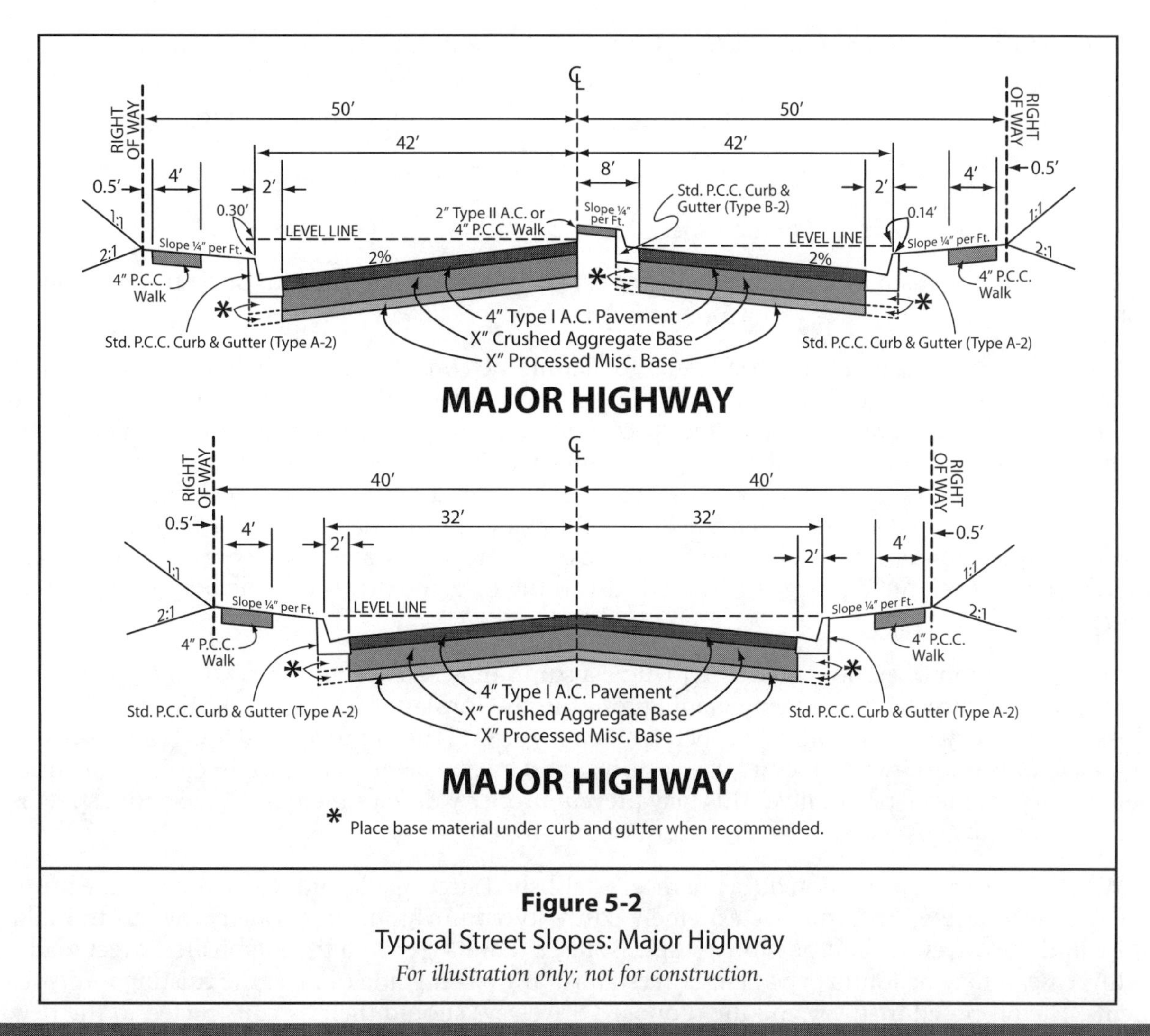

Figure 5-2
Typical Street Slopes: Major Highway
For illustration only; not for construction.

STREET DRAINAGE

Driveway construction should meet the established street grade: Where there is little or no improvement to abutting property, construct the driveway to the established street grade. In hillside streets at the site of the proposed driveway, the existing terrain may be considerably above or below the established grade.

Provide a transitional grading section between the proposed grade of the driveway and the existing street grade on either or both sides of the driveway.

This transition section should provide a smooth grade, adequate sight distance, and good drainage. Under these circumstances, the driveway construction should not create grading and drainage problems, as discussed below.

The transition grading may require the raising or lowering of existing garage floors, the remodeling or redesigning of existing buildings or buildings being designed, or the construction or reconstruction of sidewalks, walls, driveways, etc., on private property. Consult with other government offices For example:

- On drainage - Storm Drain Design Division or the Storm Drain Section of the District.
- On slopes - Bureau of Standards.

- On structures - Bridge and Structural Design Division or the Department of Building and Safety.
- On Right of Way - Street Opening and Widening Division or the Bureau of Right of Way and Land.
- On legal matters - City Attorney.
- Other involved offices - On their specialties.

Drainage problems may develop because of extensive grading, by erosion, or by the creation of a sump.

The street runoff from a small residential lot should be considered a negligible drainage factor. However, the roof, parking area, and other impervious surfaces of the lot may drain toward the street. When the driveway is constructed to concentrate and channelize the flow of water to the unimproved street, some form of drainage control must be included with the driveway construction. Driveways should not normally be designed nor permits automatically issued for this drainage situation. However, it may be determined by field inspection or design study that the driveway pavement will eliminate poor drainage conditions, such as local water pockets in a graded gutter, and will not worsen the drainage situation. If this is the case, the driveway construction should be permitted.

The elimination of accumulated water where a sump may be created occurs when the driveway construction and the accompanying grading result in the formation of a sump in the street or the blocking of drainage from abutting property. When the street is unimproved and the development is limited to only one or two properties, the costs incurred by one owner due to drainage and construction problems may be prohibitive. This may prevent property development until the entire street is improved and the costs shared by all or most of the property owners.

Where driveway construction does not meet established street grade and where most of the property has been developed, it may be extremely costly to require all of the property owners to build or rebuild the driveways and possibly garages, walls, sidewalks, etc., to the established street grade. In this case, it may be found expedient to reestablish the street grade to meet the existing improvements. The proposed roadway and the proposed driveways should then be constructed to the new grade. However, before attempting to redesign the established street grade, you should consult with the Division or District Engineer or his assistant.

Where a street grade has not been previously established, a grade should be determined for the entire street. Use the design criteria as outlined by the governing manual. The driveway grade is then superimposed on the proposed street grade profile. Adjustments should be made to the proposed street grade at the proposed driveway site to minimize the cost of the driveway construction. However, this adjustment should be made without sacrificing the City's design standards or policies, and should not be made at the expense of the existing or future development of abutting properties.

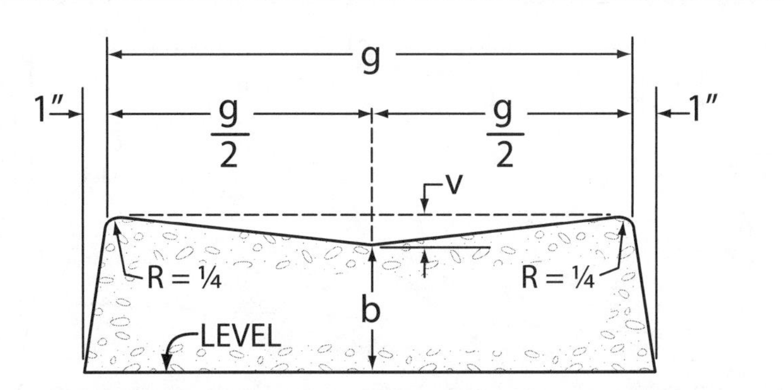

CROSS GUTTER OR LONGITUDINAL GUTTER

Concrete Use Table		
Type of Concrete Work	Class	Aggr. Type
Curb, Type A or B	B	4
Integral Curb & Gutter, Type C	B	4
Integral Curb & Pavement, Type C	A	1
Integral Curb, Gutter & Concrete Base	B	4
Integral Gutter & Concrete Base	C	1
Gutter, Cross & Longitudinal	B	4

Figure 5-3

Cross Gutter

For illustration only; not for construction.

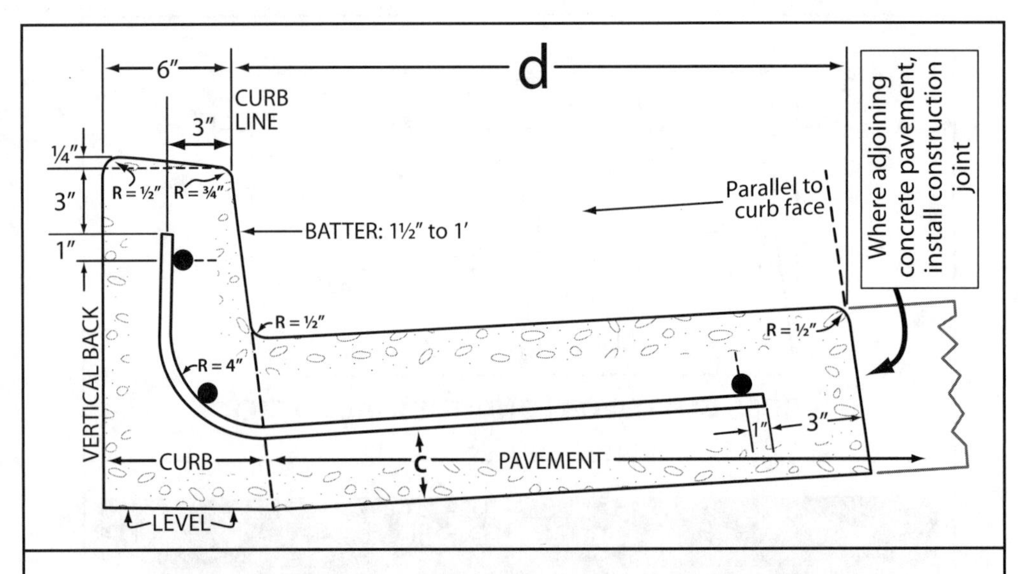

Type C - Integral Curb & Concrete Pavement

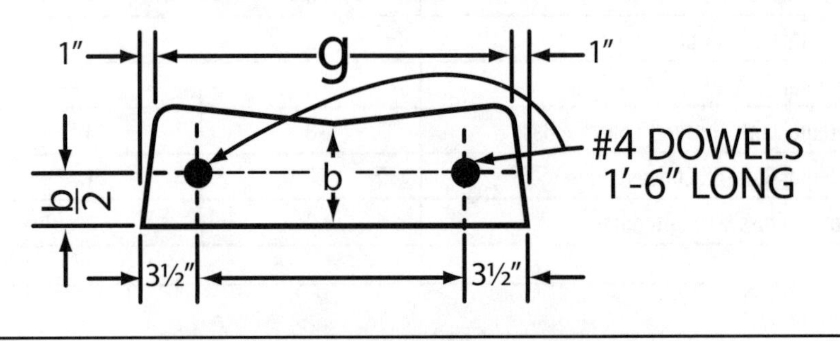

g Gutter Width	Number of Dowels
2′	2
4′	4
6′	6

Figure 5-4

Typical Curb and Gutter

For illustration only; not for construction.

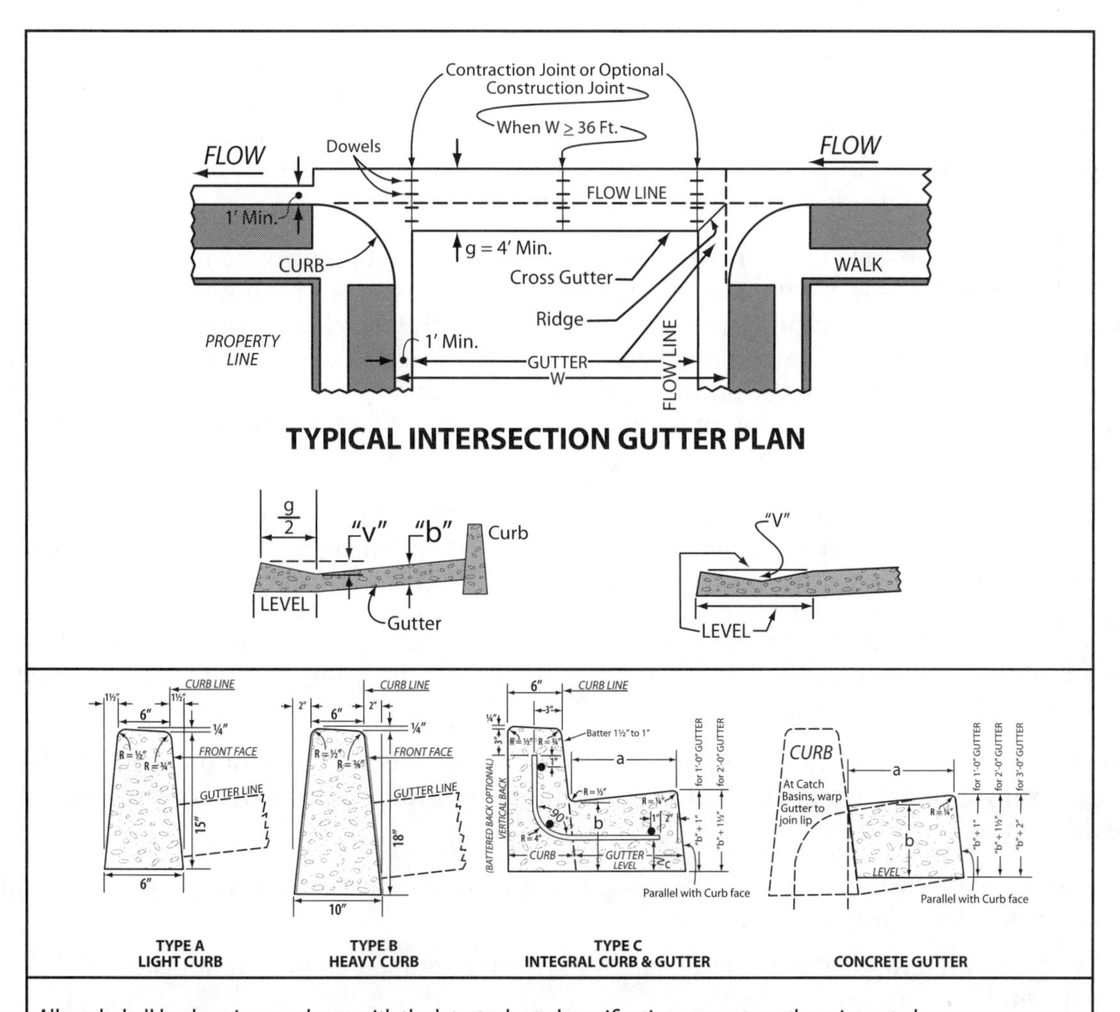

All work shall be done in accordance with the latest adopted specifications, except as otherwise noted.

Concrete surface for 4″ adjacent to all gutter flow lines shall be a steel trowel finish.

Transverse bars shall be placed 5′-0″ O.C. unless otherwise specified.

Where lap is necessary, bars shall be lapped 1′-0″.

Longitudinal bars shall terminate 2″ from all expansion joints.

Top and front face of all curbs shall be brush finish.

Expansion Joints: Preformed expansion filler 1/2″ thick shall be installed in all types of curb at the B.C. and E.C. of returns except E.C. of alley returns and at intervals of 60′ between returns. When the spacing causes expansion joint to be located in a driveway of 30′ or less in length, joint shall be installed at that end of the driveway nearest to the normal location. Similar expansion joints shall be installed in all gutter adjoining curb at the same location as constructed in the adjacent curb, and in longitudinal alley gutter only where the same joins concrete alley intersection.

Contraction Joints: Shall be installed in all types of gutter at intervals of 20′ at equal spacing between the expansion joints.

Scoring Lines 1/4″ deep shall be constructed in curbs at locations of the contraction joints in the gutter.

Tack Coat: The contact surfaces of all curb, gutters, manholes and the like shall be painted with a tack coat immediately before the adjoining asphaltic paving material is placed.

Figure 5-5

Typical Street Curbs and Gutters

For illustration only; not for construction.

At locations where curb has been or will be constructed and where there is no existing or proposed roadway pavement, the elevation of the top of the depressed curb should be 7 inches below the top of a theoretical 8-inch curb face, 6 inches below the top of a 7-inch curb face, 5 inches below a 6-inch curb face, etc. These streets are subject to erosion problems.

Generally, streets that are unimproved or partially improved have little or no drainage control, are discussed in the governing street manual. Where there is no existing or proposed curb on the street, construction of the driveway curb, which may extend into the traveled roadway, may constitute a traffic hazard. Therefore, adequate safety measures must be added to warn or divert traffic. These measures should include installation of such appurtenances as guardrail, warning rail, or guideposts; also grading, the construction of temporary berms, etc. When the street is further improved, these protective devices may be modified or removed.

If installing these protective devices is not deemed desirable, it may be advisable to use a temporary asphalt driveway. The asphalt driveway should be then constructed to the established grade. The pavement should be constructed flush with the surface, and curbs are omitted.

Where the proposed street and driveway is on private property, the grades should be adjusted to avoid any abrupt grade or alignment difference between such driveway within the public street and the existing driveway on private property.

Where this abrupt difference cannot be eliminated, a sufficient portion of the driveway located on the abutting property should be removed. A smooth transition can then be constructed to offset these differences.

The Department of Building and Safety standards for driveways on private property should be met, where possible. These standards are:

- The grade used for access to any garage or required parking area should not exceed 20 percent.
- The cross slope of the driveway should not exceed 10 percent.
- The maximum slope of a required parking space in any direction should not exceed 5 percent.
- The minimum dimensions for the above parking space should be 8 by 18 feet.
- The minimum vertical clearance for the vehicle entrance to the garage is 6 feet, 6 inches. This should be considered in changing the driveway grade or approach slab elevation. Where right of entry or easements is required for work done on driveways, the right of way sketch should show the maximum grade to which the driveways are to be reconstructed.

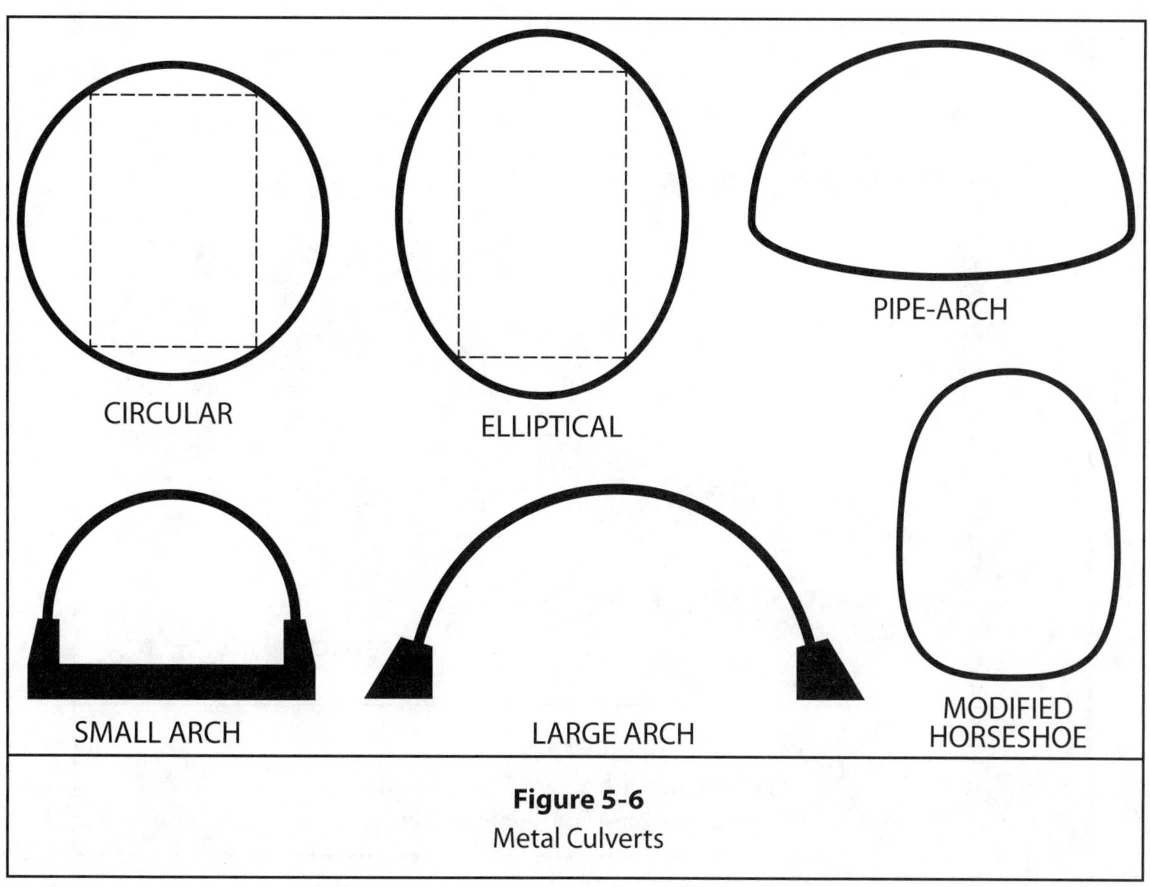

Figure 5-6
Metal Culverts

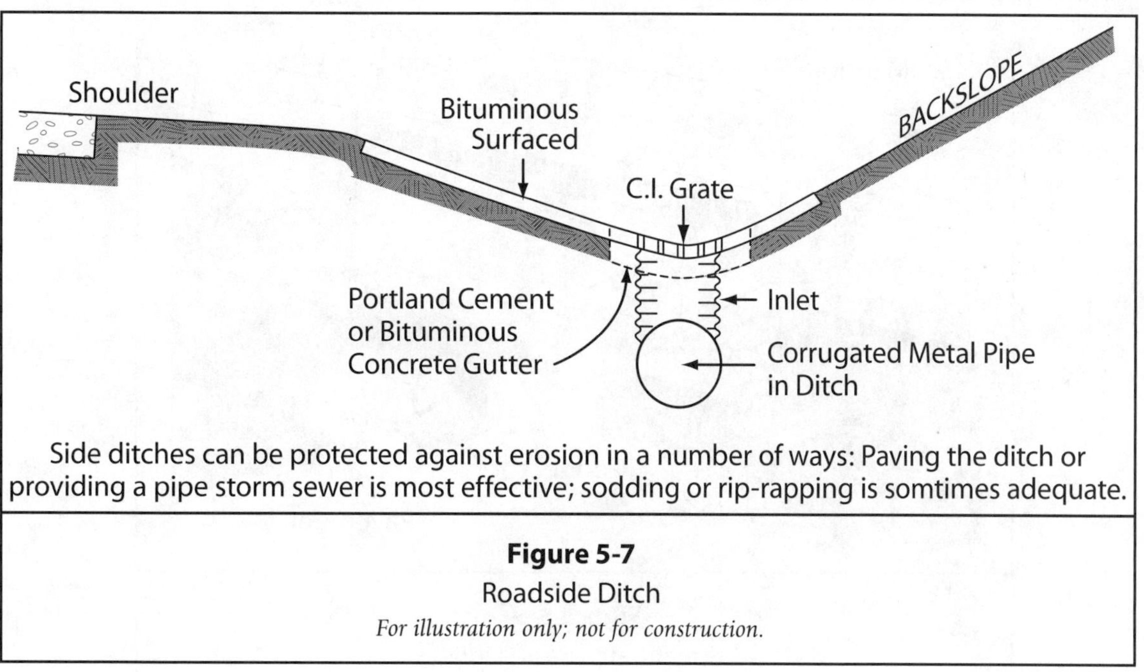

Figure 5-7
Roadside Ditch
For illustration only; not for construction.

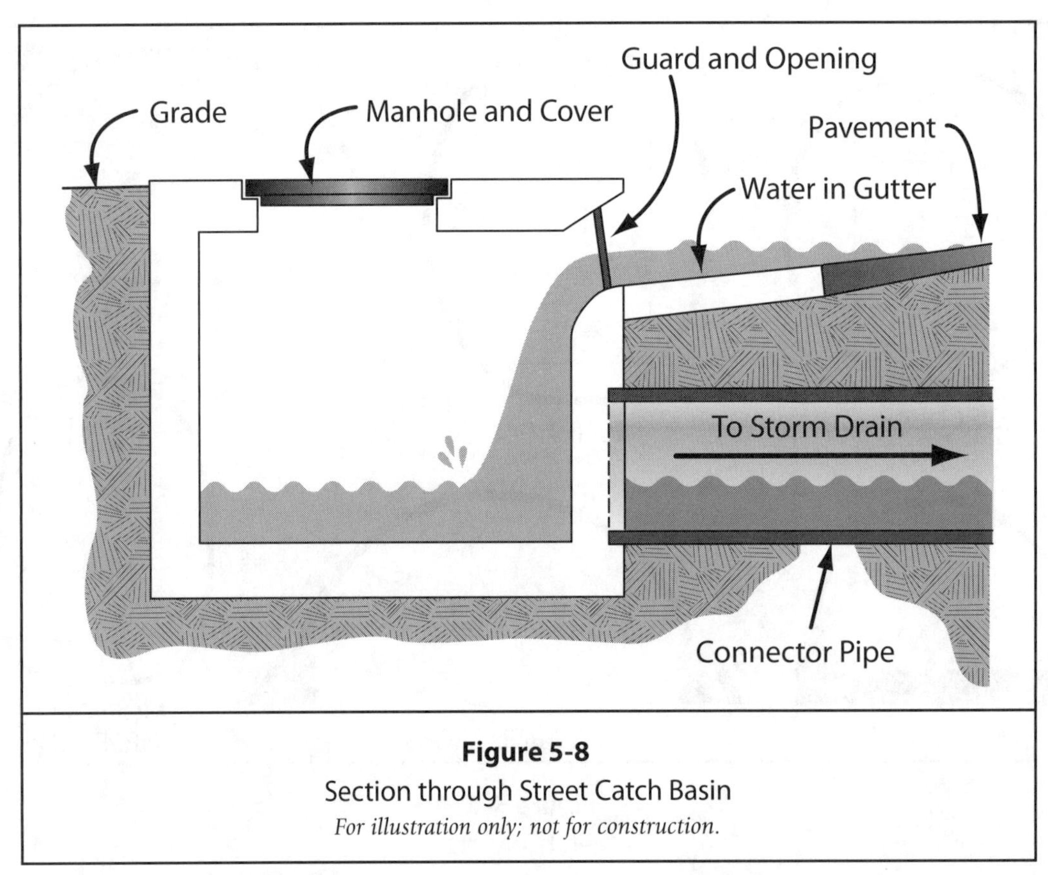

Figure 5-8
Section through Street Catch Basin
For illustration only; not for construction.

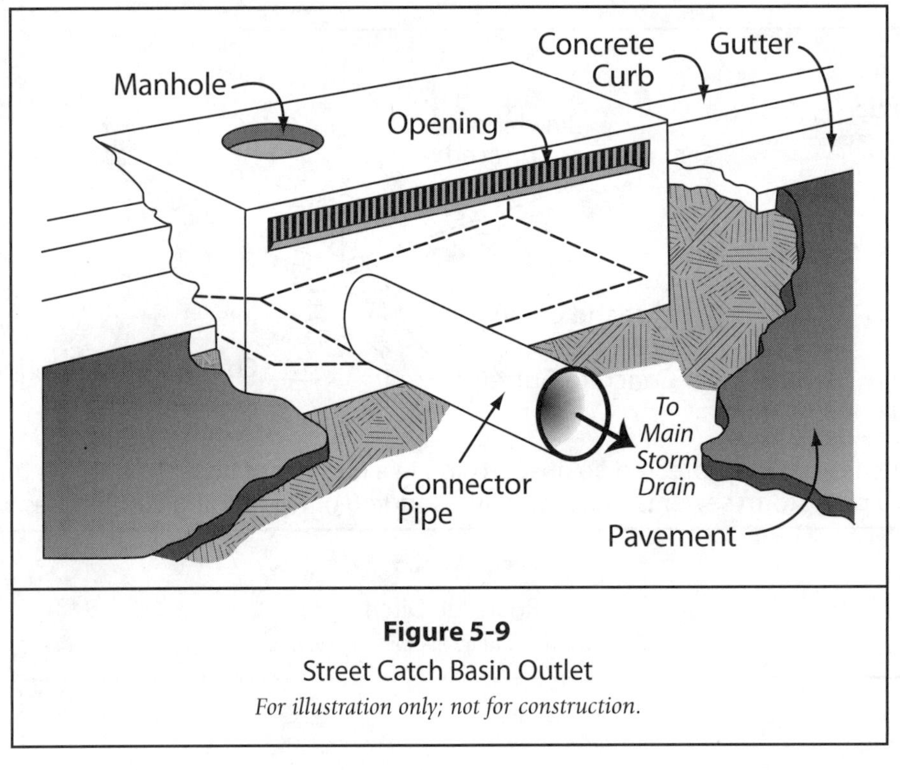

Figure 5-9
Street Catch Basin Outlet
For illustration only; not for construction.

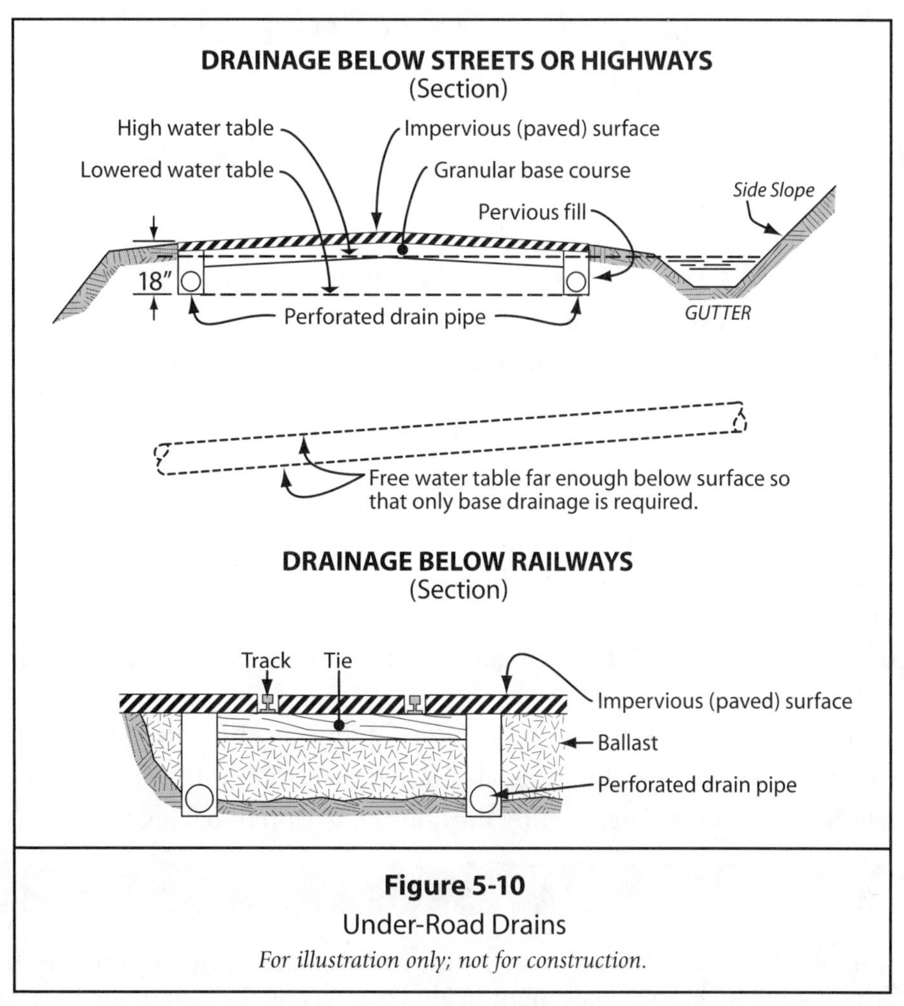

Figure 5-10
Under-Road Drains
For illustration only; not for construction.

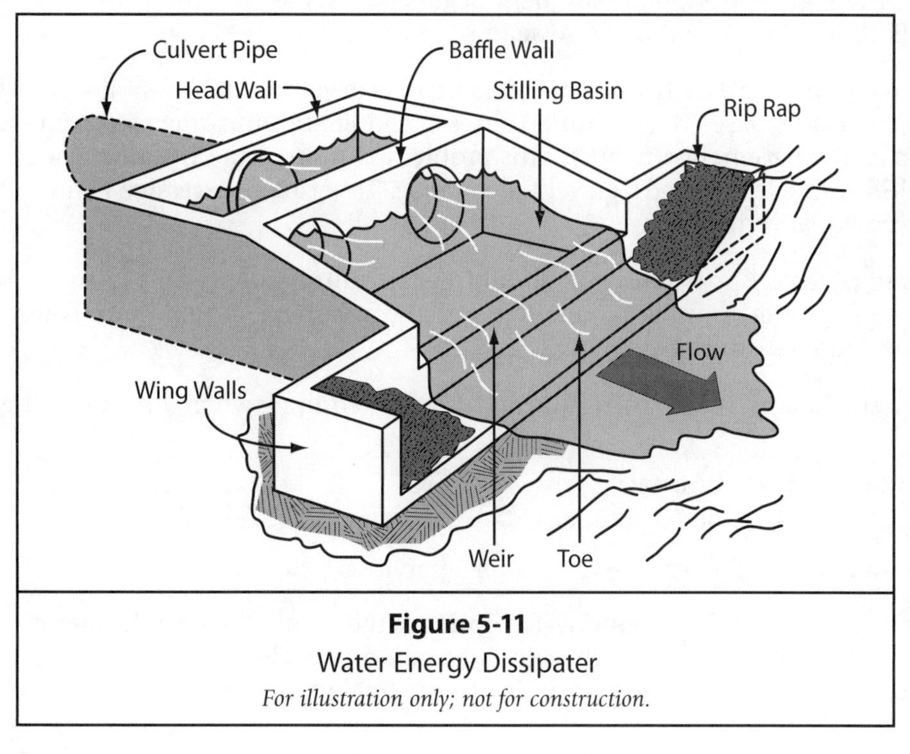

Figure 5-11
Water Energy Dissipater
For illustration only; not for construction.

SUGGESTED NOTES ON STREET CURB AND GUTTER PLANS

- Do all work in accordance with the last adopted specification except as otherwise noted hereon.
- Apply a steel trowel finish to the concrete surface for four inches adjacent to all gutter flow lines unless otherwise specified.
- Where a lap is necessary, lap bars 12 inches.
- Terminate longitudinal bars 2 inches from all expansion joints.
- Brush finish top and front face of all curbs.
- Install expansion joints of ½″ thick preformed expansion filler in all types of curbs at beginning of curb and end of curb (except of end of curb of alley returns) and at intervals of 60 ft. . When the spacing causes an expansion joint to be located in a driveway of 30 ft. or less in length.
- Install joints at the end of the driveway nearest to the normal location.
- Install similar expansion joints in all gutter and adjoining curb at the same location as constructed in the adjacent curb, and in the longitudinal alley gutter only where the same joins the concrete alley intersection.
- Install contraction joints in all types of gutter at intervals of 20 ft. at equal spacing between the expansion joints.
- Install ¼″ deep scoring lines in curb at locations of the construction joints in the gutter.
- Apply tack coat to the contact surface of all curb, gutters, manholes and the like.
- Paint the tack a coat immediately before the adjoining asphalt paving material is placed.

PUBLIC ROADS

Outside of the building lots, free water may start to percolate through joints in the road pavements and concrete gutters. Asphalt pavement begins to ravel as the sub grade becomes saturated and softens. Potholes appear and fill with water.

On roadways in hilly tracts, a hydrostatic pressure may develop sufficiently strong to lift the pavement completely off its base. This, in turn, causes cracking and, in some extreme cases, complete disintegration of the pavement structure. This problem is more acute in steep grades where water travels downhill through the road's gravel base course. This causes excessive hydrostatic pressures under the pavement at vertical curves.

Due to these potential problems, you should install sub drains under the roadway to prevent water from higher ground from accumulating under a pavement. If this water is not intercepted, subsurface water can flow into the fill areas...

In general, sub drains under roadways are usually required when the following conditions occur:

- A high water table in the area;
- Active springs or seeps beneath the pavement.
- Surface water enters the pavement through a pervious base course.

Natural sources of water are ground water, springs, and rainfall. Ground water can accumulate from sprinklers when the soil has a high water infiltration rate. This may be due to its natural properties or to improper compaction of the fill during grading operations.

Proper compaction of an earth fill involves the following steps:

- Scarifying the top 12 inches of surface to be filled;
- Watering the soil to near optimum moisture content;
- Compacting the soil to 90 to 95 percent of dry density;
- Placing fill material in thin horizontal lifts;
- Watering;
- Mixing the soil to optimum moist content; and
- Compacting the fill to 90 to 95 percent of dry density

You can do this with compactor, sheep foot roller, or wheel roller. Another possible cause of accumulation of water around a building is improper design or construction of the site drainage system. This may include:

- Lack of continuous flow lines around buildings to street or drainage devices;
- Missing concrete drainage structures in areas of concentrated flow;
- Absence of subterranean drainage devices in planting areas adjacent to building walls;
- Insufficient slope to drain surface water away from building walls; and
- Roof downspouts discharging into planter area adjacent to building.

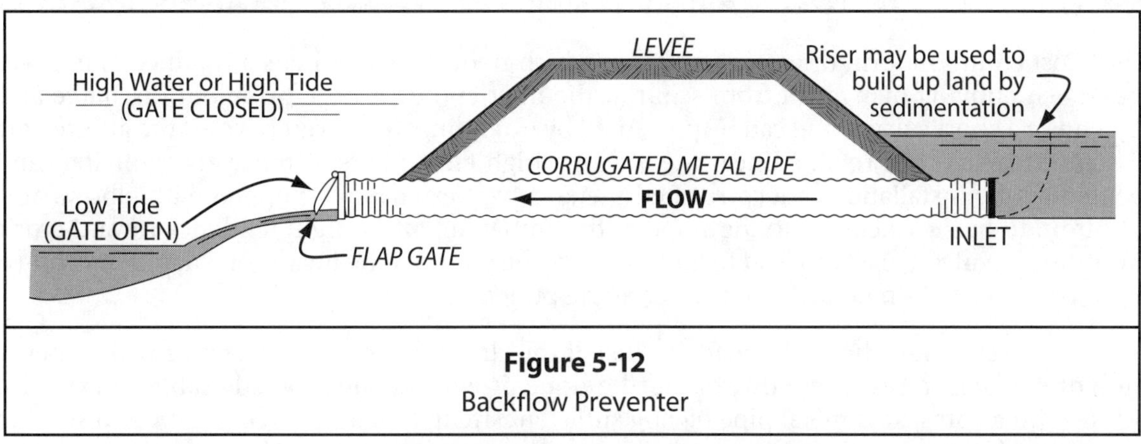

Figure 5-12
Backflow Preventer

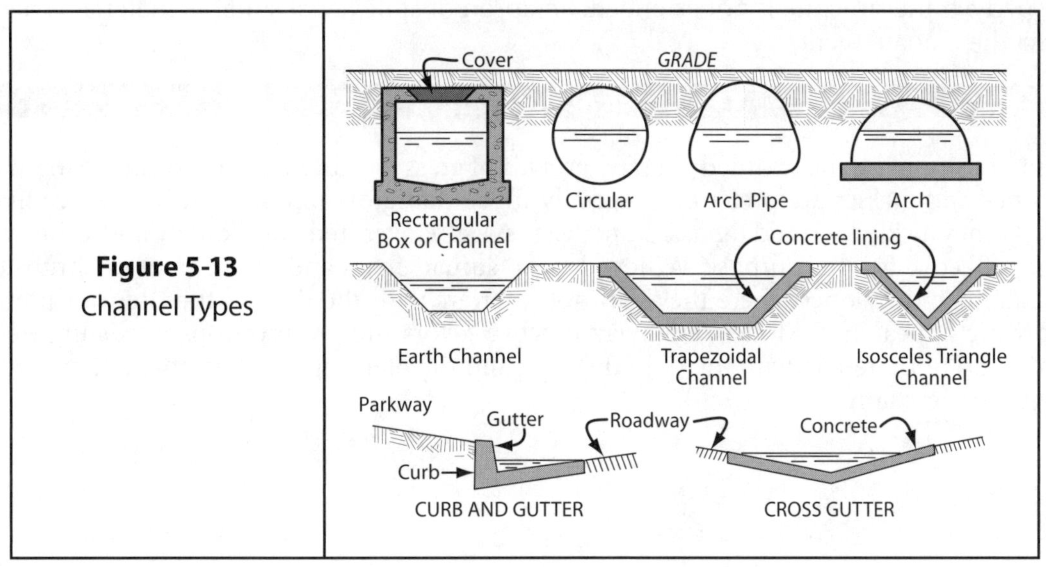

Figure 5-13
Channel Types

SOIL EROSION

One of the most destructive forces that you may be called on to combat is that of soil erosion. Erosion makes gullies on the shoulders and slopes of embankments; gouges out side ditches and endangers the road foundation as well as traffic; undermines fills and back-slopes, causing landslides; undermines bridge foundations; washes out drainage structures and other engineering works; fills and clogs ditches, culverts and other waterways with sediment.

Proper design at the time of construction and maintenance will eliminate much of the burdensome problem of maintenance.

EROSION OF SLOPES

The erosion of the side slopes of cuts and fills becomes a serious problem if any considerable amount of water flows over them. The effect of the water that falls directly on the slope may be minimized by sodding, strip-sodding or by terracing so as to reduce the velocity. It is often possible to divert the water from adjacent areas by means of a ditch along the top of the slope. Water thus diverted or collected in terraces can be concentrated and led down the slope in a pipe or flume. The water that falls on the roadway of a fill section can be confined by a berm along the shoulder and then disposed of in a similar manner.

TRANSVERSE GULLIES

The most common form of erosion and one which frequently threatens a roadbed is transverse gullies. Such gullying may result from scour at the inlet end or from a turbulent discharge at the outlet end of a short culvert, and can be prevented by extending the culvert beyond the affected area. If the culvert pierces the roadbed near the top of a high embankment, transverse gullying can be prevented by the installation of a corrugated metal culvert extension equipped with elbows at top and bottom to fit the extension to the slope of the embankment. A slight tendency toward erosion at the ends of a culvert, likely to lead to gullying, may be corrected by means of a flared or drop-type inlet, or by the use of rip rap or paved gutter at the outlet.

As a rule, every effort should be made to install the culvert so that the discharge will be beyond the toe of slope. In cases where this cannot be done, it will probably be advisable to extend the structure with a corrugated metal pipe having sufficient strength to act as a cantilever and still function properly. If the structure is not extended, an apron or spillway of suitable material is necessary to protect the embankment.

USE OF VEGETATION

Erosion of slopes can be retarded if a rank growth of grass or other vegetation can be maintained. Native vines and grasses are preferred especially those that grow rapidly and have spreading root systems. Honeysuckle, Bermuda grass, lespedeza, sweet clover, red top, kudzu and other ground covers are effective for this purpose. Willow, locust, sumac, roses and native trees or shrubs can be used to stabilize the slopes where they will not interfere with the flow in the drain or present an objectionable appearance. Modern highway practice favors the preservation of existing trees and shrubs that do not interfere with the sight distance and the planting of additional material in order to help prevent erosion.

DITCH EROSION

Aside from the use of vegetation, several systems for controlling ditch erosion are available, employed separately or in combination. Open drainage channels may be lined with rubble or be paved, the ditch may be replaced with a pipe or conduit or check dams can be installed to slow down the velocity of the water.

When open ditches are called upon to care for velocities of flow exceeding 2 ft. per second, erosion is a hazard that cannot be disregarded. The simplest method of controlling erosion is to provide a succession of ditch checks or weirs across the ditch, thereby slowing down the velocity of flow so that suspended solid matter will be deposited behind the checks and stop the erosion. The difference in elevations of the tops of successive checks should not exceed 3 ft. in order that the ditch will not be too deep.

Ditch checks are a possible hazard to traffic and they interfere with mowing and other maintenance operations. The checks may be constructed of many materials: sod, brush, logs, woven wire, wood planks; or, loose rock can be used for temporary checks. Corrugated metal, steel sheeting, bin-wall sections, rubble masonry, concrete or earth dams with drop inlet pipe culverts have all been used for permanent installations.

DITCH ENCLOSURE IN STORM SEWER

When the grade of the ditch is in excess of 5 per cent, the cost of placing ditch checks will be such that it will usually be cheaper and more satisfactory to provide a paved gutter or conduit for the entire length of the hill. Erosion of a side ditch may not only undermine the roadway but may deprive the backslopes of support and cause landslides. As an effective means of preventing pavement undercutting due to ditch erosion, a parallel drain of corrugated metal pipe can be laid in the ditch bottom and then covered over. Its continuous, flexible construction enables the pipe to resist any disjointing effect of a high velocity of flow through the pipe and to conform to a shifting or subsiding foundation soil.

PAVED DITCHES

Paved ditches are used when high velocities cannot be avoided. They may consist of Portland cement or bituminous concrete, or of rubble with dry or grout- filled joints. If the volume of water is sufficient, a storm sewer can be placed in the ditch line with frequent grating inlets.

On some highway and railroad fills, a paved or surfaced ditch is constructed along the shoulder. Inlets admit the water to pipe spillways or flumes leading down the slope. A specially designed metal inlet or "embankment protector" that has been used with much success. Inlets may also consist of wooden boxes for temporary use, or of concrete or corrugated metal pipe.

CHAPTER 6

Grading for Drainage

OBJECTIVE

The objective of this work is to provide grading which will do the following:

- Divert water away from buildings,
- Prevent standing water and soil saturation detrimental to structures and lot use to
- Provide for disposal of water from lot, except as necessary for controlled irrigation,
- Preserve desirable site features, and
- Provide grades for safe and convenient access to and around buildings and lot for their use and maintenance.

PROTECTIVE SLOPES AROUND BUILDINGS

- Slope downward from building foundation and water supply wells to lower areas or drainage swales.
- Horizontal length of slope should be a minimum 10 feet except where restricted by property line.
- Vertical fall of protective slopes should be a minimum 6 inches. However, vertical fall at high point at upper end of a swale may be reduced to 3 inches if a long slope towards the house or from a nearby high bank does not exist.
- Minimum gradient of slopes should be as follows:
  a. On concrete or other impervious surfaces, 1/16 inch per foot (1/2 percent).
  b. On pervious surfaces, 1/4 inch per foot (2 percent);
- Maximum gradient of slopes is two and one-half inches per foot (21 percent) for a minimum of 4 feet away from all building walls except where restricted by property line.

USABLE OUTDOOR AREAS

Usable outdoor areas should be of such location, size and shape as to provide for outdoor living; for outdoor service functions such as laundry drying, and for other necessary functions. Required usable outdoor area may overlap or be part of required protective slopes around buildings. Required area may be a single area, or several adjacent or separate sub-areas.

Width and length of required area or any sub-area should average at least 15 feet and be minimum 10 feet at any place. Minimum gradient of usable outdoor areas should be:

- For concrete or other impervious surfaces, 1/16 inch per foot (1/2 percent).

- For pervious surfaces in locality not subject to ground frost, 1/8 inch per foot (1 percent).
- For pervious surfaces in locality subject to ground frost, 1/4 inch per foot (2 percent).
- Maximum gradient should be five-eighths inch per foot (5 percent).

OTHER LOT AREAS

Minimum gradient should comply with above or with lesser gradients shown on exhibits accepted by the Building Official prior to building construction. To be acceptable, the lesser gradient must be adequate to drain the lot without detrimental effect upon buildings or upon essential lot use and improvements, including any individual sewage disposal system. Conditions, which would result in prolonged standing of water at any season, are not acceptable. Where surface water disposal is proposed by infiltration into the ground, technical exhibits such as soil gradation analyses and infiltration tests may be required by Building Official.

Maximum gradient, unless slope is held by satisfactory existing vegetation or rock outcropping:

- Vertical height not exceeding 30 inches; maximum 1½ feet horizontal to 1 foot vertical (1½ to 1).
- Vertical height exceeding 30 inches; maximum 2 feet horizontal to 1 foot vertical (2 to 1).

Top and bottom of banks at swales, terraces, etc., should be rounded for convenient maintenance.

STORM WATER DISPOSAL

All areas should be sloped to lower elevations off the lot or to drainage structures on the lot, except as necessary for controlled irrigation.

Unpaved drainage swales formed by intersecting slopes should have adequate depth and width. Longitudinal gradient for swale or gutter,

- For centerline within 15 feet of building or in a required outdoor area.
- For centerline in other lot areas comply with the above.

Permanence and maintenance: of off-site drainage-ways should be assured by locating in public right-of-ways, easement or by other means acceptable to Building Official. Where drain inlets or catch basins are installed, emergency surface drainage overflow should be provided to prevent possible flooding against buildings and wells in the event of failure of the underground drainage structures.

ROUGH GRADING

Natural site assets such as existing trees, shrubs, ground cover and suitable topsoil should be preserved and protected whenever practicable. The balance of the lot should be graded to the extent necessary to comply with the preceding paragraphs on grading design and the requirements below on finish grading. The subgrade should be roughly established by cut or fill, approximately parallel to proposed finish gradients and at elevations to allow for thickness of topsoil and other installations.

Filled areas other than under buildings:

- Remove all debris and other material detrimental to lot improvements.
- Remove; or cut down tree stumps 18 inches below finish grade.

- Original ground on sloping sites should be scarified and benched as necessary to provide adequate bond and prevent slippage of fill.
- Fill material should be reasonably free of debris or other detrimental material and should be placed and compacted to a density which will avoid damaging settlement to drainage structures, walks, driveways, lawns or other lot improvements.
- Fills having a depth in excess of 8 feet should be constructed in accordance with a specification determined to be acceptable to the Building Official prior to construction.

FINISH GRADING

For areas where installation of lawn or planting is required, the surface layers of soil should be workable, suitable to support plant growth and free of stones, debris, and other material detrimental to plants. For other areas, soil should be as above or soil which property owner can improve to comparable quality without its removal, or other difficult or expensive work.

Minimum compacted depth of required surface soil:

- Area to be seeded, 4 inches.
- Area to be sodded, 4 inches less thickness of sod.

Lot areas should be finish graded to the extent necessary to make the entire lot comply with preceding standards on grading design and the quality of surface soil.

The subgrade should be smoothed parallel to proposed finish grades and elevations and scarified where necessary to secure bond with surface soil. Spread surface soil uniformly to provide a smooth even surface.

Compact surface soil lightly as necessary to minimize settlement.

Seasonal Limits - Finish grading should be done when ground is frost-free and weather is favorable.

DRAINAGE STRUCTURES

The objective of this work is to provide: For the collection and disposal of surface and subsurface water to protect the dwelling and other improvements and usable lot areas.

GENERAL

- Paved gutters, drain lines and inlets or other necessary drainage structures should be installed where storm water disposal cannot be obtained without their use or where erosion cannot be prevented by finish grading and planting.
- Design, construction and installation of drainage structures should be in accordance with standard engineering practice and suitable for the use and maintenance contemplated.
- Gutters and drain lines should be connected to adequate outfalls. Permanence and maintenance of off-site drainage-ways should be assured by public rights-of-way, by easements, or by other acceptable means.
- Splashblocks should be installed at bottom of each downspout unless downspout is connected to drain line.

PAVED GUTTERS

- Construct of concrete, paving brick or other durable material.
- Gutter should have adequate depth, width and longitudinal gradient to carry water without overflow. Connect to adequate outfall.
- Install gutter in such a manner that it will provide permanent drainage with reasonable maintenance.

DRAIN INLETS

- Design, size and construction should be adequate to carry water imposed without overflow.

DRAIN LINES

- Construct of concrete, clay tile or other durable materials.
- Drain lines should be of adequate size, depth, and gradient to provide proper runoff.

Drain line for surface drainage should be sealed where necessary to prevent harmful infiltration of sand, muck and other materials. For subsurface drainage, use perforated, porous or open joint pipe with at least 9 inches gravel or crushed rock over pipe.

SPLASH BLOCKS

- Construct of concrete or other durable materials.

Size of splash blocks should be::

- Minimum width, 12 inches.
- Minimum length, 30 inches.

Splash blocks should be firmly imbedded to prevent displacement.

DRY WELLS

The minimum distance from other structures should be:

- Buildings, 10 feet
- Sewage-disposal fields or seepage pits, 20 feet
- Water-supply wells, 50 feet

Dry wells should be installed only in areas having open well-drained porous soils. Size and construction should be adequate to dispose of water.

DRIVEWAYS AND PARKING SPACES

The objective of this work is to provide the following:

- Safe and adequate on-lot parking space where a garage or carport is not provided, where other adequate, conveniently located parking space is not available,
- Safe, convenient, all-weather vehicular access to garage, carport or parking space, and
- Construction having reasonable durability and economy of maintenance.

PARKING SPACE DESIGN

- Minimum length, 20 feet.
- Width, 10 feet.
- Maximum gradient, 5/8 inch per foot (5 percent).

Crown or cross-slope:

- Minimum, 1/8 inch per foot (1 percent).
- Maximum, 5/8 inch per foot (5 percent).

DRIVEWAY DESIGN

Driveway should extend from street or alley pavement or curb line to garage, carport or parking space.

- Minimum width, 8 feet.
- Driveway entrance should have a flare or radii adequate for safe and convenient ingress and egress.
- Maximum gradient between vertical transitions, 1-3/4 inches per foot (14 percent).
- Vertical transition should prevent contact of car under-carriage or bumper with surface.
- Horizontal alignment should be safe and convenient to back car out, or an adequate turn around should be provided.

Crown or Cross-slope:

- Minimum, 1/8 inch per foot (1 percent).
- Maximum, 5/8 inch per foot (5 percent).

Ribbon driveways may be used in accordance with the following:

- Ribbons should be concrete, 5 feet O. C.
- Minimum width of ribbon, 2 feet.
- Apron at street should be flared and improved full width for at least 12 feet length, using concrete or other acceptable material.

CONSTRUCTION

Subgrade should be well drained, uniformly graded and compacted to prevent harmful differential settlement. Pavement should be concrete, bituminous pavement or other appropriate local road materials.

CONCRETE PAVEMENTS

- Minimum thickness, 4 inches nominal
- Provide an expansion joint at public walk or curb and at garage or carport slab.
- Provide contraction joints at approximate 10 feet intervals.
- Finished to provide smooth surface true to cross-section and grade.
- Concrete should be kept moist for a period of three days to insure proper curing.

BITUMINOUS PAVEMENT

- Base should be of crushed stone, gravel or other appropriate, durable road materials, properly compacted to 4 inch minimum compacted thickness.
- Wearing surface should be bituminous concrete, compacted to minimum thickness, 1½ inches.
- Other appropriate durable road materials properly placed so as to be durable under the use and maintenance contemplated.

OTHER LOT IMPROVEMENTS

Other improvements such as retaining walls and fences should be provided as needed to handle excessive grade differences, to screen unsightly views, to provide suitable access, personal safety and usable lot areas, and to protect the property. They should be structurally sound and durable.

TRACT GRADING

The following are recommendations for grading and drainage of a typical housing tract:

The objective is to provide grading that will:

- Divert water away from buildings.
- Prevent standing water and soil saturation detrimental to structures and lot.
- Provide for disposal of water from lot, except as necessary for controlled irrigation.
- Preserve desirable site features, and
- Provide grades for safe and convenient access to and around buildings and lot for their use and maintenance.

PROTECTIVE SLOPES AROUND BUILDINGS

Slope soil downward from the building foundation and water supply wells to lower areas or drainage swales. Horizontal length of slope should be a minimum 10 feet except where restricted by the property line. The vertical fall of protective slopes should be a minimum 6 inches. However, you may reduce the vertical fall at high point at upper end of a swale to 3 inches if a long slope towards the house or from a nearby high bank will not exist.

Minimum gradient

- Concrete or other impervious surfaces, 1/16 inch per foot (1/2 percent).
- Pervious surfaces, 1/4 inch per foot (2 percent).

Maximum gradient

- Two and one-half inches per foot (21 percent) for a minimum of 4 feet away from all building walls except where restricted by property line.

BASEMENTLESS SPACES (CRAWL SPACE)

Ground level shall be at least 18 inches below bottom of floor joists and 12 inches below bottom of girders. Where it is necessary to provide access for maintenance and repair of mechanical equipment located in the under-floor space, the ground level in the affected area shall be not less than 2 feet below floor joists

The interior ground level shall be above the outside finish grade unless:

- Adequate gravity drainage to a positive outfall is provided, or
- The permeability of the soil and the location of the water table is such that water will not collect in the basementless space.
- When drainage is necessary, the surface shall be properly sloped.
- Where soil and moisture conditions warrant, or when specifically required herein, surface shall be covered with a vapor barrier material.
- Remove all debris, sod, stumps and other organic materials in crawl space, and provide a reasonably smooth surface.

GRADING REQUIREMENTS

Grading regulations require that no person should conduct any grading operation for other than building site development in hillside areas.

An exception being that grading that is not connected with building site development may be conducted in hillside areas if the Grading Department finds that such work enhances the physical stabilization of property. Also, if it is not detrimental to public health, safety or welfare, and is in conformity with the approved area master plan. A tentative tract or division of land map may not be required for such exempt grading.

CONSTRUCTION REQUIREMENTS AND LIMITATIONS

In general, no structure should be constructed upon a slope steeper than two horizontal to one vertical. Exceptions are:

- Subject to approval by the Grading Department, construction may be placed upon slopes steeper than 2:1, provided reports from a soils engineer and engineering geologist recommend favorably toward construction. The reports should incorporate provisions for downhill creep in the design of footings where applicable.
- Where a minor amount of the structure is constructed on the slope, or where the construction consists of an unroofed deck or low retaining structure, the Grading Department may approve the construction without engineering and geological reports.
- Buildings should be located clear of the toe of all cut or fill slopes which exceed a slope of two horizontal to one vertical. The clearance should be one-half of the vertical height of the slope with a minimum clearance of three feet and a maximum clearance of 15 feet. Where the existing slope exceeds one horizontal to one vertical the clearance should be measured from a line where an imaginary 45-degree plane projected down from the top of the slope intersects the building pad. Where a retaining wall is constructed at the toe of the slope, the height of the slope should be measured from the top of the wall to the top of the slope.

Exceptions to this rule are:

- Second-story cantilever projections may extend a maximum of four feet into the required clearance. Where a retaining wall is constructed at the toe of the slope, the required clearance from the allowable projection should be determined from a slope height measured from the base of the wall to the top of the slope.

- Attached one-story carports, patios, porches and other similar construction which is open and unobstructed on the side facing the slope except for necessary structural members, may project into the required clearance but in no event should the clearance be reduced to less than three feet.
- For detached accessory buildings not used for living purposes, the required clearance may be reduced to three feet.

BUILDING CONSTRUCTION ON COMPACTED FILLED GROUND

In general, no building should be located within a horizontal distance of 40 feet from the top of an inclined fill slope exceeding 100 feet in vertical height measured from the toe to the top of the fill unless said building is designed to withstand the resulting total and differential settlements.

The foundation engineer should submit evidence of the anticipated settlement behavior.

MEASURE OF SETTLEMENT

Prior to permitting building on high fills, the Grading Department may require the determination of the settlement characteristics of such fills to establish that any movements have substantially ceased. In such cases, a system of benchmarks should be installed at critical points on the fill and accurate measurements of both horizontal and vertical movements should be taken for a period of time sufficient to define the settlement behavior. In no case should a period of time be less than one year; with at least four consecutive checks made at intervals of three months.

BERMS

Berms should be constructed at the top of all slopes

GUTTERS

Eave or ground gutters should be provided to receive all roof water and deliver it through a non-erosive devise to a street or watercourse.

SITE DRAINAGE

All building pads with cut or fill should slope a minimum of two per cent to an approved drainage device or to a public street. Where used, the drainage device should be an adequately designed system of catch basins and drain lines which conducts the water to a street. But, where the slope of the underlying natural ground does not exceed three percent and the compacted fill is less than three feet in depth, the slope of the pad may be reduced to one percent.

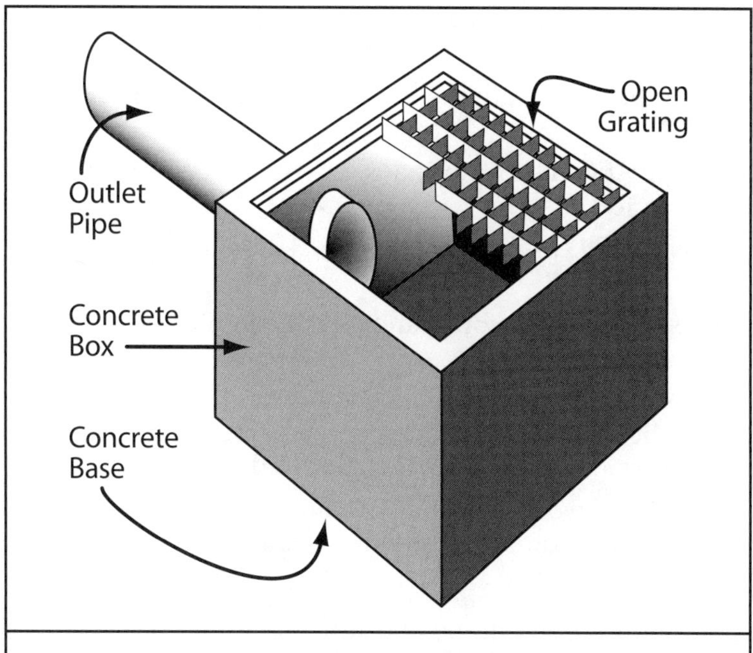

Figure 6-1

Yard Catch Basin

For illustration only; not for construction.

Yard catch basins are usually placed near the center of a flat area and the adjacent ground is sloped toward it. From here, drainage pipes carry the water away. It has a deep pit to catch sediment and prevent pipes from clogging. But the disadvantage is that if the catch basin becomes clogged, the area will be flooded. Therefore, care should be taken to keep the catch basin clear of leaves and debris. A secondary overflow is also recommended, if possible.

DRAINAGE AROUND BUILDING

On graded hillside sites, acceptable drainage devices should be installed to conduct storm water around building wherever the distance from the building to the top of any slope is less than five feet.

MAINTENANCE OF DRAINAGE

Drainage in conformance with the provisions of Grading Code should be maintained during and subsequent to construction.

GRADING VIOLATIONS

The following are a suggested list of things you should do in respect to grading to avoid violations and misdemeanor:

- Follow all orders issued by the Grading Department, all conditions imposed on grading permits, and all rules and regulations that are in effect at the time the grading permit is issued.
- Notify the Grading Department when the grading operation is ready for each of the following inspections:

- Initial Inspection: When you are ready to begin work, but before you start any grading or brushing.
- Toe Inspection: After the natural ground is exposed and prepared to receive fill, but before you place any fill.
- Excavation Inspection: After the excavation is started, but before the vertical depth of the excavation exceeds ten feet.
- Fill Inspection: After the fill emplacement is started, but before the vertical height of the lifts exceeds ten feet.
- Drainage Device Inspection: After forms and pipe are in place, but before you place any concrete.
- Rough Grading Inspection: When all rough grading has been completed. This inspection may be called for at the completion of the rough grading without the necessity of the Department having previously reviewed and approved the reports.
- Final Inspection: When all work, including installation of all drainage structures and other protective devices, has been completed and the as-graded plan and required reports have been submitted.

Issuance of Grading Certificate should be issued if upon final inspection of any excavation or fill it is found that the work authorized by the grading permit has been satisfactorily completed in accordance with the requirements of this Code. Note: On request, a separate certificate may be issued for each lot for which, building permits have been issued or applied for prior to the completion of the grading.

GRADING SPECIFICATIONS

The following are general recommendations for Grading Plans:

- Specifications have precedence over drawings.
- Keep the stamped set of plans on the job site at all times.
- Do not start any work whatsoever in or about a grading project without first notifying the grading inspector.
- Do not do grading work, import, or export, between the hours of 6:00 p.m. and 7:00 a.m. on any day and do not do work on Sunday at any time.
- Keep the construction area sufficiently dampened to control dust caused by grading and construction. Provide reasonable control of dust caused by wind.
- If a grading job extends over a period of time exceeding six months, the Governing Agency may require planting of those portions of the job where all other grading requirement have been met in order to prevent dust and erosion.
- Keep highway equipment in good operating condition and muffled as required by law.
- Control noise resulting from repair or heavy equipment after normal working hours by locating such activities as far as practicable from adjacent inhabited areas and so that such activities do not constitute a public nuisance or disturb the peace.
- Notify the Traffic Bureau of the local Police Department prior to the start of hauling.
- Detail a haul route on the plans satisfactory to the Grading Department.
- Complete the export or import of the fill material within the maximum time limit of from the start of hauling.

- Keep the fill material in each truckload low enough to prevent spillage and sufficiently wet down to prevent dust. .
- When hauling any earth, sand, gravel, rock, stone, debris, paper, or any other substance over any public street, alley or other public place, do not allow such materials to blow or spillover and upon the public street, alley or other public place or adjacent private property.
- When excavating, compacting, hauling or moving earth, sand, gravel, rock, stone, debris, or any other similar substance, cause, do not allow or permit any mud, earth, sand, gravel, rock, stone, debris or other substance to drop, be deposited, or fall from the body, tires, or wheels of any vehicle so used upon any public street or alley without immediately and permanently removing the same there from.
- Secure permission from the Department of Public Works if the trucks are loaded in the street.
- Obtain approval of the location of dumping excess soil from the grading inspector prior to starting excavation.
- Proceed brushing and scarifying of slopes only as far as periodically cleared by the grading inspector.
- Brush slopes prior to placing fill.
- Loose material should not exceed 3" in depth on a filled slope.
- Remove all debris and foreign material from the site.
- Remove and compact all loose fill.
- Locate buildings clear of the toe of all slopes, which exceed a slope of two horizontal to one vertical according to the following:
  a. A minimum of three feet provided the slope does not exceed six feet in vertical height.
  b. One-half the vertical height with a maximum of 15 feet for slopes exceeding six feet.

EXCEPTION: Attached or detached open carports or detached garages constructed a minimum of three feet clear of the toe of conforming cut or fill slopes. In the case of attached enclosed garages, all portions of the building shall comply with Subsection (b) above.

- Do not make tops of cut slopes nearer to a property line than one foot, plus one-fifth the height of the cut, but need not exceed a horizontal distance of ten feet.
- Driveway grades shall not exceed 20 per cent.

Driveways exceeding 15% should have centerline profile on plans that clearly show radius at top and bottom sufficient for vehicular access.

- Pave driveway grades exceeding 10 per cent.
- If at any stage of work on an excavation or fill the Department determines by inspecting that further work as authorized by an existing permit is likely to endanger any property or public way, the Department may require as a condition to allow the work to continue that plans for such work be amended to include adequate safety precautions.
- Maintain sanitary facilities on the site from beginning to completion of grading operations.
- Include as part of the plans all recommendations and conditions of the Soils Report by dated; the Geological Report by dated and the Board of Building and Safety Commissioners approval letter file.

- The engineering geologist, soils engineer and civil engineer shall comply with governmental requirements and provide the Department with a grading certification upon completion of the job.
- Tract civil engineers should submit an as-graded plan at a scale of 1" = 40′ with the required grading certifications to the Department upon job completion.
- Provide supervisory control during the grading operation to insure compliance with approved plans and with the Municipal Code. When necessary, and avail yourself of geological and foundation engineering services to implement your supervisory control.
- Lay subdrains under all fills placed in natural watercourses.
- Place subdrains along the watercourse flow line and along the flow line of any branches tributary thereto.
- Install additional subdrains to collect any active or potential springs or seeps that may be covered by the fill.
- Install subdrains after the watercourse has been excavated to firm material in preparation for receiving the fill.
- Show on the plans individual design and recommendations of the soils engineer and geologist to the satisfaction of the Department.

EXCAVATION - GUIDE

Height: Do not cut a slope to exceed a vertical height of 100 feet unless horizontal benches with a minimum width of 30 feet are installed at each 100 feet of vertical height.

Slope: Do not make an excavation with a cut face steeper in slope than two horizontal to one vertical.

EXCEPTION: The Grading Department, in case an appeal, may permit the excavation to be made with a cut face steeper in slope than two horizontal to one vertical if you show through investigation, subsurface exploration, analyses, and report by both a soils engineer and an engineering geologist, to the Department's satisfaction, that the underlying bedrock and the materials to be exposed on the slope have strength characteristics sufficient to produce a stable slope with a minimum factor of safety of not less than 1.5 for static loads.

Cut existing or proposed slopes to be no steeper than the bedding planes in any formation where the cut slope will lie on the dip side of the strike line.

Whenever grading at the top of any natural or man-made slope exposes soil or bedrock material that will allow the infiltration of water in a manner that would adversely affect the stability of the slope, the exposed area should be capped with a relatively, impervious compacted soil blanket seal having a minimum thickness of two feet. The soils engineer should certify in writing that the blanket seal is adequate to reduce water infiltration to permissible levels.

If the material of the slope is of such composition and character as to be unstable under the anticipated maximum moisture content, reduce the slope angle to a stable value. This requirement should be confirmed by the soils engineer's written certification following laboratory testing.

Intervening Benches: Intervening paved benches on cut slopes should have a minimum width of eight feet and shall be spaced at intervals of 25 feet measured vertically.

Tops of Cut Slopes. Do not make tops of cut slopes nearer to a property line than one foot, plus one-fifth the height of the cut, but need not exceed a horizontal distance of ten feet.

FILLS - GUIDE

The following recommendations are base on the requirements of the Grading department of the City of Los Angeles Department of Building and safety. Check with your local authority before beginning work.

Height: Do not a fill slope that exceeds a vertical height of 100 feet unless horizontal benches with a minimum width of 30 feet are installed at each 100 feet of vertical height.

Slope: Do not make a fill that creates an exposed surface steeper in slope that two horizontal to one vertical. The fill slopes abutting and above public property should be so placed that no portion of the fill lies above a plane through a public property line extending upward at a slope of two horizontal to one vertical.

EXCEPTION: The Grading Department, in case an appeal, may permit a fill to be made which creates an exposed surface steeper in slope than two horizontal to one vertical if you show through investigation, subsurface exploration, analyses, and report by both a soils engineer and an engineering geologist to the Department's satisfaction, that the fill to be used and the underlying bedrock or soil supporting the fill have strength characteristics sufficient to produce a stable slope with a minimum factor of safety not less than 1.5 for static loads. The soil engineer should verify by field-testing and observation and should certify attainment of the required strength characteristics in the fill material as specified in the approved report.

INTERVENING TERRACES

Intervening paved terraces on fill slopes should have a minimum width of eight feet and should be spaced at intervals of 25 feet measured vertically.

Compaction: All man-made fills should be compacted to a minimum of 90 per cent relative compaction as determined by ASTM method D 1557.

For slopes to be constructed with an exposed slope surface steeper than two horizontal to one vertical compaction at the exposed surface of such slope should be obtained either by over-filling and cutting back the slope surface until the compacted inner core is exposed, or by compacting the outer horizontal 10 feet of the slope to at least 92 per cent of relative compaction.

EXCEPTIONS:

Fills in non-hillside areas that do not exceed 12 inches in depth need not be compacted, but such fills should not change the existing drainage pattern.

The Department may approve uncompacted fills in self-contained areas where the fills are not to be used to support buildings or structures and no hazard will be created.

Fill material placed in areas within cemeteries used or to be used for interment sites should be compacted to a minimum of 80 percent, unless such fill is placed on a slope steeper than three horizontal to one vertical, or placed on slopes adjacent to public properties or private properties in separate ownership, or is to be used to support buildings or structures, in which cases it should be compacted to a minimum of 90 percent.

Fill slopes should be prepared for planting in one of the following ways:

a. The slope surface of fills may be prepared for planting by casting top soil over the slope surface. The top soil layer should not exceed three inches in depth.

b. The slope surface may be scarified to a depth not to exceed three inches. .

c. Loose material not to exceed three inches in depth may be left on the slope.

d. Toes of Fill Slopes. Toes of fill slopes should not be made nearer to a project boundary line than one-half the height of the fill, but need not exceed a horizontal distance of 20 feet.

e. Inspection and Control. Every man-made fill should be tested for relative compaction by a soil testing agency approved by the Department.

f. A certificate of compliance setting forth densities so determined should be filed with the Department before approval of any fill is given.

g. Old Fills. All man-made fills, whether compacted or not, which were placed prior to October 17, 1952, should be tested for relative compaction by an approved soil testing agency before any approval to build should be issued.

EXCEPTION:

Where proposed buildings are no higher than one story high and contain an area less than 1,000 square feet, the Department may waive this requirement, provided inspection shows the fill satisfactory for the proposed use.

h. Combined Cut and Fill Slopes. Where a combined cut and fill slope exceeds 25 feet in height, the required drainage bench should be placed at the top of the cut slope. The effect of surcharge of the fill upon the cut bedrock should be considered by the soils engineer and engineering geologist, and specific recommendations should be made relative to the setback between the cut and fill.

i. Fill Areas. Areas on which fill is to be placed should be investigated by the soils engineer or by the soils engineer and geologist to determine it is adequate to support the fill.

EXCEPTION: The Department may waive this investigation where it determines by inspection that the underlying material is adequate to support the proposed fill.

COMPACTION

All man-made fills should be compacted to a minimum of 90 per cent relative compaction as determined by ASTM method D 1557.

For slopes to be constructed with an exposed slope surface steeper than two horizontal to one vertical compaction at the exposed surface of such slope should be obtained either by over-filling and cutting back the slope surface until the compacted inner core is exposed, or by compacting the outer horizontal 10 feet of the slope to at least 92 per cent of relative compaction.

EXCEPTIONS:

Fills in non-hillside areas that do not exceed 12 inches in depth need not be compacted, but such fills should not change the existing drainage pattern.

The Department may approve uncompacted fills in self-contained areas where the fills are not to be used to support buildings or structures and no hazard will be created.

Fill material placed in areas within cemeteries used or to be used for interment sites should be compacted to a minimum of 80 percent, unless such fill is placed on a slope steeper than three horizontal to one vertical, or placed on slopes adjacent to public properties or private properties in separate ownership, or is to be used to support buildings or structures, in which cases it should be compacted to a minimum of 90 percent.

Fill slopes should be prepared for planting in one of the following ways:

a. The slope surface of fills may be prepared for planting by casting top soil over the slope surface. The top soil layer should not exceed three inches in depth.

b. The slope surface may be scarified to a depth not to exceed three inches.

c. Loose material not to exceed three inches in depth may be left on the slope.

d. Toes of Fill Slopes. Toes of fill slopes should not be made nearer to a project boundary line than one-half the height of the fill, but need not exceed a horizontal distance of 20 feet.

e. Inspection and Control. Every man-made fill should be tested for relative compaction by a soil testing agency approved by the Department.

f. A certificate of compliance setting forth densities so determined should be filed with the Department before approval of any fill is given.

g. Old Fills. All man-made fills, whether compacted or not, which were placed prior to October 17, 1952, should be tested for relative compaction by an approved soil testing agency before any approval to build should be issued.

EXCEPTION:

Where proposed buildings are no higher than one story high and contain an area less than 1,000 square feet, the Department may waive this requirement, provided inspection shows the fill satisfactory for the proposed use.

h. Combined Cut and Fill Slopes. Where a combined cut and fill slope exceeds 25 feet in height, the required drainage bench should be placed at the top of the cut slope. The effect of surcharge of the fill upon the cut bedrock should be considered by the soils engineer and engineering geologist, and specific recommendations should be made relative to the setback between the cut and fill.

i. Fill Areas. Areas on which fill is to be placed should be investigated by the soils engineer or by the soils engineer and geologist to determine it is adequate to support the fill.

EXCEPTION: The Department may waive this investigation where it determines by inspection that the underlying material is adequate to support the proposed fill.

SOIL CLASSIFICATION SYSTEM

The Unified Soil Classification System has been used by the Corps of Engineers and Bureau of Reclamation. It also identifies soils according to their textural and plasticity characteristics, dividing soils into course-grained soils, fine-grained soils, and organic soils.

The U.S. Department of Agriculture has combined several soil classification systems into a simple triad that places sand, silt, and clay on each side, and describes various soils as composed of percentages of the three.

Knowledge of soil types and their characteristics interpreted for land-use decisions forms an important input to grading. In many metropolitan areas, the U.S. Department of Agriculture has reinterpreted agricultural soils information for use in making urban land-use decisions.

Their soil scientists analyzed what land-related activities occurred during urbanization, such as long-term problems (landslides, flooding, erosion, water pollution, etc.), what had to be constructed (foundations, septic drain fields, utility systems, roads, drainage, etc.) and broad land use categories (industrial, residential, recreation, woodland, ecological, etc.). They then looked at soil characteristics that might affect these activities, such as drainage, depth to bedrock, soil type, etc., and mapped soils according to the following limitations.

PROBLEM SOILS

According to the U.S. Department of Agriculture, the following lists problem soils:

- Gravelly or stony.
- Less than optimum bearing capacity.
- Drainage class unfavorable; depth to seasonal water table too shallow; drainage may be needed; flooding or ponding hazard.
- Slope or topography unfavorable; imposes use or construction hazard; slippage potential.
- Depth to bedrock or impervious material is limiting.
- Frost action potential is moderate or high.
- Shrink-swell potential is moderate or high.
- Permeability (percolation rate) is restricted.
- Accessibility by equipment is restricted.
- Excavation difficult owing to consolidated materials, or clay or silty clay soil texture.
- Piping hazard.
- Thickness of suitable material is limited.
- Erosion hazard is moderate, severe, or very severe.
- Shear strength is moderate or low; slope stability is fair or poor.
- Compaction characteristics are fair or poor.
- Permeability rate is too rapid.
- Low water-holding capacity.
- Organic soil.
- Possible pollution of groundwater even though site is at least well drained.
- Soil textures permit ready sloughing of sidewalls.
- Compressibility is medium, high, or very high.
- Moist consistency is loose, firm, very firm, or extremely firm.
- Excessive fines or limited supply.

ESTIMATED SOIL PROPERTIES

- *Depth to seasonal high water table* refers to the highest level at which the groundwater stands for a significant period of time.
- *Depth from surface* indicates the thickness of significant layers of a typical profile. The thickness of the horizons differs somewhat among mapping units of the small soil series.
- *Percentage passing sieve* refers to the percentage of dry soil material that will pass sieves of the indicated sizes.

- *Permeability* refers to the rate at which water moves downward through undisturbed soil. It depends largely on the texture, structure, porosity, and density of the soil.
- *Available water capacity* represents the maximum amount of water that plants can obtain from the soil.
- *Reaction* refers to the acidity or alkalinity of the soil, expressed in terms of pH. A pH of 7.0 is neutral. Values less than 7.0 indicate acidity, and values more than 7.0 indicate alkalinity.
- *Shrink-swell potential* is an indication of the volume change that can be expected with a change in moisture content. It depends largely on the amount and type of clay in the soil. In general, soils classified as CH or A- 7 have a high shrink-swell potential; soils classified as SP, GW, or A-I have a low shrink-swell potential.
- *Corrosivity* refers to the deterioration of concrete or untreated steel pipelines as a result of exposure to oxygen and moisture and to chemical and electrolytic reactions.

On a large grading project or one involving difficult terrain, soil, types should be mapped and understood to minimize problems during the grading. The principal soil characteristics that affect grading are as follows:

- Angle of repose
- Permeability
- Erosion hazard
- Slippage potential

ANGLE OF REPOSE

Each soil has an equilibrium angle at which it win stand naturally. If this angle is exceeded, the soil will slump or slip off. This angle varies with the soil grain size and shape, the character of the material, and the proportion of water present. Any cut or fill cannot exceed this natural angle of repose if stability is expected. Generally, sandy and gravelly soils have a fairly steep angle of repose, whereas clay, silt, or loams do not. Constantly wet or completely dry soils have a shallower angle of repose than wen-drained but moist soils.

Typical angles of repose are as follows:

- Firm earth in place, 1.1
- Loose earth or humus soil, 1.5: 1
- Firm clay, 1.5:1
- Wet clay, 3:1
- Dry sand, 1.75: 1
- Wet sand, 1.25:1

These ratios vary with rainfall, moisture content, subsurface geology, plant cover, etc.

PERMEABILITY

This is particularly important when designing the drainage plan. Runoff can be disposed of in a closed system, or it can infiltrate into permeable soils as if it were a sponge. Disposing of runoff through permeable soils is good practice as it reduces cost and raises the water table. Generally, sandy and gravelly soils are more permeable than clay and loamy soils. However, many sandy soils are underlain with impermeable clay layers that reduce infiltration and may flood following severe rain. Additionally, Paving permeable soils prevents water from percolating into the soil. This is unecological in principle, and should be done only where the situation is extreme. Build on impermeable soil and drain to permeable soils.

Planting exposed banks with thick, matty, and deep-rooted plants helps stabilize the soil. Planting can reduce minor slippage, but is of little effect against a major slide.

Terrace with cross ditches to prevent excessive water from crossing the bank. Drain ditches should be placed above the bank, with terraces at intermediate intervals to reduce the quantity of water.

Subsurface drains may be drilled into the bank to relieve groundwater. This method is expensive, and it is difficult to predict the results.

Soil permeability is the important factor in locating any septic system. Obviously, gravel or sandy soils will ensure proper operation of a septic system; however, topography and underlying impermeable soils are almost more important. An impermeable subsurface or steep topography may cause effluent to flow laterally and surface at a lower elevation raising health hazards.

In pervious soils, grading below the water table usually requires dewatering or drainage of the site during construction. Drainage is accomplished by constructing a sump, well point, or deep well at the lowest level and pumping out the water.

EROSION HAZARD

All grading causes erosion. All soils erode, although some more than others. There are two principal concerns: first, topsoil, which is the lifeblood of our plant community, should not be allowed to erode. The process of building this top layer of soil is time consuming and cannot be easily duplicated by man. Generally, the top layer of organic soil should be stripped carefully and stockpiled for redistribution and use following grading. The second concern is to be careful with soils subject to erosion. Erosion occurs following the steepest terrain. It begins as rivulets and eventually becomes gullies, with sedimentation somewhere below. Erosion is part of consuming and cannot be easily duplicated by man. Generally, the top layer of organic soil should be stripped carefully and stockpiled for redistribution and use following grading. The second concern is to be careful with soils subject to erosion. Erosion occurs following the steepest terrain. It begins as rivulets and eventually becomes gullies, with sedimentation somewhere below. Erosion is part of the natural river-forming process and is quite normal as long as the time table is slow and long; however, man tends to hasten the process to the point of causing problems.

GRADING FOR EROSION CONTROL

On large projects where erosion is likely, it is becoming acceptable practice to prepare the grading plan, anticipate where erosion will occur, and devise a pre-grading plan to stop erosion. Pre-grading utilizes small check dams or holding basins below the area of major grading. The pre-grading work must be carefully carried out using small equipment (possibly a backhoe or by hand) and should be immediately seeded or planted so it will stabilize. Pre-grading adds to the overall grading costs, and benefits are not immediately realized, so there is some resistance to its use. However, grading permits are becoming more difficult to obtain, and the addition of an erosion-control plan may help, perhaps proving an immediate benefit from a client's point of view.

Grading during the dry season followed immediately by planting may reduce erosion damage. Scheduling sensitive projects should ensure that difficult grading operations are performed during the dry season. Small, robber-tired equipment or hand grading is economical in sensitive grading situations to minimize disturbance.

SLIPPAGE POTENTIAL

Slippage occurs when shear stress between layers of soil exceeds the shear strength of those soils. This can occur by increasing the stress or by decreasing the shear strength. The increased stress must be over a large area, and the slope must be relatively high and quite steep before the slip is likely to occur.

Increased soil stress is caused by removing the bottom (toe) of the bank or by increasing weight at the top of the bank. An excavation at the bottom of the bank for a building or a swimming pool is likely to cause increased stress. Steepening the slope beyond its normal angle of repose also causes stress. The addition of a building at the top of a slope with more drainage water increases the weight or pressure on the soil and is likely to cause slippage.

Reduction of shear strength is caused by a change in the soil property. Clay can be softened by the addition of water. As the surface dries, it can crack, allowing water to penetrate into the soil.

Finally, water entering the soil can create friction between layers of noncohesive soil, which may result in a buoyancy that causes the upper layers of soil to float free of the soils below. Noncohesive soils such as sands and gravels are held together purely by friction between particles, which is easily dislodged.

Prevention of slippage requires study of the geology and soil types, with careful design to solve any uncovered problems. Possible solutions include the following:

- Avoiding all impossible sites.
- Reducing the proposed gradient by extending the bank. This uses more land and exposes more soil to erosion, but sometimes can reduce the slippage potential.
- Under draining the bank to allow groundwater to flow out, rather than be trapped within the bank. Under draining is expensive and not always successful, but may help.
- Building on pile or pole foundation to avoid disturbing the ground.
- Removing an amount of soil equivalent to the weight of the proposed structure. This balances the pre and post condition, by adding no weight.
- Installing concrete retaining walls with deep footings can produce satisfactory results, but is expensive.

CHAPTER 7

Slope Drainage

RESIDENTIAL PROPERTY

Residential developments throughout the country have expanded into outlying hills and mountains. Some builders have cut building pads into the top soil to bedrock and others have cut into the bedrock and covered the pad with a relatively thin blanket fill.

A form of building defect appears when the homeowners begin to notice, a year, or two after moving in, that surface water rises around their homes and moisture penetrates the concrete floor slab of their houses of basements.

This becomes very noticeable when the flooring shows signs of deterioration or discoloration. Hardwood flooring begins to warp and separate from the floor slab. The edges of vinyl-encased hardwood discolor from mildew and resilient flooring, such as vinyl and linoleum, separate from the floor as moisture dissolves the adhesive binding. Where builders or homeowners place plastic floor mats over carpets, water accumulates under the mat. These and other symptoms indicate that free water or water vapor may have penetrated the floor slab from the soil below.

Occasionally, individual property owners, in a properly designed tract, change the contours of the land around their houses. This may void the original design. To prevent this from occurring, a governing grading ordinance may state:

"The owner of any property on which grading has been performed pursuant to a permit issued under the provisions of the code or any other person or agent in control of such property shall maintain in good condition and repair all drainage structures and other protective devices shown on the grading plans filed with the application for grading permit and approved as a condition precedent to the issuance of such permit. "

Another grading rule alerts the developer as to the hazards of ground water. This warning may appear as follows:

"In areas where ground water has risen to within 10 feet of the ground surface, difficulties can arise in subsurface structures, such as basements and vaults. This condition can be relieved by the construction of a surface or subsurface drainage, system adequate to keep the ground water down to a harmless level or by building improvements that are not harmed by such conditions."

Water enters a building mainly due to capillary action through concrete floor slab. This phenomenon is dependent upon several factors:

- Level of the top of slab in respect to, the finish grade;
- Type of sub grade;

- Depth to ground water; and
- Ductwork and piping under slab.

When a slab is below grade, the critical factors are:

- Type and thickness of base course;
- Type, location, and thickness of vapor barrier; and
- The integrity and density of the concrete slab.

Where you expose concrete slabs to atmospheric temperature changes, the soil beneath the slab may tend to become wet. They call this process *hydrogenesis.* Cool, moisture-laden air is drawn into the voids in base courses or soil during the late night or early morning. During sunlit hours, the pavement warms, as does the air trapped in the voids. The air expands and flows from the voids, but the moisture concentrates in the soil. When this process is repeated day after day, there is a net gain of moisture in the soil.

When water, under hydrostatic head, is in contact with one surface of the concrete, it penetrates the concrete through channels or openings. They call this type of flow *permeability.*

Another way that water moves is by *capillarity.* Even where there is no hydrostatic head, the movement of water may occur from a constant supply of moisture on one surface of the concrete and evaporation from the other surface. Permeability is controlled by waterproofing to stop the flow from capillarity. Moisture also tends to flow away from heat.

The homeowners notice that the grounds around the buildings are saturated even when the irrigation system is not operating. The fill slopes may begin to fail as the wet soil causes minor landslides and slump-offs. Cracks may begin to appear in buildings as their foundations settle. This can cause the soil to lose its bearing strength.

Some of the possible man-made causes for these conditions are:

- High water table;
- Lack of subterranean drains.
- Shallow blanket fill over pads cut into bedrock.
- Leaks from water pipes.
- Curbed planters constructed too close to the building foundation, and
- Excessive irrigation of landscaping.

Automatic sprinklers can discharge between 0.24 and of water per square foot per hour. Over an extended period, this rate of precipitation exceeds that of a heavy rainstorm.

Sprinkler system left on inadvertently over a weekend can cause between 12 and 28 inches of water per square foot over landscaped area.

PUBLIC ROADS

Outside of the building lots, free water may start to percolate through joints in the road pavements and concrete gutters. Asphalt pavement begins to ravel as the sub grade becomes saturated and softens. Potholes appear and fill with water.

On roadways in hilly tracts, a hydrostatic pressure may develop sufficiently strong to lift the pavement completely off its base. This, in turn, causes cracking and, in some extreme cases, complete disintegration of the pavement structure. This problem is more acute in steep grades where water travels downhill through the road's gravel base course. This causes excessive hydrostatic pressures under the pavement at vertical curves.

Due to these potential problems, you should install sub drains under the roadway to prevent water from higher ground from accumulating under a pavement. If this water is not intercepted, subsurface water can flow into the fill areas...

In general, sub drains under roadways are usually required when the following conditions occur:

- A high water table in the area;
- Active springs or seeps beneath the pavement.
- Surface water enters the pavement through a pervious base course.

Natural sources of water are ground water, springs, and rainfall. Ground water can accumulate from sprinklers when the soil has a high water infiltration rate. This may be due to its natural properties or to improper compaction of the fill during grading operations.

Proper compaction of an earth fill involves the following steps:

- Scarifying the top 12 inches of surface to be filled;
- Watering the soil to near optimum moisture content;
- Compacting the soil to 90 to 95 percent of dry density;
- Placing fill material in thin horizontal lifts;
- Watering;
- Mixing the soil optimum moist content; and
- Compacting the fill to 90 to 95 percent of dry density

You can do this with compactor, sheep foot roller, or wheel roller. Another possible cause of accumulation of water around a building is improper design or construction of the site drainage system. This may include:

- Lack of continuous flow lines around buildings to street or drainage devices;
- Missing concrete drainage structures in areas of concentrated flow;
- Absence of subterranean drainage devices in planting areas adjacent to building walls;
- Insufficient slope to drain surface water away from building walls; and
- Roof downspouts discharging into planter area adjacent to building.

Proper drainage control requires that all of the devices carry rainwater away from structures and foundations and prevent uncontrolled flow over slopes.

SLOPE FAILURES

Because of the high cost of land, builders are constructing buildings and subdivisions on hillsides or in mountainous terrains. Since costs of grading, street construction and other processes are much higher on hilly sites than on level sites, slopes are often kept at their maximum steepness to provide as many salable lots as possible. Consequently, much litigation is related to slope failures. These may be evidenced by landslides, surface failures, mud slides, or flood damage.

A problem commonly encountered with grading is getting the work done before the winter rains come. The work can include fill embankments, cut slopes, fills, drainage devices, ground cover, and rodent control. Many communities prohibit grading work during the winter months when rain can destroy uncompleted work.

The amount of governmental control and engineering required in hillside development work is far greater than that required for buildings on level sites. More agencies and professional engineers are involved, and heavier construction equipment is necessary. There are more unknowns and risks when entire hills have to be cut down and canyons filled. Some grading work may directly conflict with the natural forces of geology, erosion, drainage, and soil stability in order to provide the buildable lots and access streets.

In one case, a homeowner sued the builder, alleging that the builder had implicitly warranted his house suitable for its ordinary purpose, but that it was built in an unskillful and unworkmanlike manner. The house was built on sunken concrete piers under the foundation because the lot was on a hill with a 30-degree slope. The piers were to prevent the owner's two-story structure, with an elevated deck and concrete patio, from settling. However, within a few months after completion, the posts supporting the elevated deck were separated from the piers. This caused the deck and concrete patio to separate from the house, the roof to lift, and the floor to begin to drop out of the house. Remedial work was done by a new contractor who subsequently testified that the original work was improper in that:

- The piers were not placed on solid rock, but upon loose dirt and stone which was in turn upon rock; and
- The builder was wrong in not sinking piers to solid rock, which would have prevented any settlement. The verdict was for the homeowner. However, on appeal the court reversed, citing the fact that the owner's complaint was for breach of implied warranty and not for negligence. Therefore, any testimony that the house was built in an unskillful and unworkmanlike manner was irrelevant and possibly prejudicial. Caveat: a plaintiff should not limit causes of action against culpable parties.

Fig. 7-1
La Conchita Landslide

Fig. 7-2
Landslide

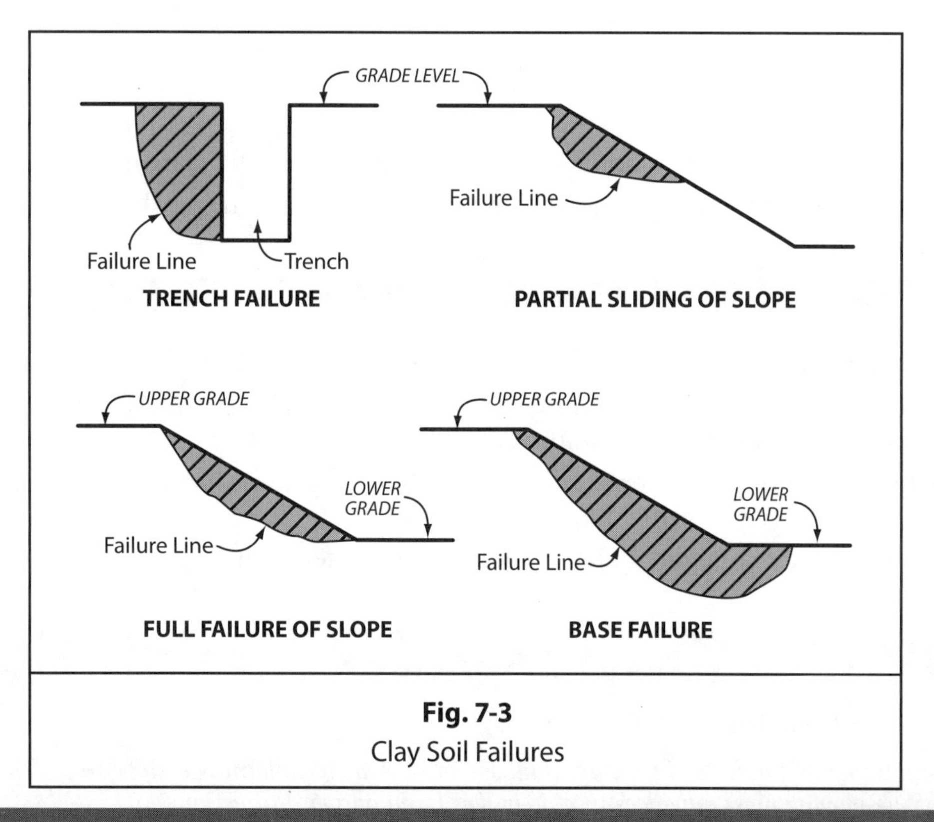

Fig. 7-3
Clay Soil Failures

STABILITY OF CUT AND EMBANKMENTS

Mass earth movements are divided into four types:

- Slow flowage;
- Rapid flowing;
- Sliding; and
- Subsidence.

Slow flowage is a slow, down slope movement of surface soil or rock debris only a few feet deep. Slow flowage of water-saturated masses from higher grounds to lower grounds is sometimes called solifluction (from Latin solum, "soil;" and fluere, "to flow").

Rapid flowing is due to the over saturation of earth or mud masses. This may or may not be accompanied by a sliding motion. These are also known as earth flows and mudflows. A landslide is a perceptible downward sliding or falling movement of relatively dry masses of earth, rock, or a mixture of both. Subsidence is a consolidation or settlement of soil due to lack of underground support.

CHECKLIST FOR INVESTIGATION OF SLOPE FAILURE

- Temporary erosion plans if work was to be performed during rainy season
- Original topographic map of site before grading started
- Final grading plan of site
- Description of fill material used if imported from another site
- Gradient of slope of cut embankments and fill embankments
- Engineering calculations substantiating stability of slopes under conditions of saturation

- Soil test data
- Compaction test reports of all compacted fills
- Report by soils engineer certifying the proper stripping of vegetation, removal of unsuitable soils, and installation of sub drains (if any) before fill started
- Name of soils engineer who provided continuous inspection during all fill and compaction operations

CAST OF CHARACTERS IN A LAND FAILURE CASE

The following describes the parties intricately connected with earthwork:

- The owner of the property who commissions the services of an architect if the project is an individual building or group of buildings.
- The architect, in order to become familiar with the topographic features and property lines of the lot, retains a civil engineer.
- Where only property lines are to be staked and monuments set, a licensed land surveyor may be contacted.

LANDSLIDES

Landslides have been defined as follows:

"A rapid displacement of a mass of rock, residual soil or settlement adjoining a slope in which the center of gravity of the moving mass advances in a downward and outward direction."

Landslides are a major phenomenon. They involve not only hillsides and mountainsides, but cuts and fills for roadways, channels and other engineering works.

The change from equilibrium and stability to the movement of vast volumes of earth or rock may require only a slight change in the balance of forces. These include the addition of water to masses of earth or to lubricating planes. Other disturbing factors could be undermining, cave-ins due to underground settlement and explosions or earthquakes.

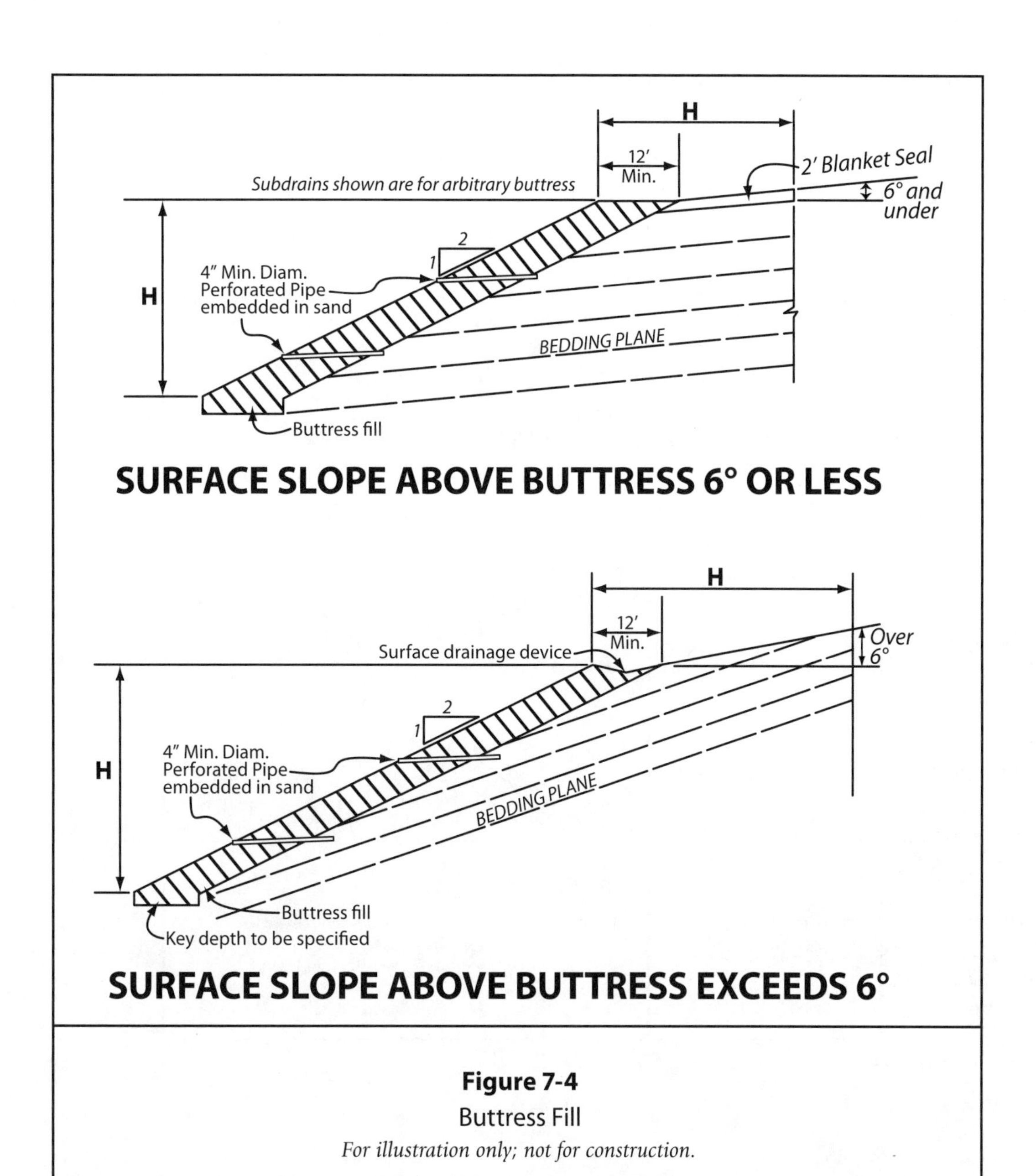

Figure 7-4

Buttress Fill

For illustration only; not for construction.

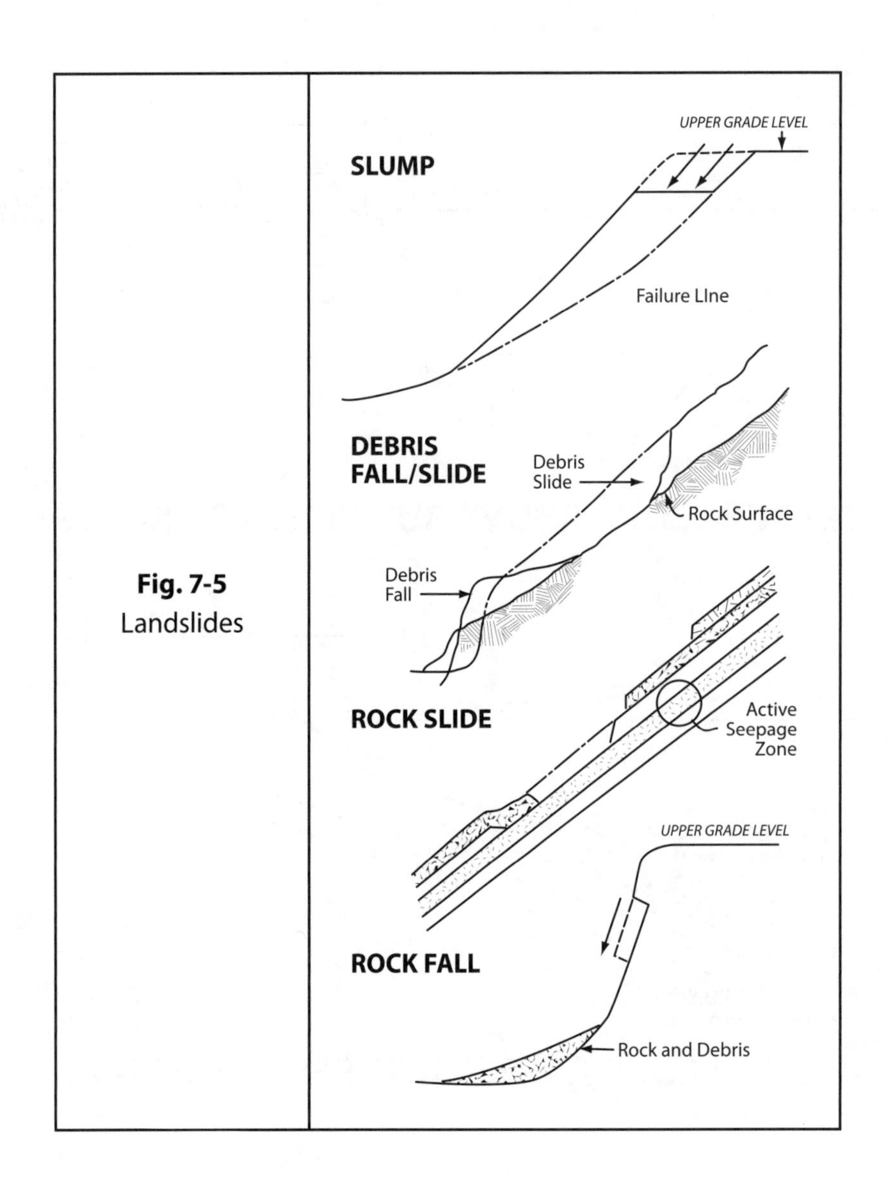

Fig. 7-5
Landslides

Fig. 7-6
Mud Slide

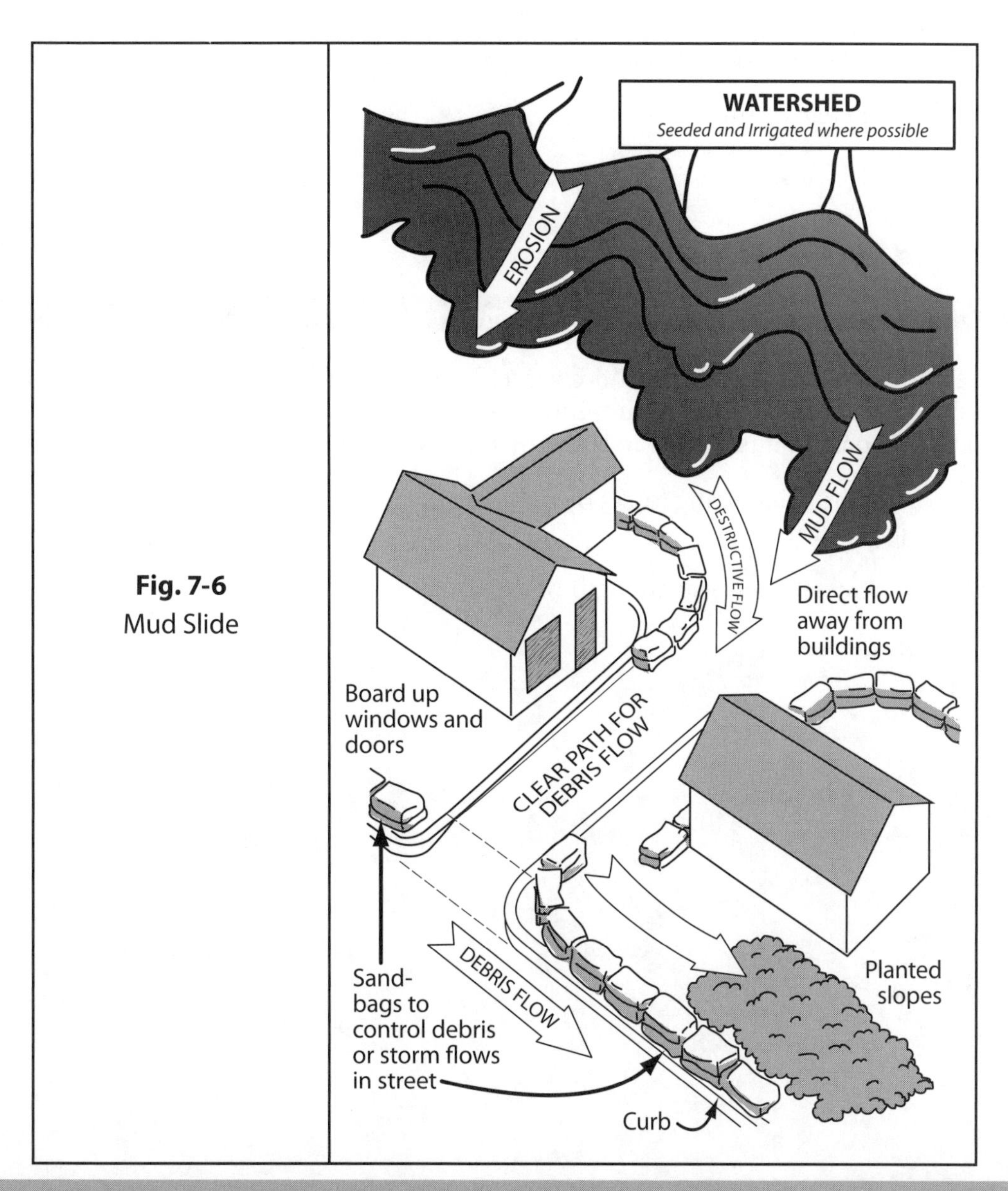

CORRECTIVE MEASURES

The analysis of landslides and a recommendation of the corrective measures to apply requires the skill of geologists and soils engineers.

Five control methods or retaining devices on highways are:

1. Buttresses
   a. Rock.
   b. Cementation of loose material at toe.
   c. Chemical treatment, flocculation at toe.
   d. Excavate, drain and backfill at toe.
   e. Relocation, raise grade at toe.
   f. Drainage of the toe.
2. Cribbing - concrete, steel or timber.

3. Retaining wall - masonry or concrete.
4. Piling - steel, concrete or timber.
   a. Floating.
   b. Fixed.
5. Tie-rodding slopes.

DRAINAGE

Since one of the basic causes of landslides is excess moisture, the fundamental way of preventing or curing them is by drainage. This requires intercepting both surface water and ground water before they can reach the mass that is subject to slides. Also, it requires draining the mass to stabilize it.

Drainage reduces the soil bearing forces by reducing the weight of the moving mass and by eliminating hydrostatic pressures.

Drainage also increases the shear strength of the soil.

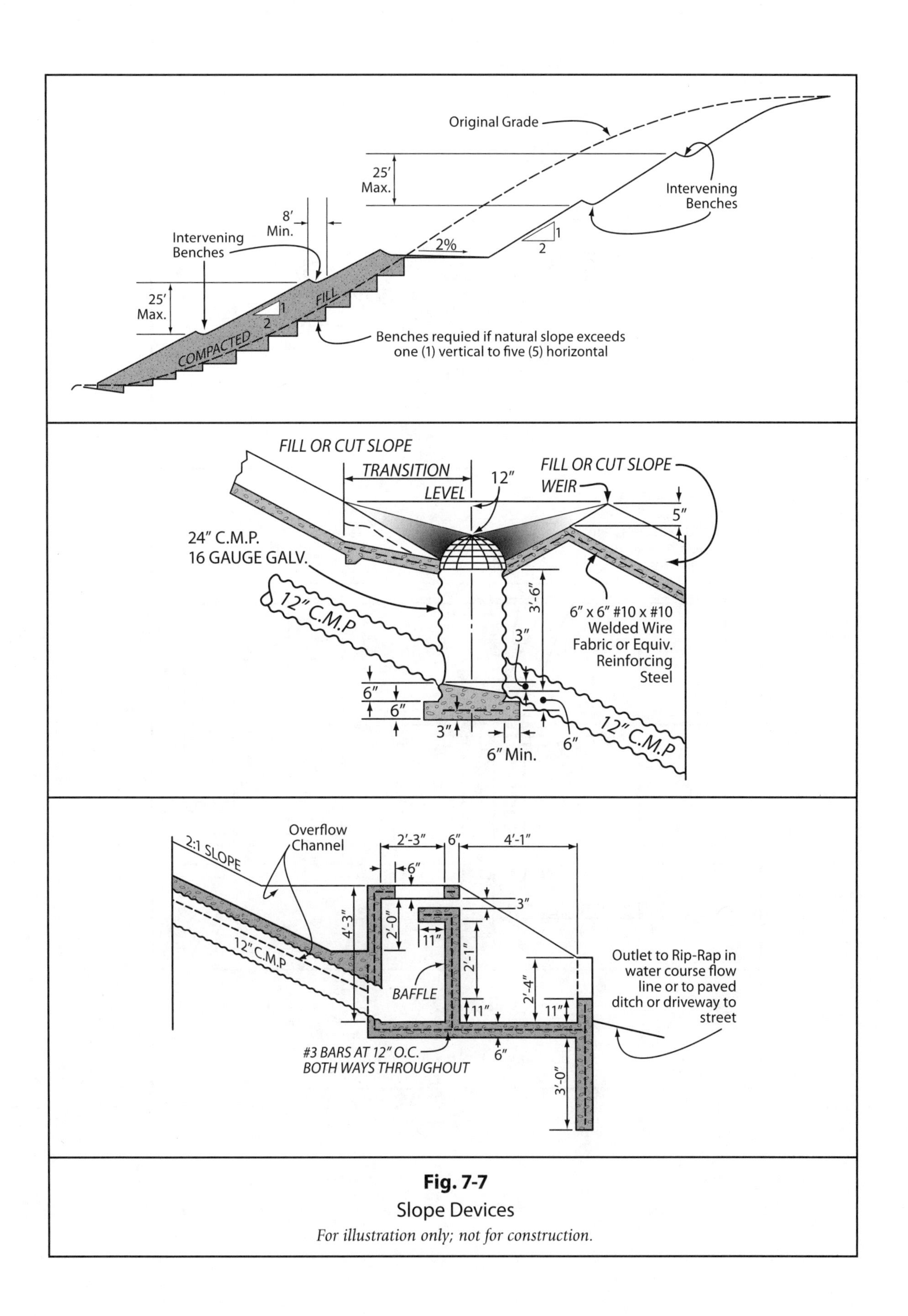

Fig. 7-7

Slope Devices

For illustration only; not for construction.

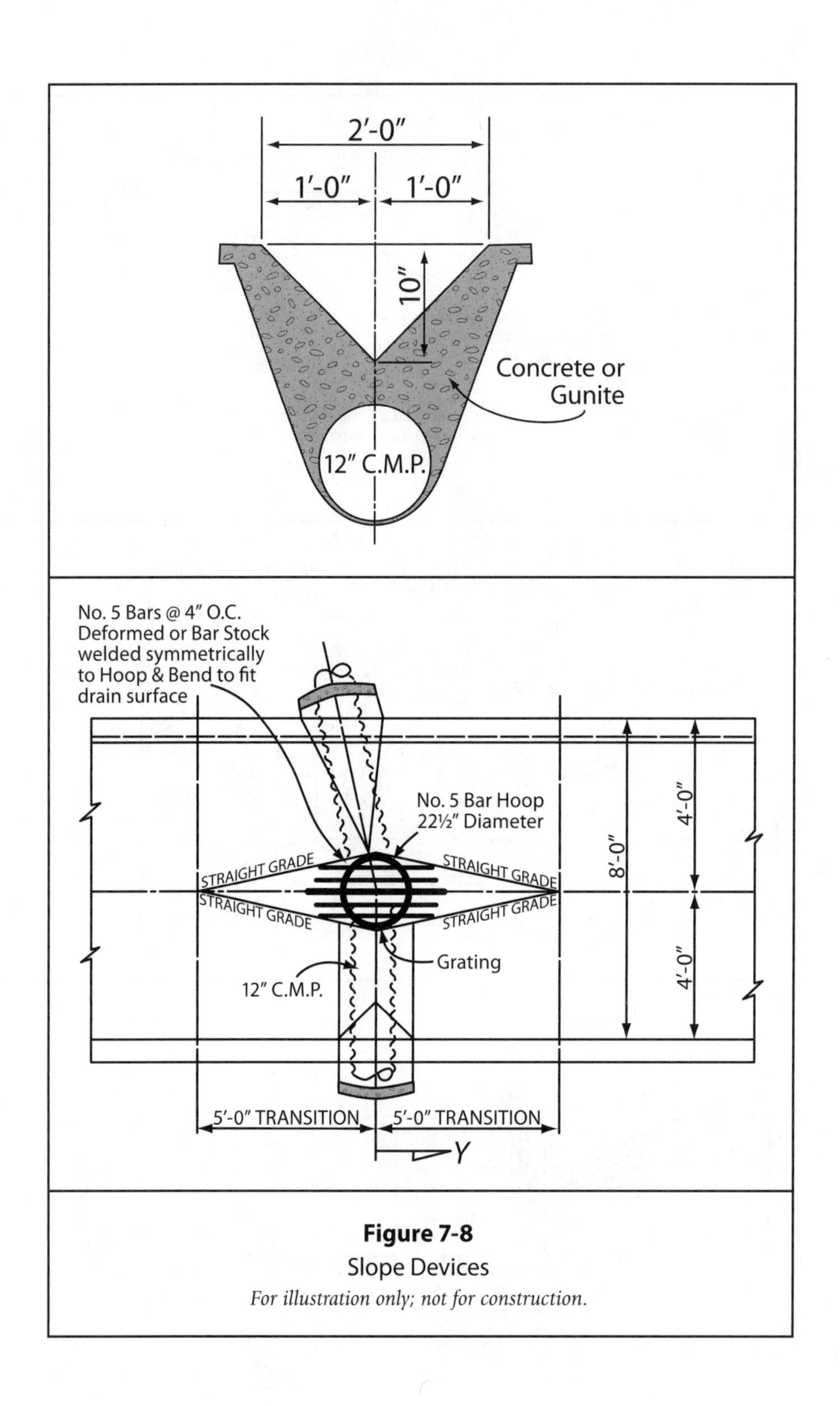

Figure 7-8

Slope Devices

For illustration only; not for construction.

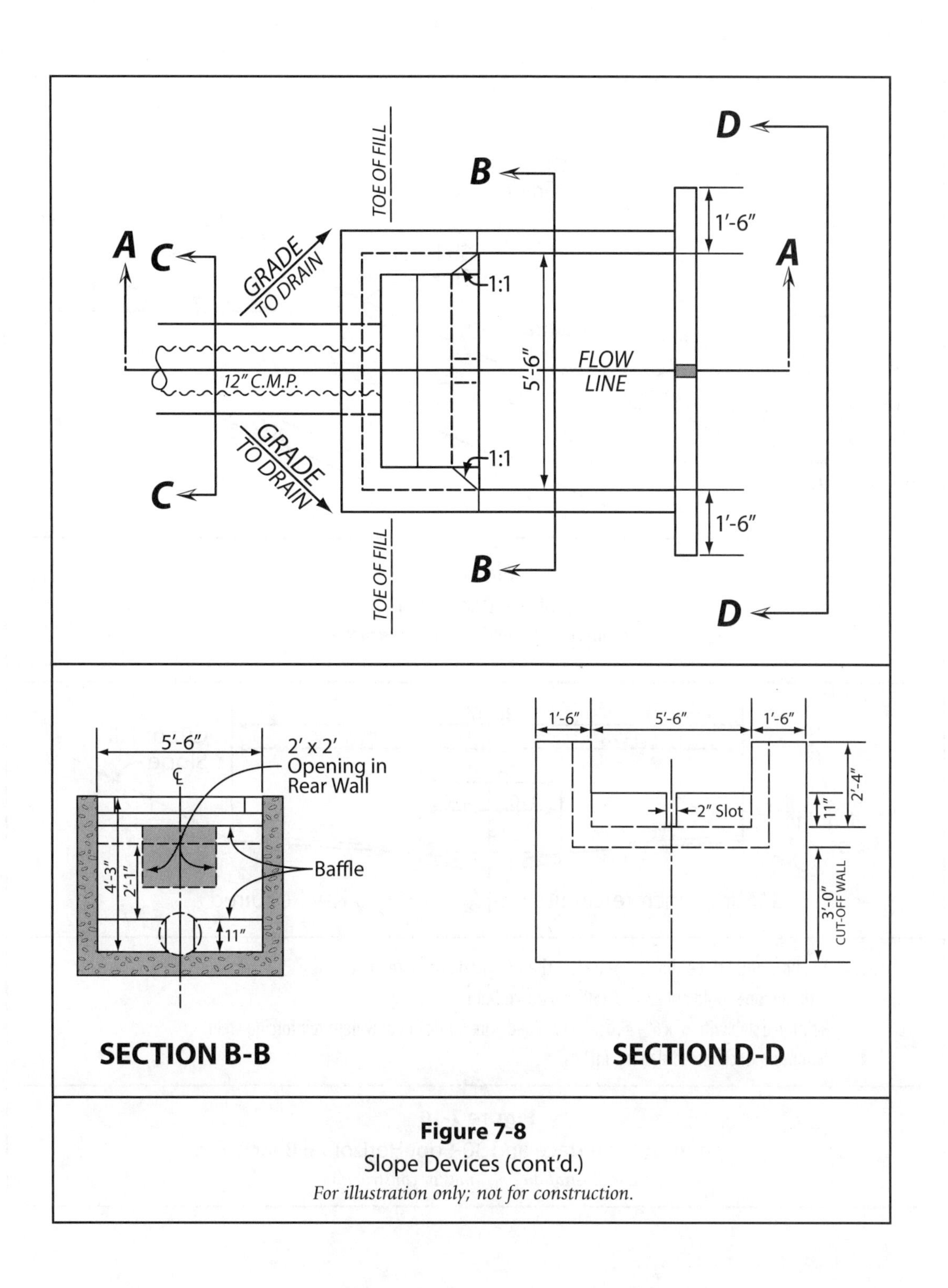

Figure 7-8
Slope Devices (cont'd.)
For illustration only; not for construction.

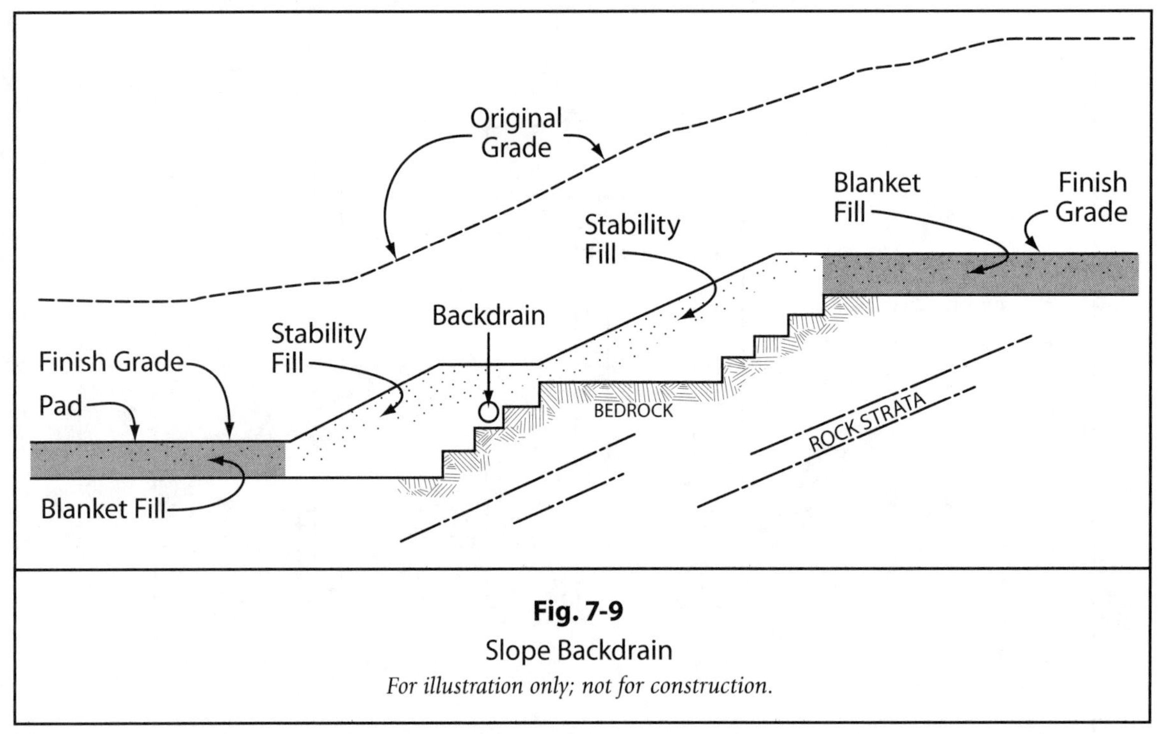

Fig. 7-9

Slope Backdrain

For illustration only; not for construction.

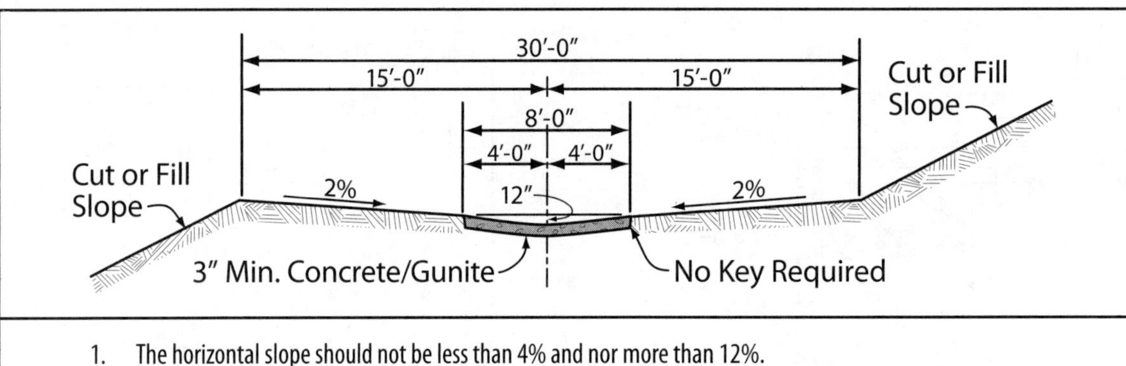

1. The horizontal slope should not be less than 4% and nor more than 12%.
2. A single run should not exceed 150′ to a downdrain.
3. Reinforce slab with 6″ x 6″ - #10 x #10 welded wire fabric or equivalent reinforcing steel.
4. Minimum thickness of slab is 3 inches.

Figure 7-10

Interceptor Terrace and 30-Foot Horizontal Bench

For illustration only; not for construction.

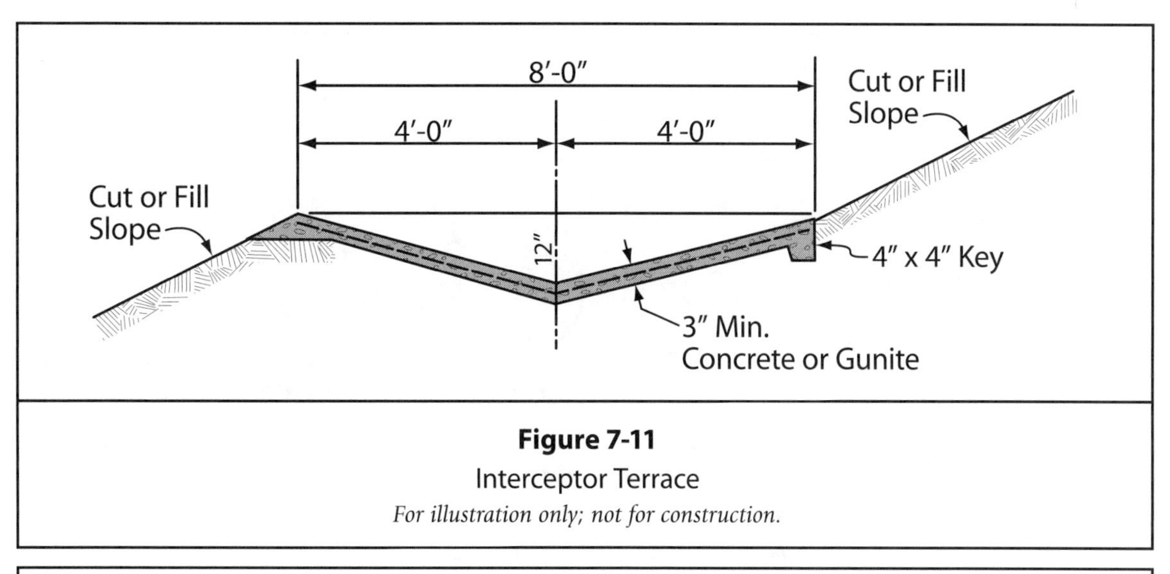

Figure 7-11
Interceptor Terrace
For illustration only; not for construction.

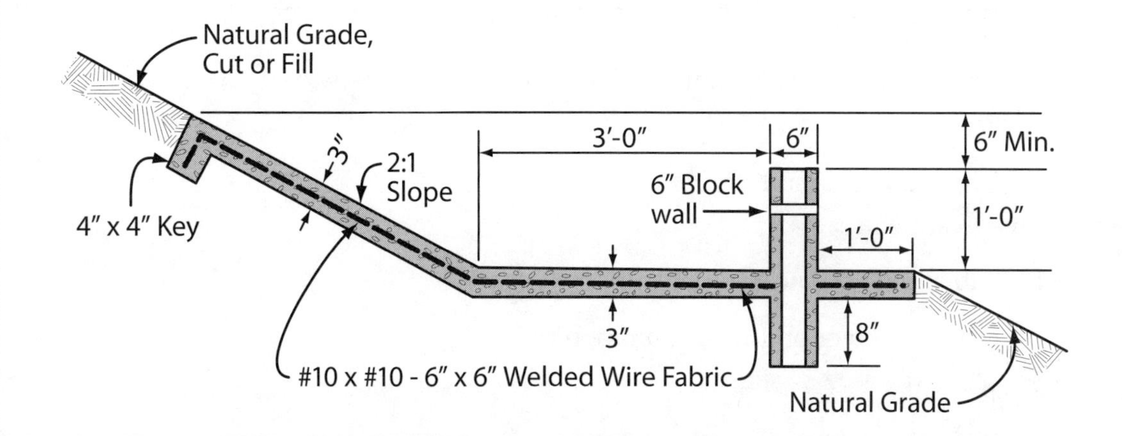

A concrete block drainage dispersal wall should be constructed as shown above, wherever it is necessary to convert channel flow to sheet flow.

1. Grout all cells and omit head joints in first course.
2. Wall to be located along contour line to establish uniform overflow or seepage.
3. Length of wall to equal length of contour line affected by grading.

Figure 7-12
Drainage Dispersal Wall
For illustration only; not for construction.

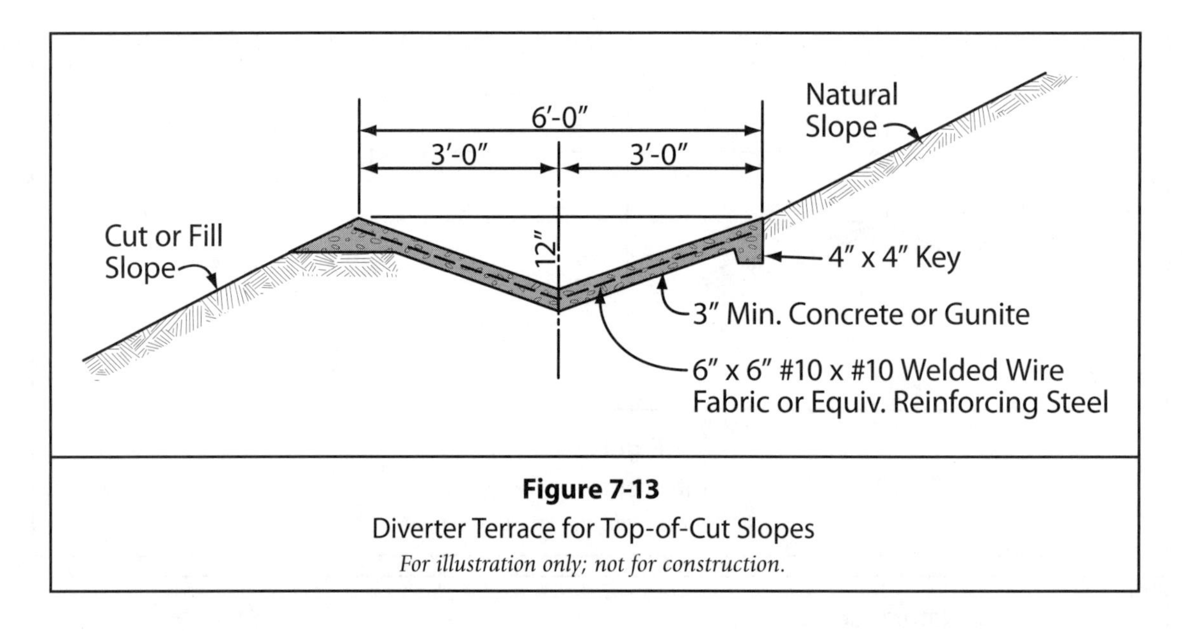

Figure 7-13

Diverter Terrace for Top-of-Cut Slopes

For illustration only; not for construction.

Corrective measures by means of drainage are subdivided as follows:

1. Surface.
   a. Reshaping landslide surface.
   b. Slope treatment.
      1. Sub-surface (French drain type).
      2. Jacked-in-place or drilled-in-place pipe.
      3. Tunneling.
      4. Blasting.
      5. Sealing joint planes and open fissures.

Drainage may not be simple nor inexpensive. The geologist who has thoroughly examined the site or who has available the results of the survey and borings can best advise the remedy. He can tell where to apply the drainage so that it will be effective and cost least.

Interception of surface water may be done by a ditch or impervious gutter. Intercepting all of the ground water before it reaches the mass is more difficult. It may be necessary to go very deep with a pipe subdrain in a pervious-backfilled trench. It may be necessary to jack a subdrain pipe or to tunnel.

Draining the mass is often of the greatest importance. It requires much skill if it is to be done effectively and economically. Because of continued soil movements and shrinkage during drying, only strongly connected pipe can be effective. Long lengths of perforated corrugated metal pipe, tightly joined, are most satisfactory for this service.

Culverts and sewers in areas are subject to sliding should also have positive joints, otherwise they will pull apart and discharge their contents into the mass, thereby aggravating the slide.

To repeat, landslides are expensive and their cure may be, too. Various control methods are available but a specialist should determine which, if any, is economical. Because drainage eliminates one of the basic causes of slides, it deserves prime consideration.

PLANTING AND IRRIGATION OF SLOPES

The following are recommendations for planting and irrigation of slopes to prevent erosion and landslides.

In general, all fill and cut slopes in designated hillside areas which are determined by the Grading Department to be subject to erosion should be planted and irrigated with a sprinkler system to promote the growth of ground cover plants to protect the slopes against erosion.

The owner is responsible for planting and maintaining all slopes.

Minimum Requirements. Low slopes to 15 feet in vertical height:

- Plant with grass or ground cover plants as recommended on the planting schedule approved by the Grading Department. Other plants recommended by a registered landscape architect may be considered for approval by the Department.
- A sprinkler system should be installed to irrigate these slopes as a part of the house plumbing installation.
- The owner should water the slopes, which have been planted with grasses and/or ground cover plants at sufficient time intervals to promote growth.

EXCEPTION:

Where the Department finds the slope is located in such an area as to make hand-watering possible, conveniently located hose bibs may be accepted in lieu of the required sprinkler system when a hose no longer than 50 feet would be necessary.

For medium slopes (15 to 38 feet in vertical height):

- Plant with grass or ground cover plants as recommended on the planting schedule approved by the Department. Other plants may be recommended by a landscape architect for approval by the Department.
- In addition to ground cover plants, approved shrubs having a one-gallon minimum size at ten feet on center in both directions on the slope when the sprinkler system is available for irrigation may be used.
- The plants and planting pattern may be varied to include trees upon the recommendation of the landscape architect and approved by the government department.
- Install an adequate sprinkler system during grading prior to planting of shrubs and trees and before grading is approved.

For high slopes (38 feet or over in vertical height):

- Plant with grass or ground cover plants as recommended on the planting schedule approved by the Department. Other plants recommended by landscape architects may be submitted to the Department for approval.
- In addition to ground cover plants, approved shrubs having a minimum one-gallon size at ten feet on center in both directions on the slope, or trees at 20 feet on center both ways may be used. A combination of shrubs and trees may be utilized. This plant and planting pattern may be varied upon the recommendation of a landscape architect and approval by the department.

Slopes exceeding a height where a drainage terrace is required should be planted with shrubs, minimum size one gallon, two feet on center, parallel to the benches, and within two feet of the uphill side. Larger varieties may be staggered on each side of the bench as an alternate.

Install an adequately designed sprinkler system prior to planting shrubs and trees and before grading is approved.

SPECIAL REQUIREMENTS FOR SPRINKLER SYSTEMS

- Plans for the sprinkler system should be submitted to and approved by the Department prior to installation.
- Sprinkler systems should be designed to provide a uniform water coverage at a rate of precipitation of not less than 1/10 inch per hour nor more than 3/10 inch per hour on the planted slope. In no event should the rate of precipitation duration of sprinkling be permitted to create a saturated condition and cause an erosion problem, or allow the discharge of excess water into any public or private street.
- A check valve and balance cock should be installed in the system where drainage from sprinkler heads will create an erosion problem.
- Adequate back flow protection should be installed in each sprinkler system as required by the Plumbing Code.
- A functional test of the sprinkler system should be performed by the installer for every sprinkler system prior to approval.

SLOPE MAINTENANCE

The scope of work involved in the permanent maintenance of flood prevention devices can easily exceed the initial cost in the development of the property. It therefore becomes incumbent upon the owner of property susceptible to water damage to maintain his or her property in a manner which will assure the continued stability of the property.

The scope of the maintenance program includes but is not limited to the following:

1. Maintain existing slope planting and irrigation system in working order;
2. Maintain paved diverter terraces, interceptor terraces, downdrains, appurtenances such as inlets, and velocity reducer structures in a clean condition and in good repair;
3. Earth berms that prevent water from flowing over slopes;
4. Prevention of standing storm water on pad directly above descending slopes, whether natural, cut or fill, as this is a major contributor to slope failure;
5. Side swales that direct water around the house should be maintained;
6. Catch basins, grates, and subsurface drainage pipes should be kept free of silt and debris;
7. Roof gutters and downspouts should be inspected periodically to assure that they are not broken or clogged;
8. All anti-erosion devices should be kept clean and in good repair;
9. As extensive landscaping or revisions to the property may seriously alter the surface drainage pattern, the owner should avoid disrupting flow patterns created when the property was originally graded;
10. Problems such as erosion should be repaired immediately in order that more serious problems may be averted;

11. Owners should prevent loose fill from being placed on the grading site, especially on slopes.

A suggested checklist for identifying the responsibilities of parties involved with catastrophic flood or mud slides is as follows:

1. What are the lot numbers of the properties involved with the damage?
2. What is the tract number?
3. What is the description of the drainage easements?
4. What is the flooding history of this property?
    a. Topographic map prior to subdivision
    b. Grading and drainage plans submitted to authorities for approval
    c. Comments and corrections made by planning department and flood control department regarding submitted drainage plan
    d. Final and approved grading and drainage plan of subdivided land
    e. Hydrological history of site
    f. Topographic map of existing condition at time of subject flood or rainstorm
    g. Description of damages due to flood or rainstorm
5. Description of protective drainage devices, including
    a. Paved diverter terraces
    b. Interceptor terraces
    c. Downdrains
    d. Appurtenances such as inlet and velocity reducer structures
6. What changes have been made to the properties since the original grading?
    a. Fill or cut
    b. Landscaping
    c. Building additions
    d. Retaining walls or other structures
    e. Drainage devices

The following are some typical laws and policies relating to flood hazards and mud slides are representative of municipal interest in such matters.

Los Angeles County Ordinance No. 447, Subdivision Ordinance Sec. 158 (90714-22-66):

LAND SUBJECT TO FLOOD HAZARD, INUNDATION, OR GEOLOGICAL HAZARD. If any portion of the land within the boundaries shown on a tentative map of a division of land is subject to flood hazard, inundation, or geological hazard and the probable use of the property will require structures thereon, the advisory agency may disapprove the map or that portion of the map so affected, and require protective improvements to be constructed as a condition precedent to approval of the map.

If any portion of a lot or parcel of a division of land is subject to flood hazard, inundation, or geological hazard, such fact and portion shall be clearly shown on the final map or parcel map by a prominent note on each sheet of such map whereon any such portion is shown.

Los Angeles County Ordinance No. 447, Subdivision Ordinance Sec. 159:

LAND SUBJECT TO OVERFLOW, OR PONDING OR HIGH GROUND WATER. If any portion of such land is subject to sheet overflow or ponding of local storm water or should the depth to ground water be less than ten feet from the ground surface, the Regional Planning Commission shall so inform the State Real Estate Commissioner.

General Grading Requirements: City of Los Angeles, Sec. 91:3007:

PLANTING AND IRRIGATION OF CUT-AND-FILL SLOPES IN HILLSIDE AREAS

a. General. All cut-and-fill slopes in designated areas that are determined by the Department to be subject to erosion shall be planted and irrigated with a sprinkler system to promote the growth of ground cover plants to protect the slopes against erosion, as required in this section. The owner shall be responsible for planting and maintaining all slopes where such is required in this section.
b. A sprinkler system shall be installed to irrigate these slopes as a part of the house plumbing system.
c. The owner shall water the slopes that have been planted with grass and/or ground cover plants at sufficient time intervals to promote growth.

In one case, the court considered whether lack of maintenance caused landslide damage to two lots in a creek-side subdivision. The subject creek meandered through the city and eventually emptied into a bay. When the subdivision was developed, the developer was required to construct storm drains to carry surface water into the creek. The city accepted the dedication of the storm drain and a drainage easement along the creek. At trial, the homeowners' experts testified that the landslide was a result of the absence of erosion control devices. The city's experts testified that a number of factors contributed to the slide, including the presence of a zone of weak clay underlying the fill which was placed by the developer.

The appellate court in review found that despite the absence of erosion control devices, the storm drain system added only a minimum amount of water to the creek and there was no evidence that the minimum addition caused the erosion, which led to the landslide.

The court found that the erosion which occurred was a natural phenomenon in all meandering creeks and that the natural flow of the creek alone caused the damage.

CHAPTER 8

Retaining Walls

RETAINING WALL BASICS

Various schemes are to support or retain earth. Depending upon the type of soil and its saturation, earth causes a horizontal force against the retaining wall. The Uniform Building Code requires that retaining walls resist this lateral pressure in accordance with accepted engineering practice.

Figure 8-1
Steel Post and Timber Beam Retaining Wall

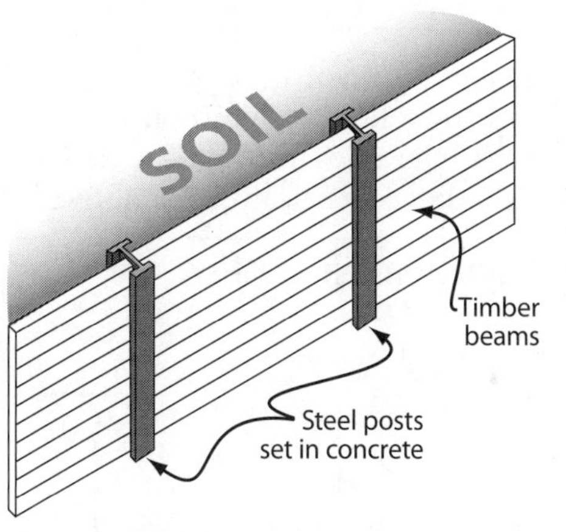

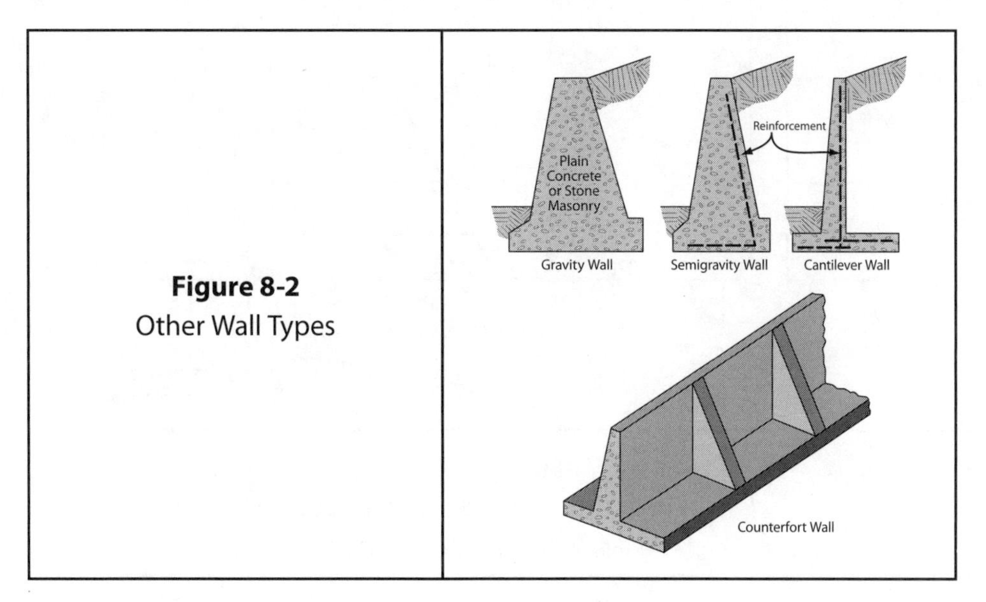

Figure 8-2
Other Wall Types

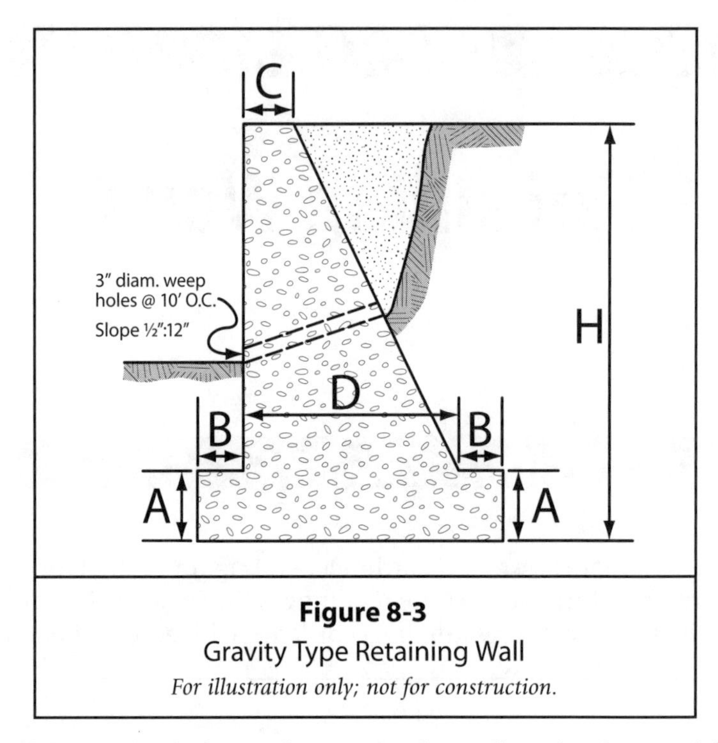

Figure 8-3

Gravity Type Retaining Wall

For illustration only; not for construction.

If well-drained earth is involved, the Code permits the walls to be designed for a pressure equivalent to that exerted by a fluid weighing not less than 30 pounds per cubic foot. Figure 8-3 above shows a small retaining wall. This wall is usually termed a cantilever type, since the stem or vertical part of the wall is designed as a cantilever above the base.

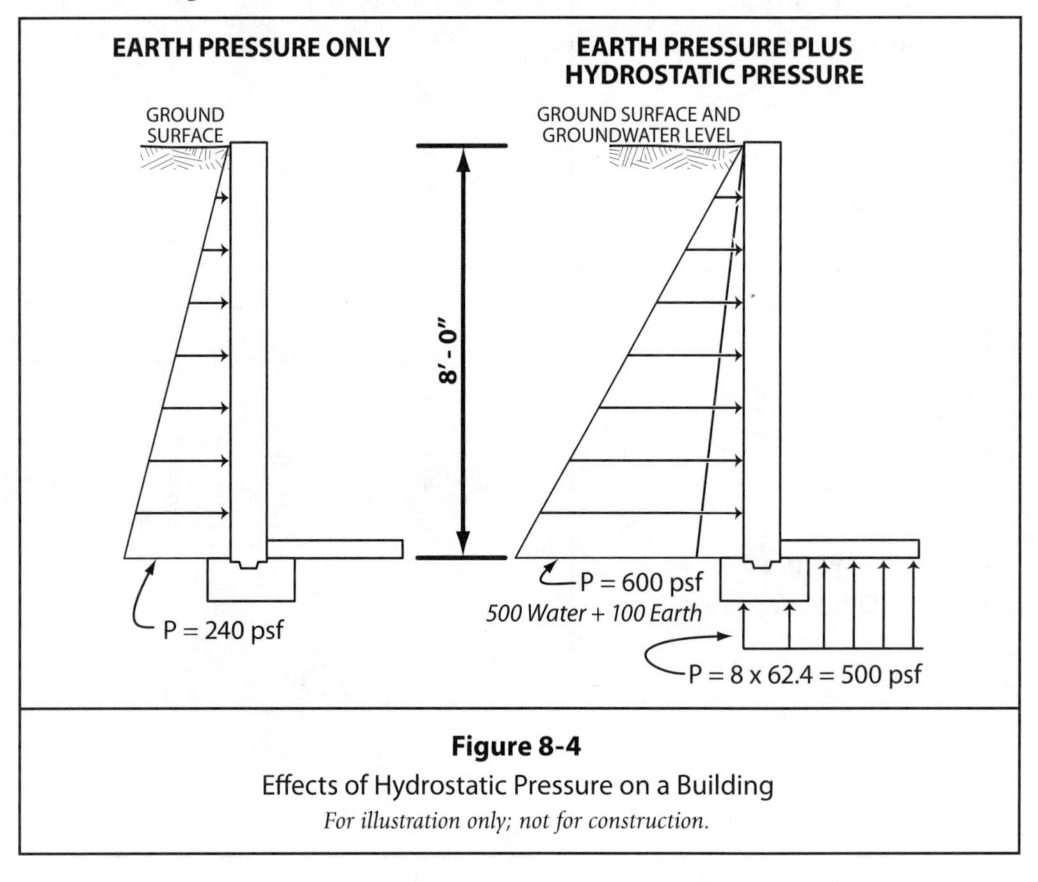

Figure 8-4

Effects of Hydrostatic Pressure on a Building

For illustration only; not for construction.

In Figure 8-5, you see the forces acting on a retaining wall. The triangular pressure diagram on the vertical stem is computed utilizing the 30-pound per cubic foot hydrostatic force.

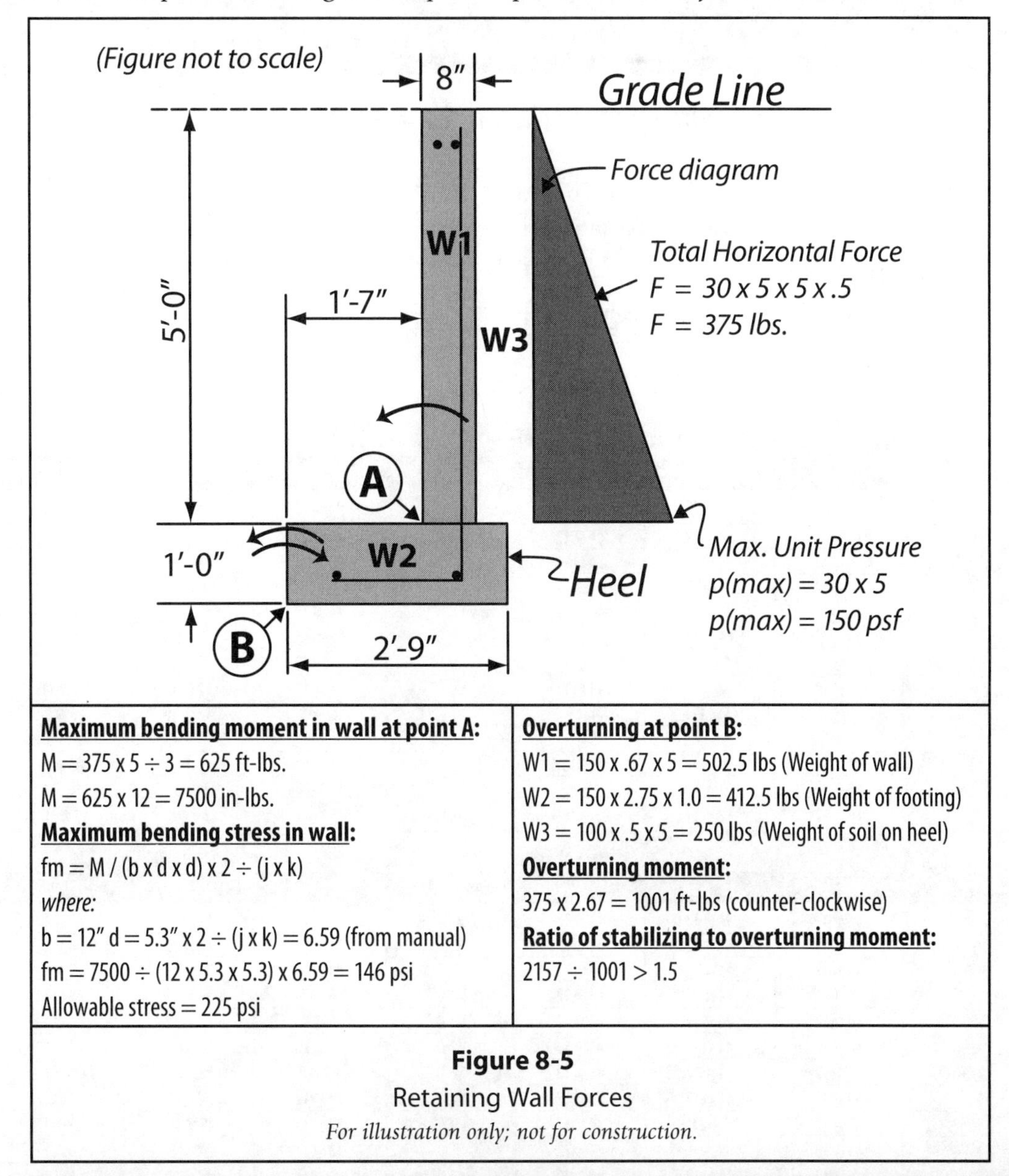

Maximum bending moment in wall at point A:	Overturning at point B:
M = 375 x 5 ÷ 3 = 625 ft-lbs.	W1 = 150 x .67 x 5 = 502.5 lbs (Weight of wall)
M = 625 x 12 = 7500 in-lbs.	W2 = 150 x 2.75 x 1.0 = 412.5 lbs (Weight of footing)
Maximum bending stress in wall:	W3 = 100 x .5 x 5 = 250 lbs (Weight of soil on heel)
fm = M / (b x d x d) x 2 ÷ (j x k)	**Overturning moment:**
where:	375 x 2.67 = 1001 ft-lbs (counter-clockwise)
b = 12″ d = 5.3″ x 2 ÷ (j x k) = 6.59 (from manual)	**Ratio of stabilizing to overturning moment:**
fm = 7500 ÷ (12 x 5.3 x 5.3) x 6.59 = 146 psi	2157 ÷ 1001 > 1.5
Allowable stress = 225 psi	

Figure 8-5
Retaining Wall Forces
For illustration only; not for construction.

TYPES OF RETAINING WALLS

There are many types of structures built to retain earth. The most simple type is probably the gravity wall, as shown in various forms in Fig. 8-2 above. It functions in the same way as the common reservoir dam, which resists the water pressure by its shear weight and the friction at the bottom of its footing. The retained earth presses on the back of the gravity retaining wall and tries to slide it or overturn it. This force is resisted by its mass of concrete.

Table 8-1
Reinforced Concrete Basement Retaining Walls

Ultimate strength design: **f'_c = 3,000 psi**

DRY EARTH
Equivalent fluid pressure 70 pcf

Horizontal Bars #4 @ 12″
2″
H
Vertical Bars *SEE TABLE*

SATURATED EARTH
Equivalent fluid pressure 70 pcf

For illustration only; not for construction.

GRADE 40 REINFORCEMENT

H	12″ wall	16″ wall	20″ wall	H	12″ wall	16″ wall	20″ wall
8′-0″	#4 @ 12	#4 @ 12	#4 @ 12	8′-0″	#4 @ 12	#4 @ 12	#4 @ 12
9′-0″	#4 @ 12	#4 @ 12	#4 @ 12	9′-0″	#4 @ 12	#4 @ 12	#4 @ 12
10′-0″	#4 @ 12	#4 @ 12	#4 @ 12	10′-0″	#4 @ 10	#4 @ 12	#4 @ 12
11′-0″	#4 @ 12	#4 @ 12	#4 @ 12	11′-0″	#5 @ 11½	#4 @ 10½	#4 @ 12
12′-0″	#4 @ 12	#4 @ 12	#4 @ 12	12′-0″	#6 @ 12	#5 @ 12	#4 @ 10½
13′-0″	#4 @ 10	#4 @ 12	#4 @ 12	13′-0″	#6 @ 9½	#5 @ 9½	#5 @ 12
14′-0″	#5 @ 12	#4 @ 11½	#4 @ 12	14′-0″	#7 @ 10	#6 @ 11	#5 @ 10
15′-0″	#5 @ 10	#4 @ 9	#4 @ 12	15′-0″	#8 @ 11	#7 @ 12	#6 @ 11½
16′-0″	#5 @ 8	#5 @ 11½	#4 @ 9½	16′-0″	#8 @ 9	#7 @ 10	#6 @ 9½
17′-0″	#6 @ 9½	#5 @ 9½	#5 @ 12	17′-0″	#9 @ 9	#7 @ 8	#7 @ 10½
18′-0″	#7 @ 11	#5 @ 8	#5 @ 10½	18′-0″	#10 @ 9½	#9 @ 11	#8 @ 11
19′-0″	#8 @ 12	#6 @ 9½	#5 @ 9	19′-0″	#10 @ 8	#10 @ 12	#8 @ 10
20′-0″	#9 @ 12	#7 @ 11	#6 @ 10½	20′-0″	#11 @ 8	#9 @ 8	#9 @ 10½

GRADE 60 REINFORCEMENT

H	12″ wall	16″ wall	20″ wall	H	12″ wall	16″ wall	20″ wall
8′-0″	#4 @ 12	#4 @ 12	#4 @ 12	8′-0″	#4 @ 12	#4 @ 12	#4 @ 12
9′-0″	#4 @ 12	#4 @ 12	#4 @ 12	9′-0″	#4 @ 12	#4 @ 12	#4 @ 12
10′-0″	#4 @ 12	#4 @ 12	#4 @ 12	10′-0″	#4 @ 12	#4 @ 12	#4 @ 12
11′-0″	#4 @ 12	#4 @ 12	#4 @ 12	11′-0″	#4 @ 11	#4 @ 12	#4 @ 12
12′-0″	#4 @ 12	#4 @ 12	#4 @ 12	12′-0″	#5 @ 12	#4 @ 12	#4 @ 12
13′-0″	#4 @ 12	#4 @ 12	#4 @ 12	13′-0″	#5 @ 10	#4 @ 9	#4 @ 12
14′-0″	#4 @ 12	#4 @ 12	#4 @ 12	14′-0″	#5 @ 8	#5 @ 11′	#4 @ 9½
15′-0″	#4 @ 12	#4 @ 12	#4 @ 12	15′-0″	#7 @ 12	#5 @ 9	#5 @ 12
16′-0″	#5 @ 12	#4 @ 11½	#4 @ 12	16′-0″	#7 @ 10	#5 @ 7½	#5 @ 12
17′-0″	#5 @ 10	#4 @ 9	#4 @ 12	17′-0″	#8 @ 11	#6 @ 9	#6 @ 12
18′-0″	#6 @ 12	#5 @ 12	#4 @ 10	18′-0″	#8 @ 9	#7 @ 10	#6 @ 10
19′-0″	#6 @ 10	#5 @ 10	#4 @ 8½	19′-0″	#10 @ 12	#8 @ 11	#6 @ 8
20′-0″	#7 @ 10½	#6 @ 12	#5 @ 11	20′-0″	#10 @ 10	#9 @ 12	#7 @ 9½

Another common type of wall is the reinforced concrete retaining wall consisting of a stem wall and a footing. This is called a "cantilever wall" as the stem is cantilevered from its foundation or footing. The footing may project both in front and back of the wall, only in the front, or only in the back. You call the portion in the back "heel" and the part in front the "toe."

Selection of the type of cantilever retaining usually depends on the location of the property line, obstruction, or access for work.

If the wall height is not too great, you can use a reinforced concrete block stem, or concrete masonry stem, economically. See Figure 9-2.

You may choose a counterfort retaining wall for heights in excess of 20 feet. It consists of brackets, which strengthen the junction of the stem to the base slab. When these brackets are at the back of the stem, they act in tension and are called "counterforts." When they are in front of the stem, they act in compression, and are called "buttresses."

Often roads built in mountainous terrain are too close to the edge of the embankment for a conventional retaining wall. In this case, you can use steel posts and timber planks, as shown on Fig. 8-1.

On the high side of a mountain road, you can retain the slope by building a "bin-type" retaining wall. These may be made of precast concrete or steel planks or preformed metal modules, as discussed in Chapter 10.

RETAINING WALL DESIGN

You usually build a retaining wall of stone, masonry, or reinforced concrete to sustain the lateral pressure of the earth. Such walls depend for their stability on their own weight (gravity walls) or on their own weight in addition, an additional weight of the laterally supported material resting on the heel of the footing.

Fig. 8-5 represents a wall supporting an earth fill. There is a tendency for a portion of the earth next to the wall to break away along some such line to settle downward and move forward. This tends to slide and overturn the wall. The internal frictional resistance of the retained soil also includes a cohesive quality. If cohesion is absent, the surface of rupture may be assumed an inclined plane, while if cohesion is present, the surface of rupture stands more nearly vertical at the surface of the ground than at the base of the wall.

Usually this curved direction of the surface of rupture is influenced by a variation in moisture content in the retained soil. The more moisture, the less cohesion, and hence the flatter the angle of rupture. It should be seen at the outset that the removal of all possible moisture from the supported material reduces greatly the size of the prism of earth actually effective in pressing against the wall and this increases the stability of the wall.

Earth-pressure theories generally assume a granular mass possessing no cohesion, since this assumption is on the safe side of the usual conditions of soil possessing some cohesion. Probably the most important rule in the design of retaining walls is to provide adequate drainage.

RETAINING WALL NOTES

- Design is based on weight of earth 100 p.c.f., and angle of repose assumed 33° and no surcharge. Designed for 2000-lb. controlled concrete, low water-cement is recommended for permanency.
- The resultant pressure on walls above is at the outer edge of middle third of the wall.
- Alternate vertical rebar in Types I, II, and III walls may be cut at 1/2H. See Figures 8-11 through 8-14.
- Expansion joints in walls should not be over 75′ -0″ on center, construction joints at 30′ 0″ on center.

RETAINING WALL FOUNDATION LAYOUT

The first step in laying out a foundation is to surround the work with batter boards. Keep the batter boards clear of the foundation excavation. The top of the batter board is usually at the same elevation as the top of the foundation wall. Drive a single nail into the batter board at the extension of the outer lines of the foundation wall. When you connect these nails with a tight string, you have outlined the foundation.

You can check if the structure is square by measuring the diagonals between opposite corners. The diagonals should be equal lengths. You can also check whether corners are ninety degrees by using a triangle with a 3-foot, 4-foot, and 5-foot side. Remember the rule for right triangles – the sum of the square of the sides is equal to the square of the hypotenuse.

Use the elevation of the top of batter boards to measure down to the bottom of footing using a leveling rod and level. This establishes the depth of excavation. After the excavation is done, you can start building the formwork for the footings. Now place the concrete into the forms.

When the concrete in the footings has set, remove the forms and build the wall forms. Use the nails in the batter boards to locate the wall forms. You usually use the outside surface of the foundation walls as a horizontal control of the building.

When you are about to place concrete for an elevated slab on forms and shoring, you should locate all sleeves, openings, and embedded items by surveying. Mark the plywood form with centerlines of these items. Place the sleeves and forms for openings before placing the reinforcement and concrete. Use the outside surface of the perimeter walls as the base of the survey.

Table 8-2
Estimating Concrete Retaining Walls

- Material costs for concrete placed directly from chute are based on 2,000 PSI concrete, 5.0 sack mix, with 1" aggregate, including 5% waste.
- Pump mix costs include an additional cost of $10.00 per CY (cubic yard) for pump.
- Labor costs are for placing only.
- Add the cost of excavation, formwork, steel reinforcing, finishes and curing.
- Square foot costs are based on SF (square foot) of wall measured on one face only.
- Costs do not include engineering, design or foundations.
- Craft crew B1 is one laborer and one carpenter
- Craft crew B3 is two laborers and one carpenter

For Concrete Footings, Grade Beams and Stem Walls, use the figures below for preliminary estimates. Concrete costs are based on 2,000 PSI, 5.0 sack mix with 1" aggregate placed directly from the chute of a ready-mix truck. Figures in parentheses show the cubic yards of concrete per linear foot of foundation (including 5% waste). Costs shown include concrete, 60 pounds of reinforcing per CY of concrete, and typical excavation using a 3/4 CY backhoe with excess backfill spread on site.

For Concrete Footings and Grade Beams: Cast directly against the earth, no forming or finishing required. These costs assume the top of the foundation will be at finished grade. For scheduling purposes estimate that a crew of 3 can layout, excavate, place and tie the reinforcing steel and place 13 CY of concrete in an 8-hour day. Use $900.00 as a minimum job charge.

	Craft@Hrs	Unit	Material	Labor	Total
Using concrete (before 5% waste allowance)		CY	76.90		76.9C
4" thick walls (1.23 CY per CSF, or 100 square feet)					
To 4' high, direct from chute	B1@.013	SF	.99	.34	1.33
4' to 8' high, pumped	B3@.015	SF	1.12	.38	1.50
8' to 12' high, pumped	B3@.017	SF	1.12	.44	1.56
12' to 1 6' high, pumped .	B3@.018	SF	1.12	.46	1.58
16' high, pumped	B3@.020	SF	1.12	.51	1.63
6" thick walls (1.85 CY per CSF)					
To 4' high, direct from chute	B1@.020	SF	1.49	.53	2.02
4' to 8' high, pumped	B3@.022	SF	1.68	.56	2.24
8' to 12' high, pumped	B3@.025	SF	1.68	.64	2.32
12' to 16' high, pumped	B3@.027	SF	1.68	.69	2.37
16' high, pumped	B3@.030	SF	1.68	.77	2.45
8" thick walls (2.47 CY per CSF)					
To 4' high, direct from chute	B1@.026	SF	1.69	.69	2.38
4' to 8' high, pumped	B3@.030	SF	2.24	.77	3.01
8' to 12' high, pumped	B3@.033	SF	2.24	.84	3.08
12' to 16' high, pumped	B3@.036	SF	2.24	.92	3.16
16' high, pumped	B3@.040	SF	2.24	1.02	3.26
10" thick walls (3.09 CY per CSF)					
To 4' high, direct from chute	B1@.032	SF	2.50	.85	3.35
4' to 8' high, pumped	B3@.037	SF	2.80	.95	3.75
8' to 14' high, pumped..	B3@.041	SF	2.80	1.05	3.85
12' to 16' high, pumped	B3@.046	SF	2.80	1.18	3.98
16' high, pumped	B3@.050	SF	2.80	1.28	4.08
12" thick walls (3.70 CY per CSF)					
To 4' high, placed direct from chute	B1@.040	SF	2.99	1.06	4.05
4' to 8' high, pumped	B3@.045	SF	3.36	1.15	4.51
8' to 12' high, pumped	B3@.050	SF	3.36 .	1.28	4.64
12' to 16 high, pumped	B3@.055	SF	3.36	1.41	4.77
16' high, pumped	B3@.060	SF	3.36	1.54	4.90

ESTIMATING WALL FORMS

Table 8-3
Estimating Wall Forms

Formwork over 6' high includes an allowance for a work platform and handrail built on one side of the form for use by the concrete placing crew.

	Craft@Hrs	Unit	Material	Labor	Total
Heights to 4', includes 1.1 SF of plyform, 1.5 SF of lumber and allowance for nails, ties and oil per SFCA					
1 use	F5@.119	SFCA	2.54	4.81	7.35
3 uses	F5@.080	SFCA	1.40	3.24	4.64
5 uses	F5@.069	SFCA	1.17	2.79	3.96
Add for 1 side battered	F5@.016	SFCA	.25	.65	.90
Add for 2 sides battered	F5@.024	SFCA	.51	.97	1.48
Heights over 4' to 6', includes 1.1 SF of plyform, 2.0 SF of lumber and allowance for nails, ties and oil					
1 use	F5@.140	SFCA	2.86	5.66	8.52
3 uses	F5@.100	SFCA	1.56	4.04	5.60
5 uses	F5@.080	SFCA	1.29	3.24	4.53
Add for 1 side battered	F5@.016	SFCA	.29	.65	.94
Add for 2 sides battered	F5@.024	SFCA	.57	.97	1.54
Heights over 6' to 12', includes 1.2 SF of plyform, 2.5 SF of lumber and allowance for nails, ties and oil					
1 use	F5@.160	SFCA	3.30	6.47	9.77
3 uses	F5@.110	SFCA	1.78	4.45	6.23
5 uses	F5@.100	SFCA	1.47	4.04	5.51
Add for 1 side battered	F5@.016	SFCA	.33	.65	.98
Add for 2 sides battered	F5@.024	SFCA	.66	.97	1.63
Heights over 12' to 16', includes 1.2 SF of plyform, 3.0 SF of lumber and allowance for nails, ties and oil					
1 use	F5@.180	SFCA	3.62	7.28	10.90
3 uses	F5@.130	SFCA	1.94	5.26	7.20
5 uses	F@@.110	SFCA	1.60	4.45	6.05
Add for 1 side battered	F5@.016	SFCA	.38	.65	1.01
Add for 2 sides battered	F5@.024	SFCA	.72	.97	1.69

STEEL FRAMED PLYWOOD FORMS

Rented steel framed plywood forms can reduce forming costs on many jobs. Where rented forms are used 3 times a month, use the wall forming costs shown for 3 uses at the appropriate height but deduct 25% from the material cost and 50% from the labor manhours and labor costs. Savings will be smaller where: layoffs change from one use to the next, when form penetrations must be made and repaired before returning a form, where form delivery costs are high, and where non-standard form sizes are needed.

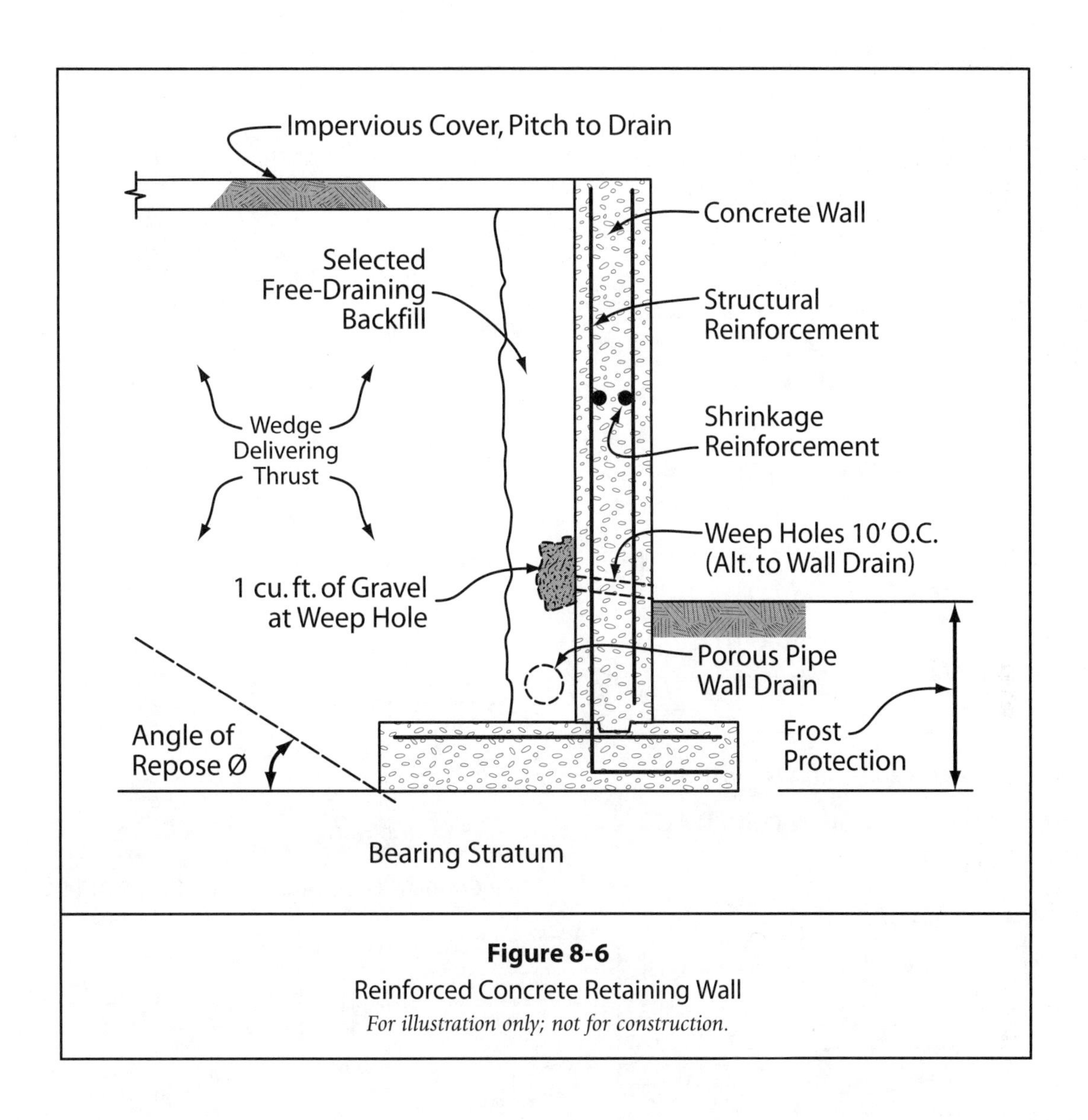

Figure 8-6

Reinforced Concrete Retaining Wall

For illustration only; not for construction.

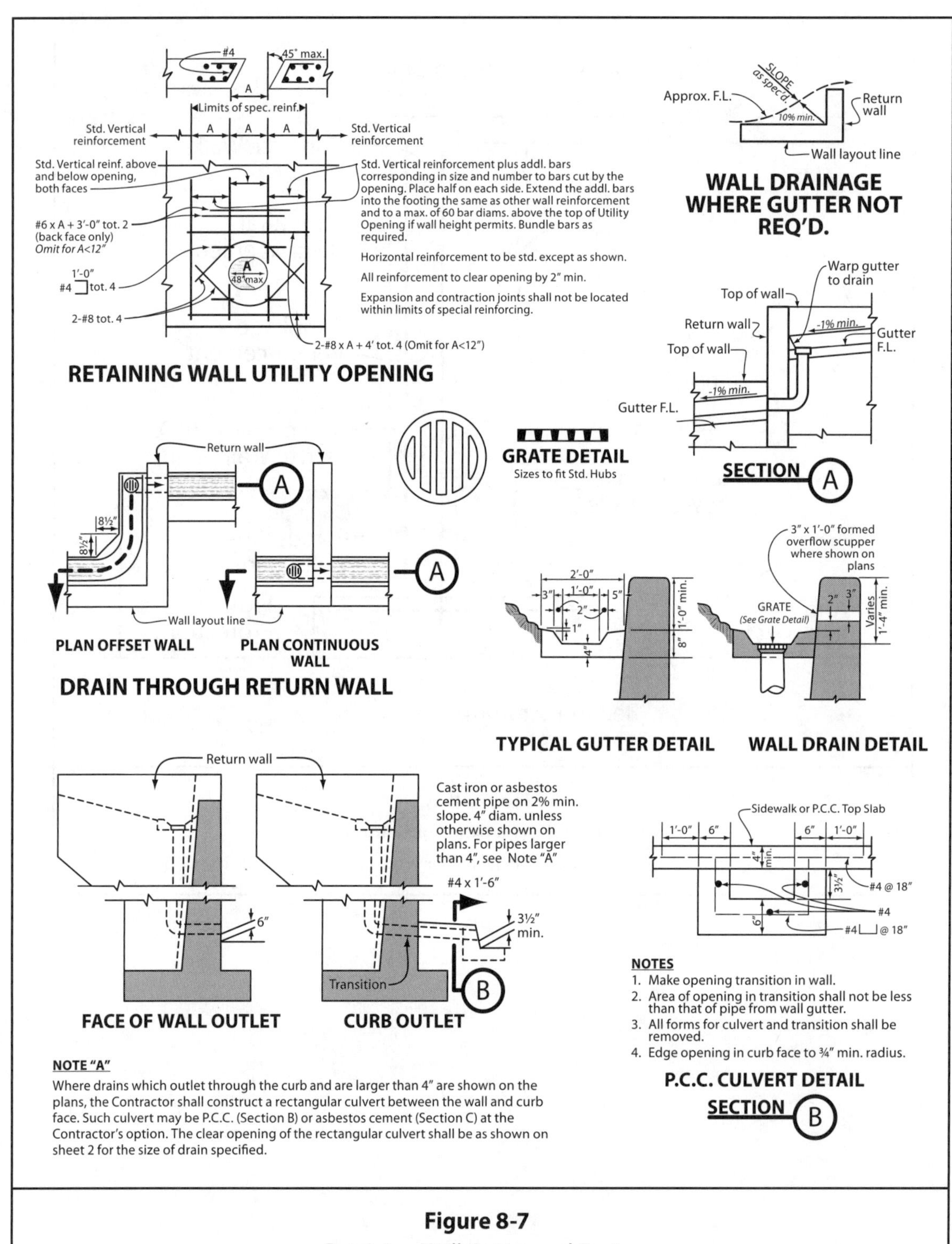

Figure 8-7

Retaining Wall Gutter and Drain

For illustration only; not for construction.

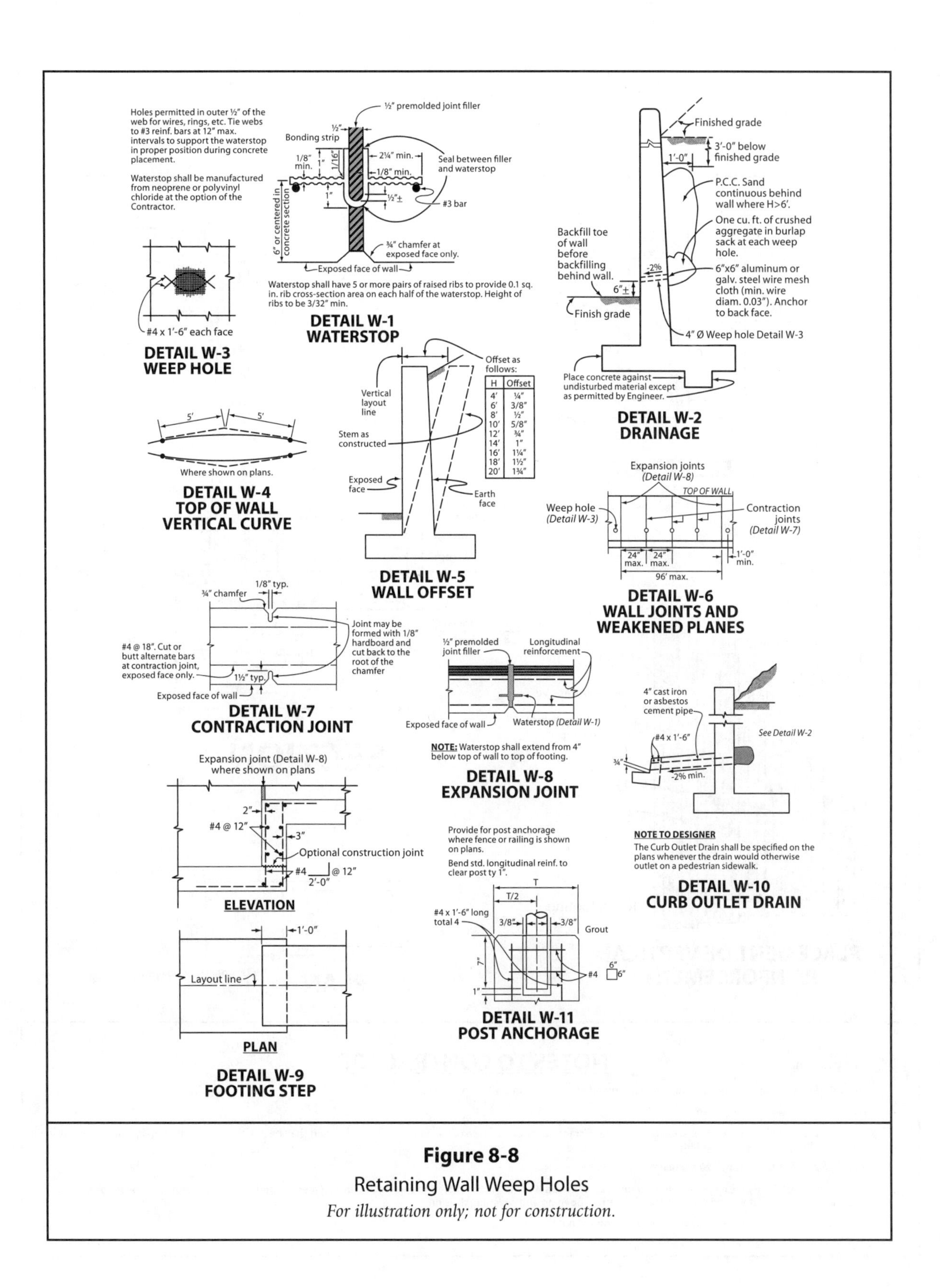

Figure 8-8

Retaining Wall Weep Holes

For illustration only; not for construction.

Figure 8-9

Retaining Wall Design

For illustration only; not for construction.

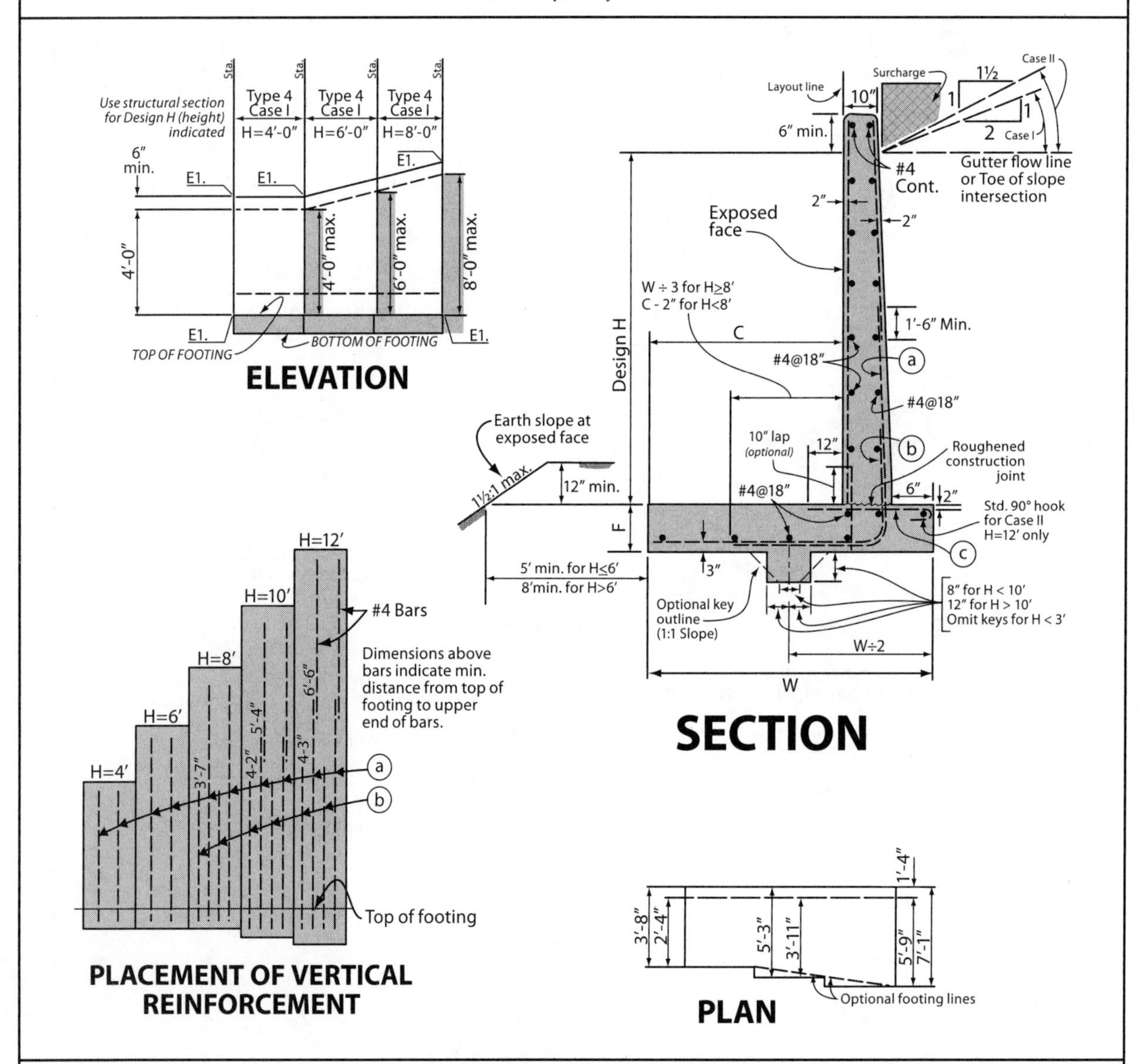

NOTES TO CONTRACTOR

1. All bar reinforcement shall conform to the requirements of ASTM Designation A615, Grade 60.
2. All concrete shall be Class 6.0-B-3250.
3. Bar spacings are center to center of bars. Bar cover is clear distance between surface of bar and face of concrete and shall be 2" unless otherwise noted. Reinforcement shall terminate 2" from concrete surfaces unless otherwise noted.
4. Longitudinal bars shall be lapped 20 bar diameters at splices.
5. The contractor shall notify the engineer when footing excavations are completed and prior to placing of reinforcement to permit an inspection of the bearing soil by a representative of the geology and soils section of the street opening and widening division.
6. Compaction of the backfill material by ponding or jetting will not be permitted.

Figure 8-10

Retaining Wall Reinforcement

For illustration only; not for construction.

REINFORCING STEEL, DIMENSIONS AND DATA										
	Case I Level Backfill + 2′ Surcharge or 2:1 Backfill					Case II 1½:1 Backfill				
Design H	4′	6′	8′	10′	12′	4′	6′	8′	10′	12′
W	3′-8″	5′-3″	7′-1″	9′-4″	11′-9″	3′-8″	5′-3″	7′-1″	9′-4″	11′-10″
F	0′-10″	0′-10″	0′-10″	0′-11″	1′-1″	0′-10″	0′-10″	0′-10″	1′-0″″	1′-3″
C	2′-4″	3′-11″	5′-9″	8′-0″	10′-5″	2′-4″	3′-11″	5′-9″	7′-11″	10′-3″
T	0′-10″	0′-10″	0′-10″	0′-10″	1′-0″	0′-10″	0′-10″	0′-10″	0′-11″	1′-1″
Bar (a)	4 @ 18	5 @ 15	4 @ 15	5 @ 13½	6 @ 11½	4 @ 18	5 @ 14½	4 @ 16	5 @ 12	6 @ 10
Bar (b)			6 @ 15	7 @ 13½	7 @ 11½			7 @ 16	7 @ 12	7 @ 10
Bar (c)	4 @ 18	4 @ 18	4 @ 18	4 @ 18	4 @ 18	4 @ 18	4 @ 18	4 @ 18	4 @ 18	4 @ 18
Soil Pressure (psf)	630	650	660	660	700	490	560	610	680	750

Figure 8-11

Retaining Wall Types: TYPE I

For illustration only; not for construction.

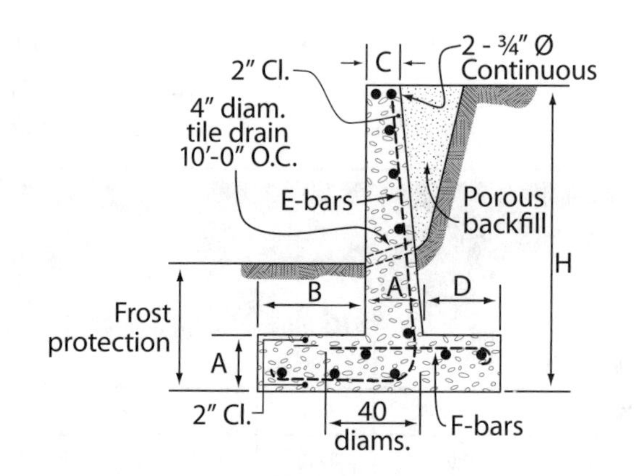

TYPE I

H (ft.-in.)	A (in.)	B (ft.-in.)	D (ft.-in.)	C (in.)	Toe Pressure p.s.i.	E-Bars (in.)	F-Bars (in.)
5-0	8	1-2	0-6	8	765	#3 – 12 O.C.	#3 – 18 O.C.
6-0	8	1-5	0-8	8	865	#3 – 12 O.C.	#3 – 18 O.C.
7-0	8	1-8	0-10	8	930	#3 – 9 O.C.	#3 – 18 O.C.
8-0	12	1-11	0-9	8	1125	#3 – 11½ O.C.	#3 – 18 O.C.
9-0	12	2-2	1-0	8	1230	#3 – 7½ O.C.	#3 – 18 O.C.
10-0	12	2-5	1-2	8	1315	#4 – 10 O.C.	#3 –12 O.C.
11-0	12	2-8	1-5	8	1420	#5 – 10½ O.C.	#3 – 12 O.C.
12-0	12	2-11	1-8	8	1515	#6 – 12 O.C.	#4 – 14 O.C.
13-0	12	3-2	1-11	10	1630	#7 – 12 O.C.	#4 – 12 O.C.
14-0	12	3-5	2-2	10	1735	#8 – 12½ O.C.	#4 – 10 O.C.
15-0	14	3-8	2-3	12	1895	#8 – 12½ O.C.	#4 – 12 O.C.
16-0	15	3-11	2-4	12	2010	#8 – 11½ O.C.	#4 – 10 O.C.
17-0	16	4-2	2-6	12	2130	#9 – 13 O.C.	#4 – 9½ O.C.
18-0	17	4-4	2-7	12	2260	#9 – 11½ O.C.	#4 – 9½ O.C.
20-0	19	4-10	2-11	12	2510	#9 – 10 O.C.	#5 – 10½ O.C.
22-0	21	5-4	3-3	12	2750	#9 – 8 O.C.	#6 -11½ O.C.
24-0	24	5-10	3-5	12	3020	#9 – 7½ O.C.	#6 – 11½ O.C.
26-0	26	6-4	3-8	12	3240	#9 – 6 O.C.	#6 – 11 O.C.
28-0	28	6-10	4-0	12	3500	#9 – 5½ O.C.	#6 – 9 O.C.
30-0	31	7.3	4-2	12	3780	#9 – 5 O.C.	#6 – 9 O.C.

NOTE:

- Design of retaining wall is based on earth 100 p.c.f angle of repose assumed 33 degrees, and no surcharge.
- Designed for 2000-lb. controlled concrete.
- A low-water-cement ratio recommended for permanency.
- The resultant pressure on wall above the outer edge of the middle third.
- Expansion joints in wall should not be over 75 ft. on center.
- Construction joints 30 ft. on center.

Figure 8-12

Retaining Wall Types: TYPE II

For illustration only; not for construction.

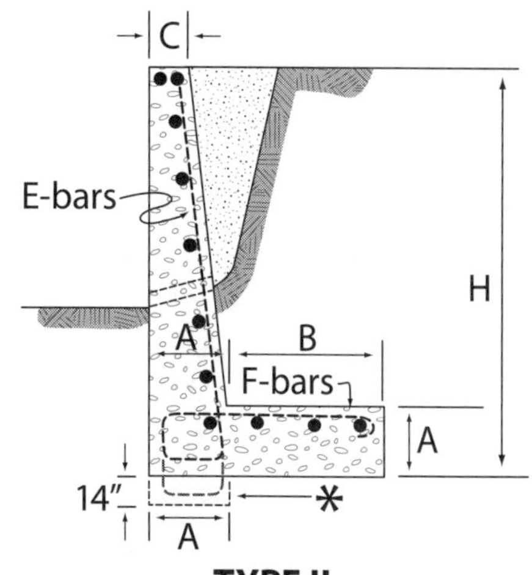

H (ft.-in.)	A (in.)	B (ft.-in.)	C (in.)	Toe Pressure p. s. i.	E-Bars (in.)	F-Bars (in.)
5-0	8	2-1	8	1170	#3 – 12 O.C.	#3 – 12 O.C.
6-0	8	2-7	8	1380	#3 -12 O.C.	#3 – 12 O.C.
7-0	8	3-2	8	1610	#3 – 9 O.C.	#3 – 8 O.C.
8-0	12	3-7	8	1825	#3 – 11½ O.C.	#3 – 8 O.C.
9-0	12	4-0	8	2030	#4 – 4 O.C.	#4 – 10½ O.C.
10-0	12	4-7	8	2230	#4 – 10 O.C.	#5 – 11 O. C.
*11-0	12	5-2	8	2440	#5 – 10½ O.C.	#6 – 12 O.C.
*12-0	12	5-9	8	2640	#6 – 12 O.C.	#7 – 12 O.C.
*13-0	12	6-4	10	2850	#7 – 12 O. C.	#7 – 9½ O.C.
*14-0	12	6-11	10	3050	#7 – 9½ O.C.	#8 – 10 O.C.
*15-0	14	7-3	12	3290	#8 – 12½ O.C.	#8 – 10½ O.C.
*16-0	15	7-8	12	3510	#8 – 11½ O.C.	#9 – 12 O.C.
*17-0	16	8-3	12	3730	#9 – 13 O.C.	#9 –10 O.C.
*18-0	18	8-6	12	3950	#9 – 12½ O.C.	#9 – 10 O.C.
20-0	20	9-6	12	4390	#9 – 10 O.C.	#9 – 8 O.C.
22-0	22	10-5	12	4820	#9 – 8½ O.C.	#9 – 7 O.C.
24-0	25	11-3	12	5270	#9 – 7½ O.C.	#9– 6 O.C.
26-0	28	12-2	12	5690	#9 – 7 O.C.	#9 – 5½ O.C.
28-0	30	13-1	12	6160	#9 – 6 O.C.	#9 – 4½ O.C.
30-0	33	13-10	12	6580	#9 – 5 ½ O.C.	#9 – 4 O.C.

NOTE:

- Design of retaining wall is based on earth 100 pcf angle of repose assumed 33 degrees, and no surcharge.
- Designed for 2000-lb. controlled concrete.
- A low-water-cement ratio recommended for permanency.
- The resultant pressure on wall above the outer edge of the middle third.
- Expansion joints in wall should not be over 75 ft. on center.
- Construction joints 30 ft. on center.

Figure 8-13

Retaining Wall Types: TYPE III

For illustration only; not for construction.

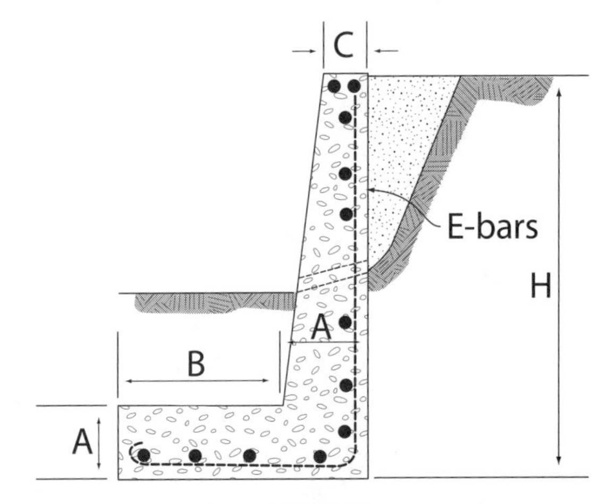

H (ft.-in.)	A (in.)	B (ft.-in.)	C (in.)	Toe Pressure p.s.i.	E-Bars (in.)
5-0	8	1-8	8	572	#3 – 12 O.C.
6-0	8	2-5	8	545	#3 – 12 O.C.
7-0	8	3-3	8	525	#4 – 14 O.C.
8-0	12	3-2	8	720	#3 – 10 O.C.
9-0	12	4-1	8	694	#4 – 12 O.C.
10-0	12	5-0	8	675	#5 – 12½ O.C.
11-0	12	6-0	8	657	#6 – 13 O.C.
12-0	12	7-1	8	640	#6 – 10 O.C.
13-0	12	7-9	10	677	#7 – 11 O.C.
14-0	12	9-0	10	657	#8 – 11 O.C.
15-0	14	8-10	12	800	#8 – 11 O.C.
16-0	15	9-6	12	838	#9 – 12 O.C.
17-0	16	10-3	12	875	#9 – 11 O.C.
18-0	18	10-9	12	955	#9– 10½ O.C.
20-0	20	12-2	12	1030	#9 – 9 O.C.
22-0	22	13-8	12	1104	#9 – 7½ O.C.
24-0	25	14-10	12	1225	#9 – 6½ O.C.
26-0	28	16-2	12	1328	#9 – 6 O.C.
28-0	30	17-5	12	1425	#9 – 5 O.C.
30-0	33	18-7	12	1545	#9 – 4½ O.C.

NOTE:

- Design of retaining wall is based on earth 100 pcf angle of repose assumed 33 degrees, and no surcharge.
- Designed for 2000-lb. controlled concrete.
- A low-water-cement ratio recommended for permanency.
- The resultant pressure on wall above the outer edge of the middle third.
- Expansion joints in wall should not be over 75 ft. on center.
- Construction joints 30 ft. on center.

Figure 8-14

Retaining Wall Types: TYPE IV

For illustration only; not for construction.

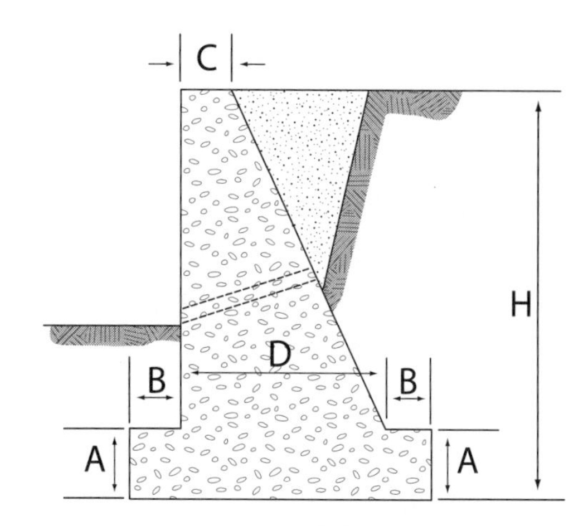

TYPE IV

H (ft.-in.)	A (ft.-in.)	B (in.)	D (ft.-in.)	C (in.)	Toe Pressure
5-0	1-0	6	1-4	8	1046
6-0	1-0	6	1-8	8	1360
7-0	1-0	6	2-4	8	1540
8-0	1-0	6	2-10	8	1760
9-0	1-0	6	3-4	8	2040
10-0	1-0	6	3-10	12	2300
11-0	1-0	6	4-4	12	2560
12-0	1-0	6	4-10	12	2800
13-0	1-0	6	5-5	12	3040
14-0	1-0	6	5-11	12	3140
15-0	1-0	6	6-6	12	3500
16-0	1-0	6	7-0	16	3780
17-0	1-0	6	7-6	16	4040
18-0	1-0	6	8-0	16	4340
20-0	1-0	6	9-4	16	4750

NOTE:

- Design of retaining wall is based on earth 100 pcf angle of repose assumed 33 degrees, and no surcharge.
- Designed for 2000-lb. controlled concrete.
- A low-water-cement ratio recommended for permanency.
- The resultant pressure on wall above the outer edge of the middle third.
- Expansion joints in wall should not be over 75 ft. on center.
- Construction joints 30 ft. on center.

CHAPTER 9

Concrete Masonry Retaining Walls

An easy way to build a retaining wall is by using prefabricated Concrete Masonry Units (CMUs), also called Concrete Block, and Concrete Hollow Units. Some units are made to simulate stone, bricks, and special architectural finishes as described below. The Concrete Masonry Association publishes manuals and information of the many types of concrete building units on the market.

These units are grouped according to the wall thickness in which they are used. These range from 4, 6, 8 and 12 inches. Some are 16 inches square and function as pilaster blocks that serve to carry concentrated loads from trusses and roof girders.

Fig. 9-1

Terms Relating to Concrete Masonry Units

A. Scoring
B. Stretcher End (Mortar Groove)
C. Breaker
D. Sash Groove
E. Plain End
F. Rectangular Core
G. Pear Core
H. Split Face
I. Fluted ("8-Rib" shown)
J. Bullnose ("Double" shown)

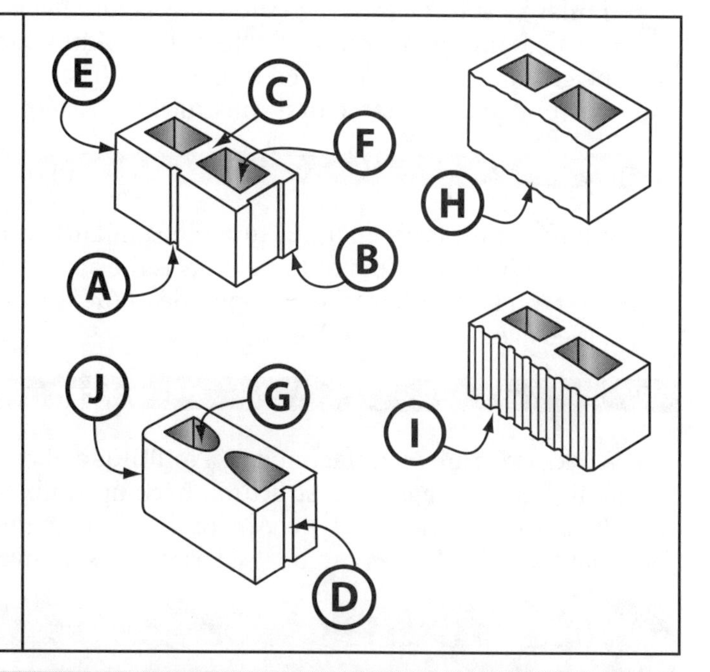

DIMENSIONS

Common practice in describing masonry units is to give the width first, the height second, and the length last, followed by descriptive names. For example an 8 x 4 x 16 Standard Block is 8″ wide, 4″ high and 16″ long. These sizes are in modular dimensions and actual sizes are 3/8″ less.

All pertinent dimensions and minimum shell thicknesses are shown. Rather than repeat dimensions for each unit, the Standard Block in each group is dimensioned, and these dimensions apply to other units in the group. Where there are variations in the units applicable dimensions are shown for the individual unit.

ARCHITECTURAL FEATURE UNITS

The ever changing pattern of the concrete masonry feature unit is the result of demands by the Architect for flexibility in scale and design.

As the architectural features of buildings change, the materials used must also change. The manufacturers of Concrete Masonry Units will be happy to consult with you for a feature unit to expand the horizons of your imagination.

Some of the categories of Architectural Feature Units are outlined as follows:

CAP OR PAVING UNITS

Cap or paving units are manufactured in a variety of sizes. These are used as capping units for parapet, garden and retaining walls and are also used for stepping stones, patios, fireplaces, barbecues, veneering, etc. Many integral colors are available. They may be used both structurally and non- structurally. When used in reinforced walls the reinforcing steel is generally laid in grout space between wythes.

SLUMPED UNITS

Similar to split-faced units (shown in Figure 9-1), slumped units are available in standard and special sizes and in a variety of colors. The widths vary with the amount of surface projection but the heights are held to specified dimensions. They are used to give special architectural effect both in structural and non-structural construction.

SPLIT-FACED UNITS

Split-faced units are manufactured in standard and special sizes and in a variety of textures and colors. The special feature of these units is the architectural appearance of one side. This makes possible a standard block texture on one side of the wall and a special stone-like texture on the exposed side. (See Figure 9-1 above.)

VENEER UNITS

Veneer units are manufactured in a multitude of colors and as many textures. They are non-structural and are laid against a structural back up wall. The width is approximately 3 1/2 inches, but the height varies with the type. Any of the architectural units which have been mentioned and the standard 4″ wide blocks can be used as veneers to give the effect which is desired by the designer.

SPECIAL SHAPED UNITS

Concrete Masonry Units are flexible in that they are capable of being adapted, modified or molded; responsive to changing conditions. Special Shaped Units provide this flexibility. These are available for structural and non-structural purposes and in standard and supplementary sizes, including 12″ high units and design face units. Your Manufacturer's literature should be consulted.

SCREEN BLOCKS

The newest member of the concrete masonry family is the most dramatic. Screen wall units are manufactured in standard face and sculptured designs. The sizes have a range to meet nearly every decorative need from a 4 x 4 x 4 to the giant 16″ square. These units in addition to complete fire safety are used in areas to screen out street noises as well as augment the already efficient insulating qualities of the other concrete masonry. The natural gray color can be painted to blend or complement other materials and colors. Consult the manufacturers' literature for additional information.

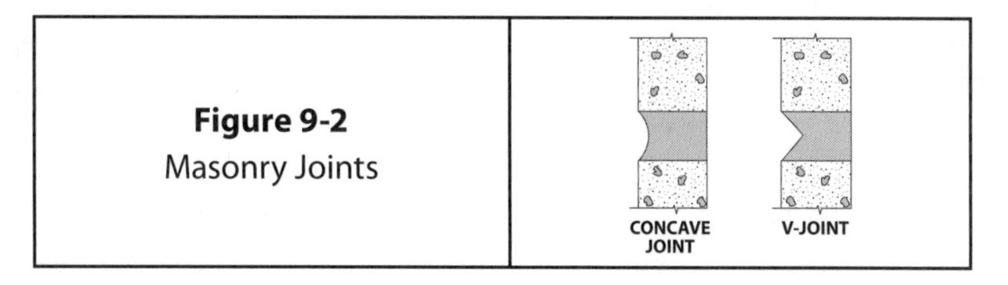

Figure 9-2
Masonry Joints

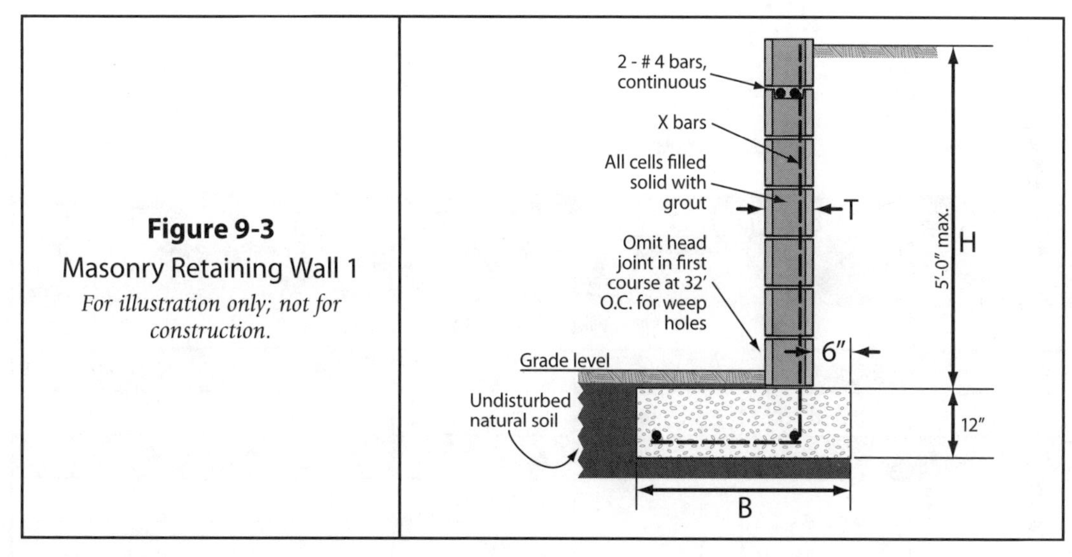

Figure 9-3
Masonry Retaining Wall 1
For illustration only; not for construction.

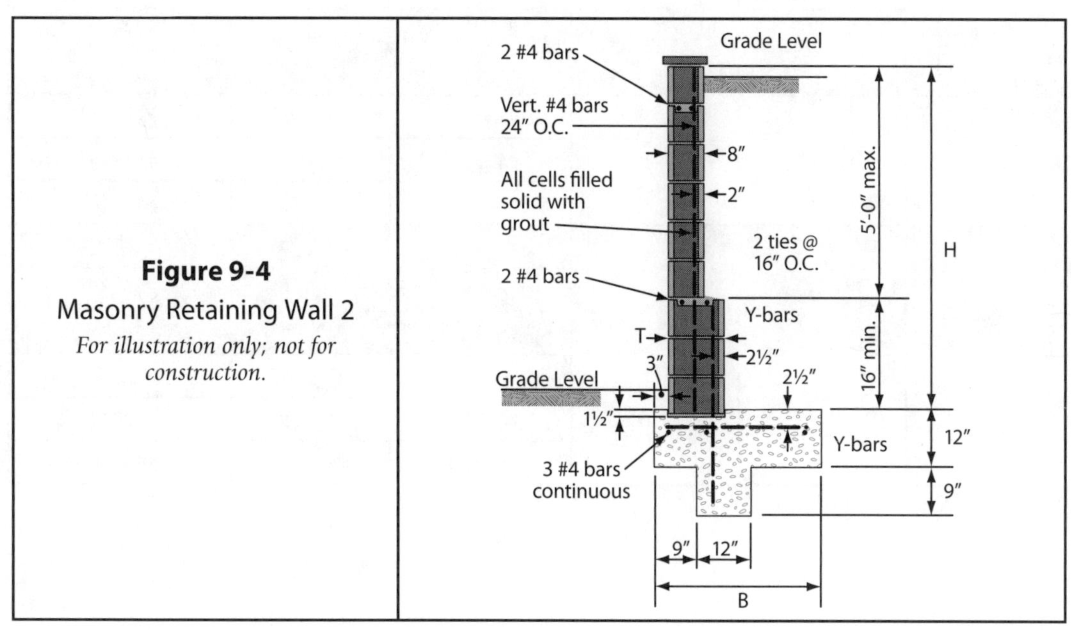

Figure 9-4
Masonry Retaining Wall 2
For illustration only; not for construction.

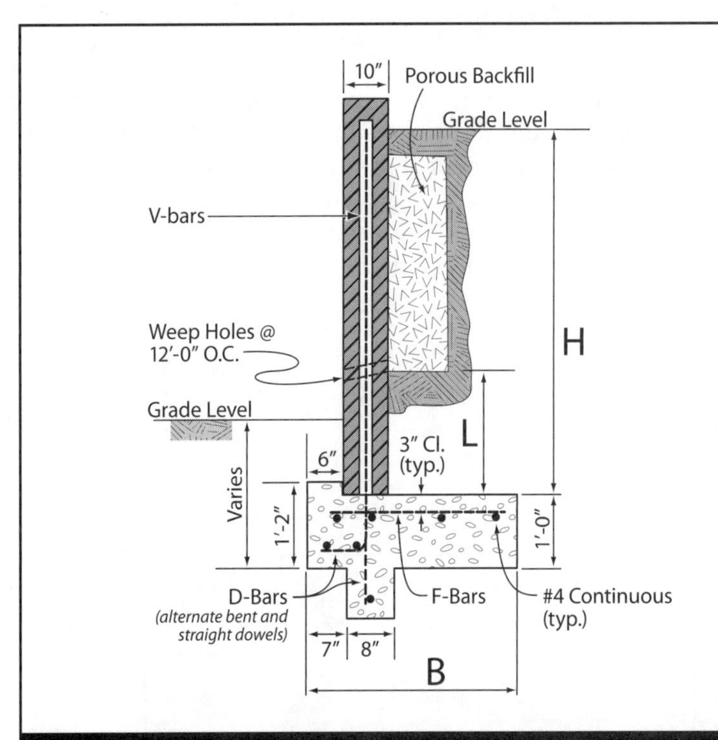

Figure 9-5
Masonry Retaining Wall Rebar
Section through typical low, reinforced masonry retaining wall

For illustration only; not for construction.

STEEL REQUIRED FOR LOW REINFORCED MASONRY RETAINING WALLS

H	B	L	D Bars	V Bars	F Bars
2′-0″	1′-9″	1′-10″	#3 @ 40″		#3 @ 40″
2′-6″	1′-9″	2′-4″	#3 @ 40″		#3 @ 40″
3′-0″	2′-0″	2′-10″	#3 @ 40″		#3 @ 40″
3′-6″	2′-0″	3′-4″	#3 @ 40″		#3 @ 40″
4′-0″	2′-4″	1′-4″	#3 @ 27″	#3 @ 27″	#3 @ 27″
			#4 @ 40″	#3@40″	#3 @ 40″
4′-6″	2′-8″	1′-6″	#3 @ 19″	#3 @ 38″	#3 @ 19″
			#4 @ 35″	#3 @ 35″	#3 @ 35″
5′-0″	3′-0″	1′-8″	#3 @ 14″	#3 @ 28″	#3 @ 14″
			#4 @ 25″	#3 @ 25″	#3 @ 25″
			#5 @ 40″	#4 @ 40″	#4 @ 40″
5′-6″	3′-3″	1′-10″	#3 @ 11″	#3 @ 22″	#3 @ 11″
			#4 @ 20″	#4 @ 40″	#3 @ 20″
			#5 @ 31″	#4 @ 31″	#4 @31″
6′-0″	3′-6″	2′-0″	#3 @ 8″	#3 @ 16″	#3 @ 8″
			#4 @ 14″	#4 @ 28″	#3 @ 14″
			#5 @ 20″	#5 @ 40″	#4 @ 20″

MODULAR DIMENSIONS

The width, height, and length of all concrete masonry units ends in 5/8″. For example, a block 15-3/8″ long plus 5/8″ mortar joints equals 16″ length. The object of modular coordination is to take advantage of economy of having all building materials apply to 4″ increments including masonry materials.

Modular dimension lines are from the center of a joint to the center of a joint. For this reason, the thickness of one mortar joint should be subtracted from the modular dimension to give the exact length. For opening dimensions the thickness of one mortar joint should be added to the modular dimensions to get exact opening dimensions. For simplicity it is usually advisable to use modular dimensioning for the floor plan and block layout plan as all fractions are eliminated. Exact dimensions can be used but they will be in fractions of an inch. Actual dimensions should be considered when dimensioning the foundation plan; however, modular dimensions may also be used.

8" AND 4" MODULAR LAYOUT

For design simplicity and economy of construction plan dimensions should be determined by using multiples of 8″ (8″ modular dimension). This not only means a total wall length should be a multiple of 8″ but walls between openings as well as openings themselves should conform to this 8″ modular dimension.

Sometimes it is impractical to adhere to the 8″ module and a 4″ module must be used. Masonry opening widths of 1′-8″, 3′-0″ and 3′-8″ and wall space between openings of 2′-4″ require 12″ long blocks (three quarter unit). The use of this unit has little effect on wall patterns and presents no great problems with cell alignment.

Corners involve variations from the regular wall layout system and in some cases involve variations in dimensional layout illustrate the common methods of corner layout). Because the 6″ wide and 12″ wide units used in combination with other widths are not adaptable to the 8″ modular standard, some cutting of units at the corners is necessary.

A block layout plan shows the blocks at the first course is important as it not only relieves dimensions from the floor plan but will save block laying time for the mason.

Vertical core alignment is important as steel reinforcement must be continuous from the foundation to the bond beam and the layout plan should provide for this alignment. Also, when the first block course is properly laid the rest of the block laying time is greatly simplified.

Foundation plan should accurately locate the dowel steel. Other items such as plumbing and electrical which are to be installed with or before concrete work can be shown on the foundation plan or this work can be properly noted and referred to other drawings which will show their location.

BUILDING CODE CONSTRUCTION REQUIREMENTS

All reinforced hollow unit masonry should be built to preserve the unobstructed vertical continuity of the cells to be filled. Walls and cross webs forming such cells to be filled should be full bedded in mortar to prevent leakage of grout. All head (or end) joints should be solidly filled with mortar for a distance in from the face of the wall or unit not less than the thickness of the longitudinal face shells. Bond should be provided by lapping units in successive vertical courses or by equivalent mechanical anchorage.

Vertical cells to be filled should have vertical alignment sufficient to maintain a clear, unobstructed continuous vertical cell measuring not less than two inches by three inches (2″ x 3″).

Cleanout openings should be provided at the bottoms of all cells to be filled at each lift or pour of grout where such lift or pour of grout is in excess of four feet (4′) in height. Any overhanging mortar or other obstruction or debris should be removed from the insides of such cell walls. The cleanouts should be sealed before grouting, after inspection.

Vertical reinforcement should be held in position at top and bottom and at intervals not exceeding 192 diameters of the reinforcement.

All cells containing reinforcement should be filled solidly with grout. Grout should be poured in lifts of eight feet (8′) maximum height. All grout should be consolidated at time of pouring by puddling or vibrating and then reconsolidated by again puddling later, before plasticity is lost.

When total grout pour exceeds eight feet (8′) in height, the grout should be placed in four-foot (4′) lifts and special inspection during grouting should be required. Minimum cell dimension should be three inches (3″). Special inspection at time of grouting should not be considered as special inspection.

When the grouting is stopped for one hour or longer, horizontal construction joints should be formed by stopping the pour of grout one and one-half inches (1½″) below the top of the uppermost unit.

COLD WEATHER CONSTRUCTION

No masonry should be laid when the temperature of the outside air is below 40° F., unless approved .methods are used during construction to prevent damage to the masonry. Such methods should include protection of the masonry for a period of at least 48 hours where masonry cement or Type I Portland cement is used in the mortar and grout and for a period of at least 24 hours where Type III Portland cement is used. Materials to be used and materials to be built upon should be free from ice or snow.

Wood members should not be used to resist horizontal forces contributed by masonry or concrete construction in buildings over one story in height. Exception: Wood floor and roof members may be used in horizontal trusses and diaphragms to resist horizontal forces imposed by wind, earthquake, or earth pressure, provided such forces are not resisted by rotation of the truss or diaphragm.

SPECIFICATIONS FOR CONCRETE MASONRY CONSTRUCTION

- The Concrete Masonry Contractor should examine all drawings and specifications and note all conditions that may affect his work and performance in fulfilling the contract.
- Where any deviation is to be made from the plans and specifications, the Engineer or Architect should be notified and his written approval obtained before proceeding with the work.
- The work should include the furnishing of all the necessary labor, materials and appliances necessary to complete the execution of the concrete masonry construction shown on the drawings and specifications.
- All preparations and all masonry work necessary to receive and adjoin other work. Others should furnish all inserts and attachments as noted in the plans and specifications, but these should be installed under this section.

- Cooperation with all other trades in laying out his work. Giving the work his personal supervision and keeping a competent foreman on the job at all times. Arranging for adequate bracing, forming, and shoring required in conjunction with and in the course of constructing the concrete masonry and not provided for under other sections.
- Advising the General Contractor as to the position of all dowels to masonry. The General Contractor should be responsible for the placement of dowels in adjoining construction.
- Placing of all reinforcing steel and furnishing all reinforcing steel for concrete masonry not provided for under other sections.
- Arranging for the necessary storage space and protection for materials at the job site.
- Calling for all inspections as required in the course of his work by the Engineer, Architect and/or Building Department.
- Arranging for furnishing test specimens and samples of materials as may be required.
- All materials should conform to the following current standards:
- Hollow Load Bearing Masonry Units should be Grade A units conforming to the ASTM. Designation C90 and in addition the requirements of the Quality Control Standards of the Concrete Masonry Association.
- Solid Load Bearing Masonry Units should be Grade A units conforming to the ASTM Designation C145.
- Hollow Non-Load Bearing Masonry Units should conform to A.S. T.M. Designation C129.
- Concrete Building Brick should be Grade A units conforming to ASTM Designation C53.
- Masonry units having a dry density of not more than 105 pounds per cubic foot of concrete should be known as lightweight masonry units.
- All masonry units should have a minimum net tensile strength of not less than 125 psi., when tested in accordance with the methods set forth in the Quality Control Standards of the Concrete Masonry Association.
- Masonry units should have cured for not less than 28 days when placed in the structure.
- All masonry units should have a maximum linear shrinkage of .06 of 1 % from the saturated to the oven dry condition, when tested in accordance with the methods set forth in the Quality Control Standards of the Concrete Masonry Association.
- Cement should be Type I (or Type II, Type III or Type V) Portland Cement conforming to A.S. T .M. Designation C150.
- Plastic Cement should have less than 12% of the total volume in approved types of plasticizing agents and should conform to all of the requirements for Portland Cement in A.S. T .M. Designation C150, except in respect to the limitations on insoluble residue, air-entrainment and additions subsequent to calcination.
- Mortar should be freshly prepared and uniformly mixed in the ratio of 1 part Portland Cement, 14 parts minimum to ___ part maximum lime putty or hydrated lime, damp loose sand not less than 2½ and not more than 3 times the sum of the volumes of the cement and lime used, and should conform to ASTM Designation C270.
- Grout for pouring should be of fluid consistency and mixed in the ratio by volumes, 1 part Portland Cement, 2¼ parts minimum to 3 parts maximum damp loose sand where the grout space is less than 3 inches in its least dimension.
- Grout for pouring should be of fluid consistency and mixed in the ratio by volumes, 1 part Portland Cement, 2 parts minimum to 3 parts maximum damp loose sand, 2 parts coarse aggregate where the grout space is 3 inches or more in its least dimension.

- Grout for pumping should be of fluid consistency and should have not less than seven sacks of cement in each cubic yard of grout. The mix design should be approved by the Engineer or Architect.
- Fluid consistency should mean that consistency as fluid as possible for pouring without segregation of the constituent parts.
- Hydrated lime should conform to ASTM Designation C207.
- Quicklime should conform to ASTM Designation C5. Quicklime should be slaked and then screened through a 16-mesh sieve. After slaking, screening, and before using, it should be stored and protected for not less than 10 days. The resulting product should weigh not less than 83 pounds per cubic foot.
- Aggregate should be clean, sharp, and well graded, and free from injurious amounts of dust, lumps, shale, alkali, surface coatings and organic matter.
- Aggregate for mortar should conform to ASTM Designation C144.
- Aggregate for grout should conform to ASTM. Designation C404.
- The use of admixtures should not be permitted in mortar or grout unless substantiating data is submitted to and approved by the Engineer or the Architect.
- The use of admixtures should not be permitted in mortar without reducing the lime content. Proportions of admixture should be as approved by the Engineer or Architect.
- Inert coloring pigments may be added but not to exceed six percent by weight of the cement. The formula should be approved by the Engineer or the Architect.
- The use of uncontrolled fire clay, dirt, and other deleterious materials is prohibited.
- Water should be free from deleterious quantities of acids, alkalies, and organic materials.
- Steel bar reinforcement should conform to ASTM Designation A615.
- Wire reinforcement should conform to ASTM Designation A82.
- Reinforcement should be clean and free from loose rust scale and any coatings that reduce bond.
- Masonry work should not be started when the horizontal or vertical alignment of the foundation is a maximum of one inch total in error.
- All masonry should be laid true, level, plumb, and neatly in accordance with the plans.
- Units should be cut accurately to fit all plumbing ducts, openings, electrical work, etc., and all holes should be neatly patched.
- Extreme care should be taken to prevent visible grout or mortar stains.
- No construction supports should be attached to the wall except where specifically permitted by the Architect or Engineer.
- Masonry units should be sound, dry, clean, and free from cracks when placed in the structure.
- All masonry units should be stored on the job so that they are kept off the ground and protected from rain. Wetting the units should not be permitted except when hot dry weather exists causing the units to be warm to the touch, and then the surface only may be wetted with a light fog spray.
- Proper masonry units should be used .to provide for all windows, doors, bond beams, lintels, pilasters, etc., with a minimum of unit cutting.
- Where masonry unit cutting is necessary, all cuts should be neat and true.

- Mortar should be mixed by placing one-half of the water and sand in the operating mixer. Then add the cement, lime and the remainder of the sand and water. Mortar should be re-tempered with water as required to maintain high plasticity. Re-tempering on mortar boards should be done only by adding water within a basin formed with the mortar and the mortar reworked into the water. Any mortar which is unused after one and one-half hours from the initial mixing time should not be used.
- After all ingredients are in the batch mixer they should be mechanically mixed for not less than three minutes. Hand mixing should not be employed unless specifically approved.
- For bonding the masonry to the foundation the top surface of the concrete foundation should be clean and with laitance removed and aggregate exposed before starting the masonry construction.
- Where no bond pattern is shown, the wall should be laid up in straight uniform courses with regular running bond.
- Intersecting masonry walls and partitions should be bonded by the use of steel ties at 24″ O.C. maximum. Corners should have a standard masonry bond by over-lapping units and should be solid grouted.
- Where stack bond is indicated on the plans, approved horizontal reinforcing should be provided at 24″ O.C. maximum.
- Veneer should be bonded to the wall in an approved manner.
- Columns, beams, joists and similar structural members should be anchored to the walls with anchor bolts or their equivalent. Anchors should be fully and solidly grouted in place. Embedment should not be less than two-thirds of the wall thickness unless otherwise noted.
- All longitudinal joints (grout space) in two or more wythe masonry should be solidly grouted.
- The starting joint on foundations should be laid with full mortar coverage on the bed joint except that the area where grout occurs free from mortar so that the grout will contact the foundation.
- Mortar joints should be straight; clean, and uniform in thickness and should be tooled as shown on the plans.
- All walls should have joints tooled with a round bar (or v-shaped bar) to produce a dense, slightly concave surface well bonded to the block at the edges, unless specifically detailed.
- Tooling should be done when the mortar is partially set but still sufficiently plastic to bond.
- All tooling should be done with a tool, which compacts the mortar, pressing the excess mortar out of the joint rather than dragging it out.
- Raked joints should be not more than one-half inch deep and where exposed to the weather should be tooled.
- Where walls are to receive plaster the joints should be struck flush.
- Where joints are to be concealed under paint, these joints should be filled flush and then sacked to produce a dense surface without sheen.
- Joints which are not tight at the time of tooling should be raked out, pointed, and then tooled.
- Unless otherwise specified or detailed on the plans, in hollow unit masonry the horizontal and vertical mortar joints should be 3/8″ thick with full mortar coverage on the face shells and on the webs surrounding cells to be filled.

- Vertical head joints should be buttered well for a thickness equal to the face shell of the unit and these joints should be shoved tightly so that the mortar bonds well to both units. Joints should be solidly filled from the face of the block to at least the depth of the face shell.
- If it is necessary to move a unit after it has been once set in place, the unit should be removed from the wall, cleaned and set in fresh mortar.
- Concrete building bricks should be laid with full head and bed joints.
- The tops of unfilled cell columns under a horizontal masonry beam should be covered with metal lath or special units should be used to confine the grout fill to the beam section.
- All bolts, anchors, etc., inserted in the wall should be solid grouted in place.
- Spaces around metal door frames and other built-in items should be filled solidly with grout or mortar.
- In hollow unit masonry low-lift grouting, the structure should be grouted in heights of less than four feet. In two wythe masonry low-lift grouting, the wall should be grouted in heights of less than eight inches or six times the grout joint thickness, whichever is lesser.
- In two wythe masonry walls one wythe may be carried up 16″ maximum before grouting. The grout joint should be at least one inch wide and should be filled solidly with grout.
- Cleanout holes in high-lift grouting should be provided at the bottom of all cores containing vertical reinforcement in hollow unit masonry and in two wythe masonry should be provided by omitting alternate units on the first course of one wythe.
- Mortar projections and mortar droppings should be washed out of the grout space and off the reinforcing steel with a jet stream of water as required to clean the space.
- All grout should be consolidated at time of pouring by puddling or vibrating and then reconsolidated by later puddling before the plasticity is lost.
- The minimum dimension of the grout space should be three inches.
- Two wythe masonry should cure at least three days and hollow unit masonry should cure at least 24 hours before grouting.
- Grout should be poured to not more than four foot depths, then wait approximately one hour and pour another four foot depth. The full height in each section of 'the wall should be poured in one day.
- Vertical grout barriers or dams should be built across the grout space of two wythe masonry the entire height of the wall to control the flow of the grout horizontally. These barriers should be less than 25 feet on center.
- All reinforcing steel should be inspected in place before grouting and there should be continuous inspection during the grouting operation.
- In two wythe masonry, wire ties consisting of #9 wire rectangles, should connect the wythes and should be spaced not more than 12″ O.C. vertically for stacked bond, not more than 24″ O.C. vertically for running bond, and not more than 32″ O.C. horizontally.
- Waterproofing should consist of a minimum of two coats with at least 24 hours between applications.
- Walls should be waterproofed as frequently as recommended in the guarantee by the waterproofing manufacturer. The first coat of waterproofing should be a filler coat unless otherwise specified.
- All basement walls below grade should be waterproofed on the exterior surface extending from the foundation pad to above grade.

- Waterproofing should be either asphalt conforming to ASTM D449, Type A, or coal-tar pitch conforming to ASTM D450, Type B, or as otherwise approved.
- Surfaces to be waterproofed should be clean, dry and should be given either a priming coat of creosote oil conforming to ASTM D43 and two mop coats of hot coal-tar pitch or a priming coat of asphalt primer conforming to ASTM D41 and two mop coats of hot asphalt. Mop coats should be applied uniformly using not less than 20 lb. of tar or asphalt per 100 sq. ft. per coat and should provide a continuous impervious coating, free from pinholes or other voids.
- Where known water is present membrane waterproofing should be used as specifically noted on the plans.

CHAPTER 10

Other Types of Retaining Walls

There are many other types of retaining walls beside the common cantilever L-shape and inverted T-shape walls. In fact, the ancient master builders constructed great retaining walls standing many stories high. An example is the wall that surrounds the Temple Mount in Jerusalem. The most common is the *gravity wall,* also called the *buttress fill.* This is similar to the dam that retains water in a reservoir. The levee is another example. These are made of a combination of earth and a core of rock.

The construction of the railroad and highways in America required excavation in mountainous areas to hold back steep slopes. This need developed into a simple form of retaining wall called *cribbing.* This is similar to the *cofferdam,* used to retain water.

Originally, they built cribbing of locally cut timbers stacked together at a steep angle and filled with rock. Cribbing made of steel and reinforced concrete followed. More recently, they used bin-type retaining walls for railroad and highway right-of-ways. The main advantage is that they transported the materials to the site by truck or railcar.

For very steep embankments below the roadway, they installed wood or steel sheet piling. This system requires the least space but needs a drop hammer or pile driver to pound the individual piles into the ground.

BUTTRESS FILLS

A buttress fill is a designed compacted earth fill used to provide lateral support to an unstabilized rock mass. The buttress fill should comply with the following.

- You should investigate the ability of the foundation soil to support the buttress. Otherwise, provide additional benching for ordinary fills.
- You should provide specifications for keying of the base of the buttress and for bonding the buttress to the natural ground.
- The minimum base width of a buttress fill should not be less than 12 feet nor less than one-half its height, whichever is the greater. The width of a buttress fill may vary uniformly to a top width of not less than 12 feet.
- The exposed surface of a buttress fill should not exceed a slope of two horizontal to one vertical.
- Provide sub drains which blanket the entire back face of the buttress or which occur at intervals to prevent build-up of hydrostatic pressure. You should provide details of sub drains.

- Provide blanket seals of relatively impervious material on cut pads above buttress fills where grading exposes the strata to infiltration of water. The blanket should be of two-foot minimum thickness or of greater dimension as specified by the foundation engineer.
- The maximum height of a buttress fill should be limited to 30 feet unless the foundation engineer provides substantiating calculations to justify a height above 30 feet.

For design purposes, use a maximum value of 75-psf cohesion and an angle of internal friction of six degrees to determine the resistance of the bedding plane. You may use greater values, if substantiated by tests taken along the probable slip plane under conditions simulating the worst possible field conditions. The method of performing these tests should be included in the foundation engineer's report. Some basic guidelines are:

- Assume the mass of earth to be retained to extend a minimum distance from the top face edge of the buttress equal to the vertical height of the buttress when the surface slope above does not exceed six degrees, or 100 feet when surface slope above the buttress exceeds six degrees.
- Specify the type, percent compaction, cohesion, and angle of internal friction of the materials to be placed in the buttress.
- Design the buttress fill for a minimum safety factor of 1.50 based upon the smaller value of yield or ultimate shear strength of the fill material.
- Upon justification by the foundation engineer, deviations from the above requirements may be approved by the governing authorities.

STEEL SHEETING

You may need sheeting at the edge of a roadway in the mountains where the embankment is too close for a conventional concrete retaining wall. The sheeting may be of wood, heavy steel sections or cold-formed sheets.

The driving of sheeting requires skilled operators, particularly if satisfactory and economical results are to be obtained. This is true for any type sheeting: wood, heavy steel sections or cold-formed sheets.

The importance of proper driving equipment and the proper use of it must be stressed, The American Society of Civil Engineers in its Manual of Engineering Practice on Pile Foundations and Pile Structures has the following recommendation on driving load-carrying piles, and they apply very well to steel sheet piles. They use steel sheet piling at the abutments of bridges where the roadway is elevated

The type, length, dimensions and material having been chosen for the pile, in view of the soil conditions and the load to be carried, the type, size, weight and velocity at, impact of the hammer (or of equivalent gravity fall) must be chosen subject to the conditions that it deliver sufficient-energy without injury to the pile. Selection may be made based on weight of striking part (ram) and foot-pounds delivered per blow.

The matter resolves itself into a practical decision, since it is not feasible to compute the various resistances that must be overcome. Good judgment, based upon knowledge of the local conditions and upon experience in pile driving, is essential, and advice of specialists would be of value.

Moist, silty soils or silty-clay soils are relatively easy to penetrate. Driving in clay soils will vary from easy to difficult or impossible. The same is true of sand. Confined quicksand is very difficult to penetrate by driving and will require the use of a water jet. Gravel-sand mixtures may contain large-size stones that will prevent the penetration of the piling.

No definite recommendations can be given on the size and kind of driving equipment that should be used unless you know all the conditions.

The character of the soil mass through which the piling is to be driven, the moisture content of the soil, the length and weight of the sheeting and whether excavation is done ahead of driving are facts that must be known in order to select the proper driving equipment for handling sheeting.

A hand maul or a light pneumatic hammer may be satisfactory for pushing sheeting in a trench where the bottom can be excavated ahead of driving and when the earth loads on the sheeting are light. If the sheeting is to be driven in advance, of excavation or the side pressures are heavy, then heavier equipment such as a drop hammer or, a pneumatic or steam pile driver will be needed. The use of heavy driving equipment will usually show faster driving with less injury to the sheeting for any given condition. Light equipment tends to batter the top edge and slow down the driving.

The driving equipment must be capable of supplying ample foot-pounds of energy to move the sheeting easily. A driver, which strikes a heavy blow with a low velocity at impact, will do the most work with the least amount of damage to the sheeting. Long, heavy sheet piles require more energy to start them moving than do short light sections. In this respect, the heavy hot rolled sections will require heavier drivers than those for lightweight sheeting. The friction of the soil on the sheeting surfaces and force required for penetration of the leading edge are factors that are very hard to evaluate. In order to select the proper driving equipment, it is essential to know local conditions and have experience with various types of driving equipment in the soil formations that will be encountered.

Regardless of the type of equipment used, it is essential that the sheeting be held firmly in place during driving. In addition, you must hold the driver on the top of the sheeting so that all blows are axial to the sheeting and squarely centered on the top surface. You should provide a good set of leads that will hold the sheeting firmly in place and a driver in proper position during driving...

In trench work where the sheeting is held by wales, it is sometimes possible to hold the hammer on top of the sheeting by means of a crane or a tripod, but the hammer tends to bounce around and cause bending and bruising of the pile. Light pneumatic equipment must be held so that the blows are axial. For driving sheeting, as opposed to "pushing" it, the hand tools strike such a light blow that the driving is very slow, resulting in bending and bruising of the top of the pile.

Medium size single or double-acting hammers will give a much more satisfactory performance when held in leads than when swung on a cable. The leads can be short and need not be as elaborate as would be required for driving long foundation piling but they must be supported to hold the driver in line with the sheeting, and they should be guyed to hold the sheeting in line. A light to medium weight gravity hammer (1200- to 2500-lb.) may give good results if properly guided and used with a 2- to 5-ft drop.

Driving heads, such as can be furnished with sheeting, are designed to spread the load over a larger area of the top of a sheet than would be the case if the driving equipment were used without the head. Therefore, the head should be used only when driving with hand tools or the light pneumatic or gasoline drivers. The driving head is unnecessary and may even be detrimental when used in conjunction with gravity hammers or single or double-acting pile hammers. These drivers have a wide flat surface that will cover the entire top end of one-and in some cases two or more-sheet piles. The driving head, if used in this case, would concentrate the blow on a smaller portion of the pile than would be the case if the driver were centered and held squarely on the top end of the sheeting and the blow struck through the wide base of the driver. If proper driving equipment is used, the special driving head is seldom necessary.

Jets are generally necessary when driving into fine, well-compacted sand or quicksand. They are used in conjunction with driving equipment and the water flow must be controlled to displace just enough material to allow the sheeting to be driven.

Driving a long row of interlocking sheeting requires careful operation to keep the top from progressively leaning toward the undriven end of the row. Carefully line and plumb each piece before starting to drive it. Hold the section firmly in place until it is driven to full penetration. A very satisfactory method is to set up the individual sections well in advance and then drive the row in a stair-step fashion, driving each piece 1 or 2 ft. at a time. Closing the interlock on the bottom leading edge of the sheeting will frequently prevent small pebbles from entering the interlock and interfering with the driving of the following section. Crawling out of plumb can be partly avoided by placing one or two 20-penny spikes in the fold at the top of the interlock of the last piece of sheeting driven, after the adjacent piece has been placed for driving. Each section of sheeting should be straight and free from kinks before it is placed for driving.

CRIB WALLS, OR CRIBBING

Retaining walls are sometimes formed of stacked rectangular elements to form cells, which are filled with soil. Their stability depends on the weight of the cell units and their filling, and on the strength of the filling material. Crib walls are relatively inexpensive. They are usually made of reinforced concrete, although timber and fabricated metal crib walls are not uncommon.

Fig. 10-1
Cribbing

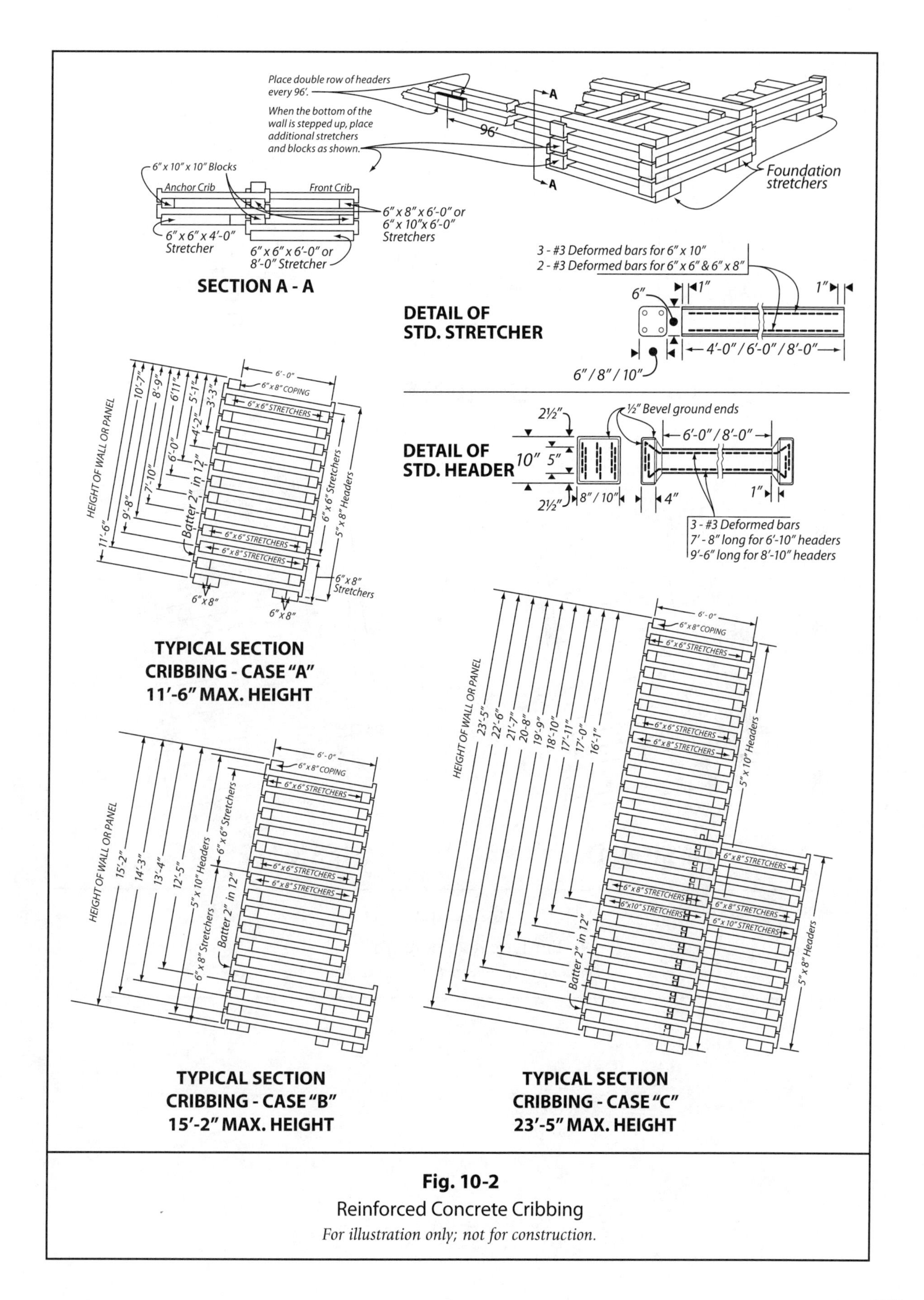

Fig. 10-2

Reinforced Concrete Cribbing

For illustration only; not for construction.

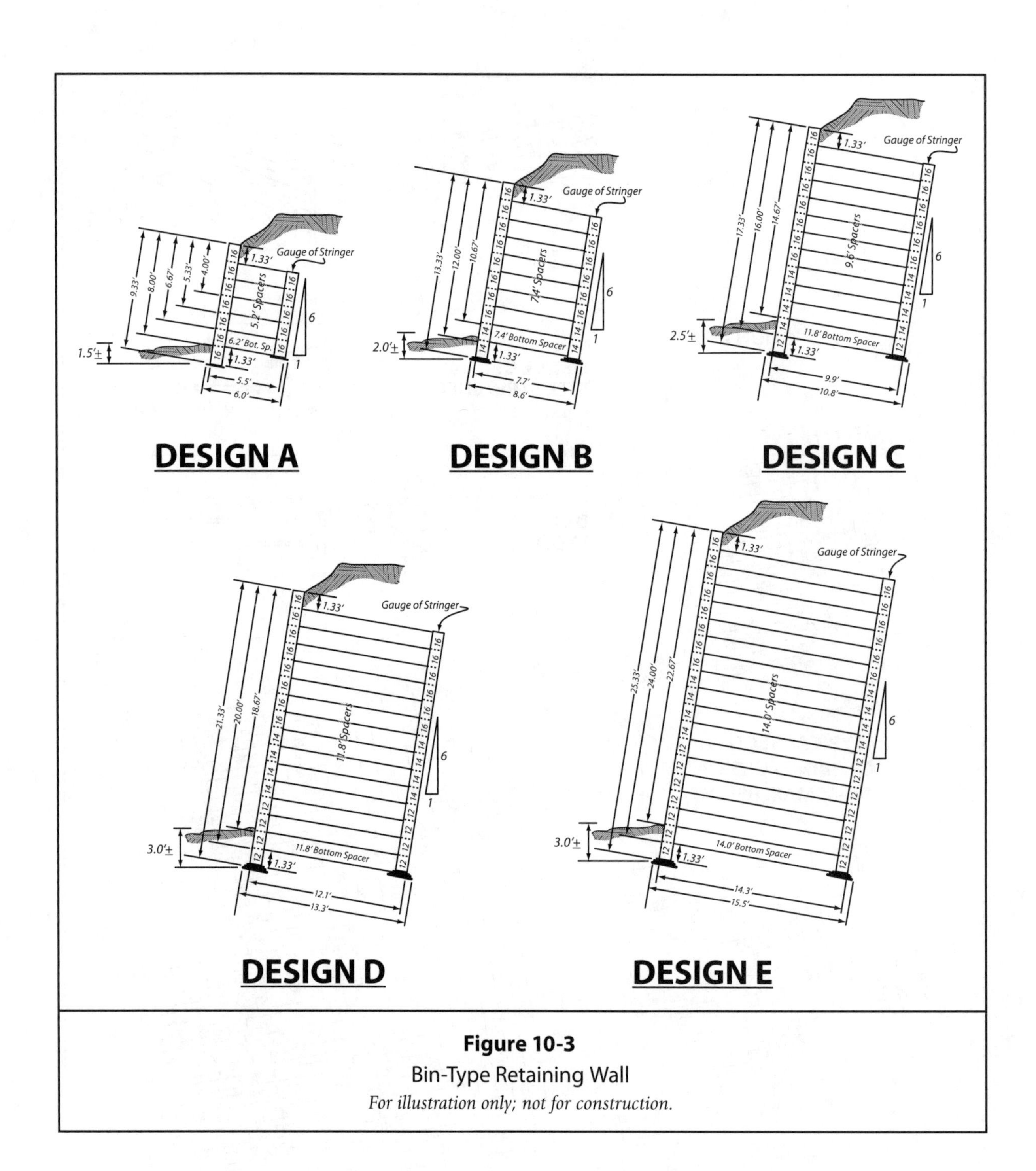

Figure 10-3

Bin-Type Retaining Wall

For illustration only; not for construction.

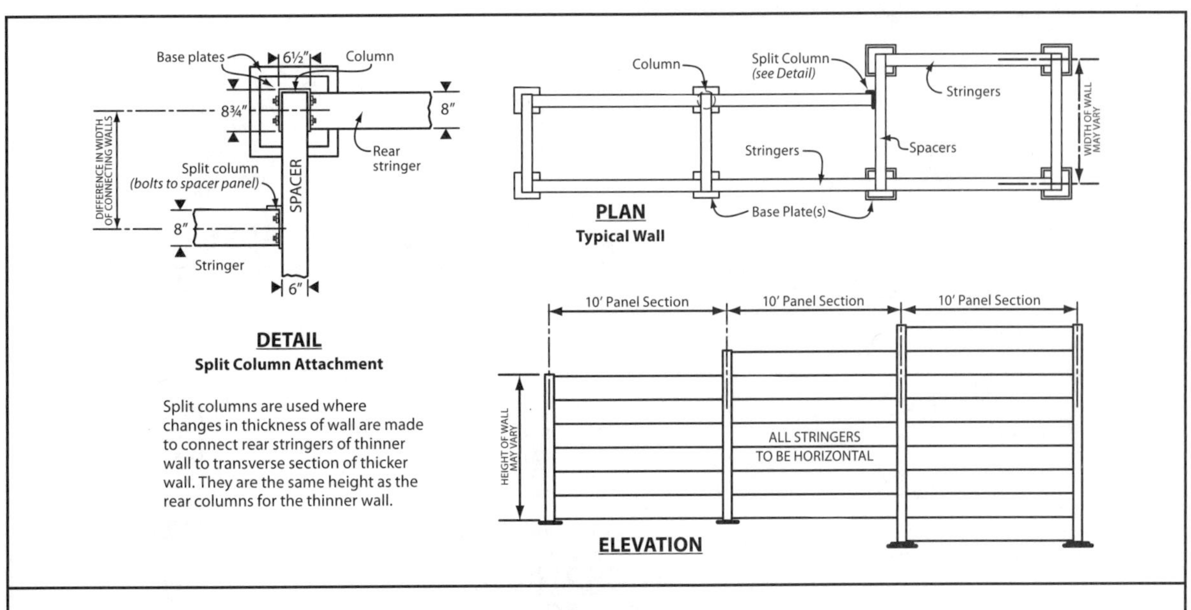

Fig. 10-4
Layout of Bin-Type Retaining Wall
For illustration only; not for construction.

Fig. 10-5
Bin-Type Retaining Walls

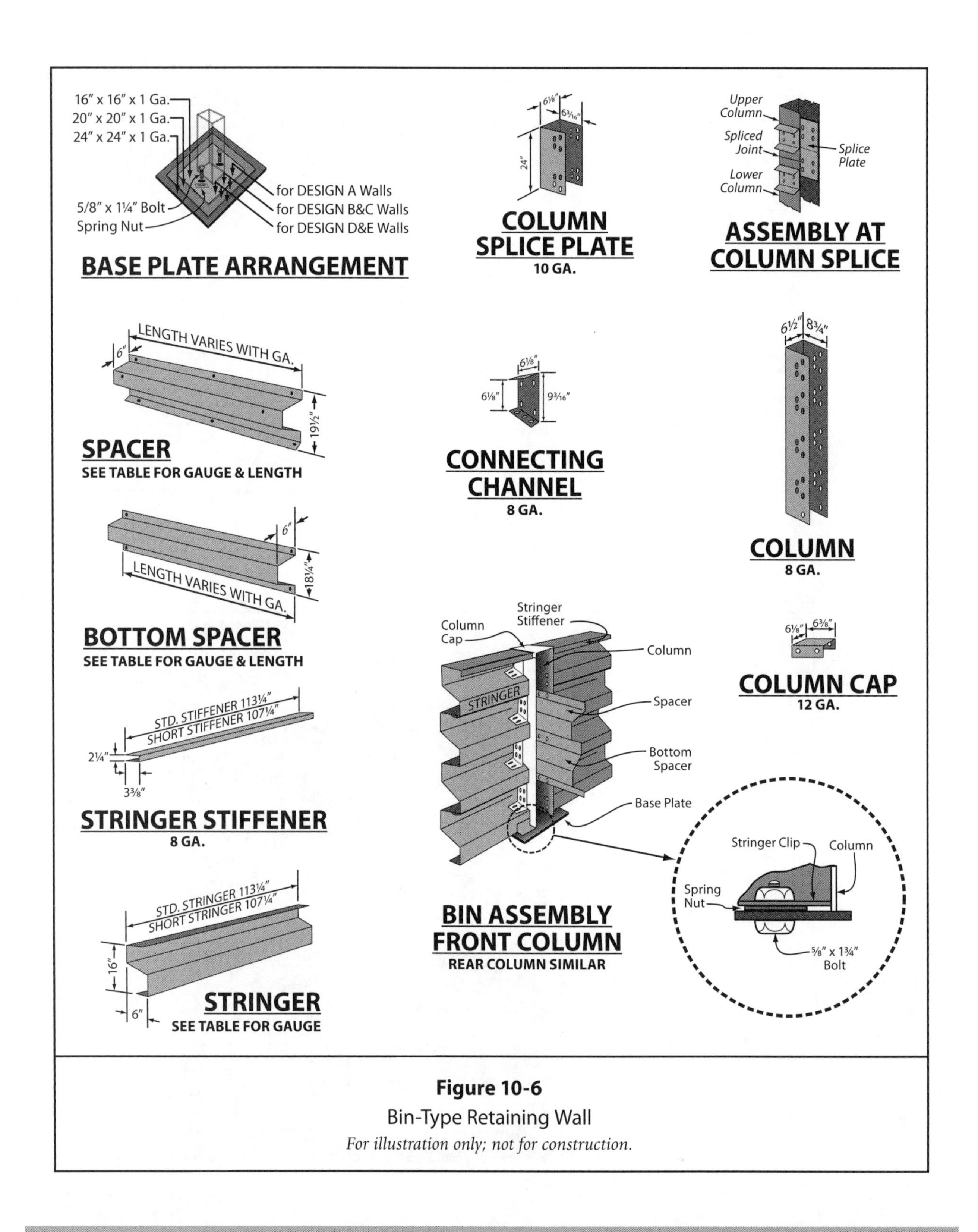

Figure 10-6

Bin-Type Retaining Wall

For illustration only; not for construction.

WOOD BULKHEADS OR STEEL POSTS AND WOOD PLANKING

Another type of wood retaining wall is the bulkhead, which is essentially posts spaced at six feet intervals with horizontal wood 3″ x 12″ planking. The posts may be of wood or steel structural

	Units Required Per Transverse Section																										Units Required Per Panel Section					
	Wall Ht.		Bearing Plates			Front Column Ht. in Feet				Rear Column Ht. in Feet							Spacers Gauge & Length					Bottom Spacer Gauge & Length					Stringers					
																	16	16	14	12	12	16	16	14	12	12	16	14	12	10		
DESIGN	Feet	Feet & In.	16"x16"	20"x20"	24"x24"	1st Lift	2nd Lift	3rd Lift	Total Ht.	1st Lift	2nd Lift	3rd Lift	Total Ht.	Total Col. Lg. in Ft.	Column Splice	Column Cap	5.2'	7.4'	9.6'	11.8'	14.0'	5.2'	7.4'	9.6'	11.8'	14.0'	9.5'	9.5'	9.5'	9.5'	Stringer Stiffener	Wall Ht. in Ft..
A	4.00	4'-0"	2			4.00			4.00	1.33			1.33	5.00		1	0					1					4				1	4.00
	5.33	5'-4"	2			5.33			5.33	2.67			2.67	6.00		1	1					1					6				1	5.33
	6.67	6'-8"	2			6.67			6.67	4.00			4.00	10.67		1	2					1					8				1	6.67
	8.00	8'-0"	2			8.00			8.00	5.33			5.33	13.33		1	3					1					10				1	8.00
	9.33	9'-4"	2			9.33			9.33	6.67			6.67	16.00		1	4					1					12				1	9.33
B	10.67	10'-8"	2	2		10.67			10.67	8.00			8.00	18.67		1		5					1				14				1	10.67
	12.00	12'-0"	2	2		12.00			12.00	9.33			9.33	21.33		1		6					1				14	2			1	12.00
	13.33	13'-4"	2	2		8.00	5.33		13.33	10.67			10.67	24.00	1	1		7					1				14	4			1	13.33
C	14.67	14'-8"	2	2		8.00	6.67		14.67	12.00			12.00	26.67	1	1			8					1			14	6			1	14.67
	16.00	16'-0"	2	2		8.00	8.00		16.00	8.00	5.33		13.33	29.33	2	1			9					1			14	8			1	16.00
	17.33	17'-4"	2	2		12.00	5.33		17.33	8.00	6.67		14.67	32.00	2	1			10					1			14	8	2		1	17.33
D	18.67	18'-8"	2	2	2	12.00	6.67		18.67	8.00	8.00		16.00	34.67	2	1				11					1		14	8	4		1	18.67
	20.00	20'-0"	2	2	2	12.00	8.00		20.00	12.00	5.33		17.33	37.33	2	1				12					1		14	8	6		1	20.00
	21.33	21'-4"	2	2	2	12.00	9.33		21.33	12.00	6.67		18.67	40.00	2	1				13					1		14	8	8		1	21.33
E	22.67	22'-8"	2	2	2	12.00	10.67		22.67	12.00	8.00		20.00	42.67	2	1					14					1	14	8	10		1	22.67
	24.00	24'-0"	2	2	2	12.00	12.00		24.00	12.00	9.33		21.33	45.33	2	1					15					1	14	8	12		1	24.00
	25.33	25'-4"	2	2	2	12.00	8.00	5.33	25.33	12.00	10.67		22.67	48.00	3	1					16					1	14	8	14		1	25.33

This table applies to standard panel sections and includes units for both front and rear of one 10' section of wall.

Figure 10-7

Bin-Type Retaining Walls

For illustration only; not for construction.

shapes and held in position by means of a steel rod anchored to the original ground. In some cases, the steel posts are driven into the ground by a pile driver. These walls can hold street fills up to 12 feet high.

Wood posts and planks are made of creosoted Construction Grade Douglas fir. Post size varies from 6″ x 6″ to 10″ x 10″ depending on the height of retained earth and spacing of posts.

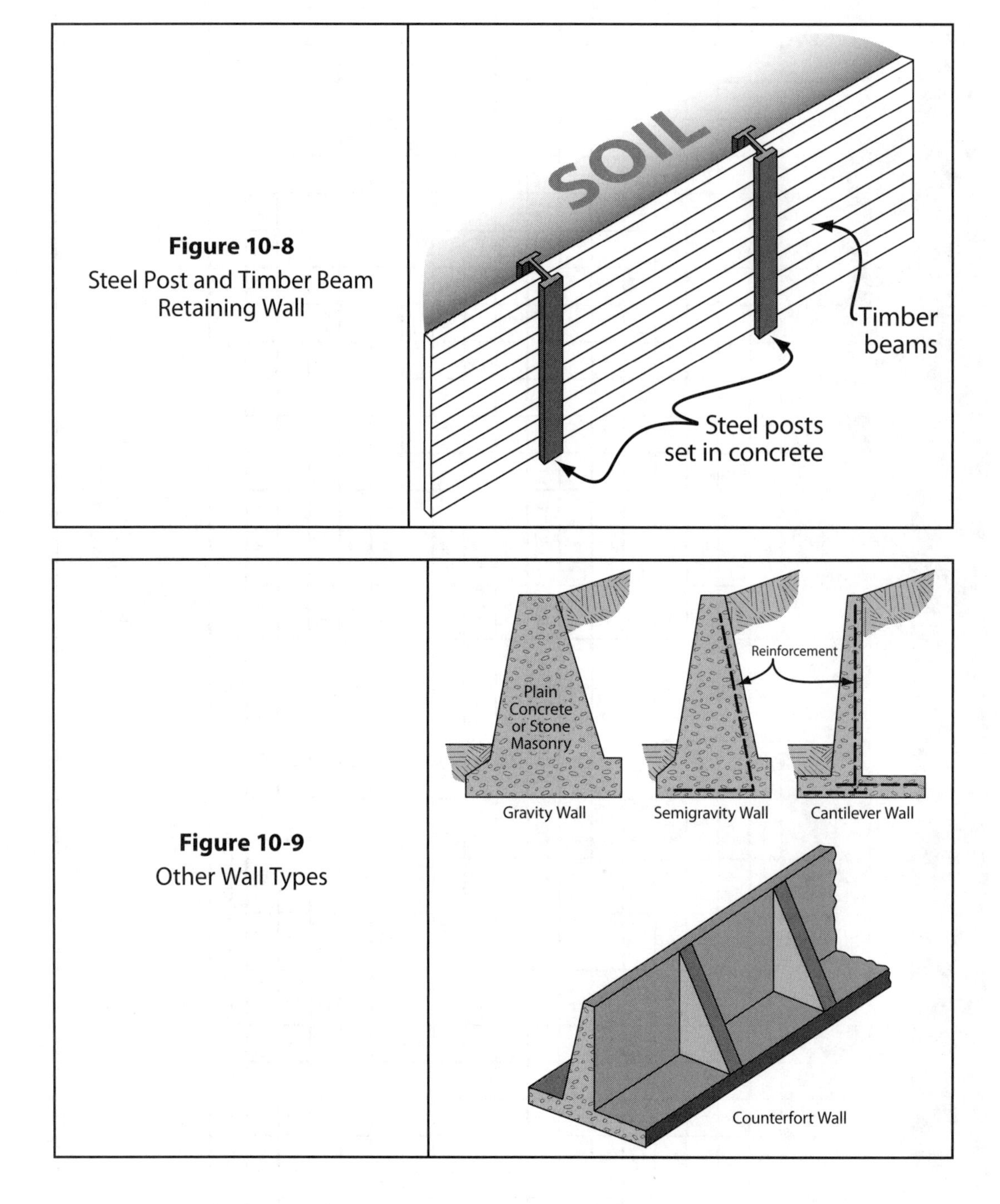

Figure 10-8
Steel Post and Timber Beam Retaining Wall

Figure 10-9
Other Wall Types

CHAPTER 11

Glossary of Drainage and Retaining Wall Terms

AASHO: American Association of State Highway Officials, an organization that has developed standards for roadway construction.

ACCELERATOR: An admixture used to increase early strength gain in concrete.

ACCEPTED ENGINEERING PRACTICE: Design and construction that conforms to accepted principles, tests, or standards of nationally recognized technical and scientific authorities.

A.D.: Air-dried.

ADDENDA: Documents issued before the bid opening which clarify, correct, or change bidding documents or contract documents.

ADMIXTURE: Material other than water, aggregate, or cement, used as an ingredient of concrete.

AGGREGATE: Granular material such as sand, gravel, crushed rock and iron blast furnace slag, which, when used with Portland cement, forms concrete.

AGGREGATE, COARSE: A concrete component with a particle size over 1/4 inch.

AGGREGATE, FINE: A concrete component with a particle size less than 1/4 inch.

AGGREGATE, LIGHTWEIGHT: An aggregate with a dry loose weight of less than 70 pounds per cubic foot.

AGGREGATE/CEMENT RATIO: Weight of aggregate divided by weight of cement.

AGREEMENT: Document signed by an owner and contractor covering the work to be performed.

AIR-DRIED (AD) LUMBER: Lumber used in formwork that has been stored in an open area to dry naturally. Also, see Seasoning.

AIR-DRY WEIGHT: Unit weight of a lightweight concrete specimen cured for 7 days with no moisture loss or gain.

ALLOWABLE STRESS: Amount of force per unit area permitted in a structural member.

ALLOWABLE STRESS INCREASE: Percentage increase in the allowable stress based on the length of time that the load acts on the member.

ANCHOR: Device used to secure formwork to previously placed concrete of adequate strength, normally embedded in the concrete during placement.

ANCHOR BOLTS: Steel bolts embedded in concrete that hold a building or structure to the foundation.

APA: American Plywood Association. They represent most of the plywood manufacturers for the purpose of research, quality, and promotion.

AQUIFER: an underground geological formation of sand, gravel, and fractured rock that transmits large amounts of water.

ARCHITECT: Person licensed by the state and charged with the design and specifications of a building.

AREA DRAIN: a drain used to collect water from a specific area and connected directly to an underground pipe.

ASTM: American Society of Testing and Materials.

AXIAL FORCE: A push (compression) or pull (tension) acting along the length of a member, usually measured in pounds.

AXIAL STRESS: Load or force divided by the cross-sectional area of a member, usually expressed in pounds per square inch (psi).

BACKFILL: Material used to fill excavations around completed foundations or piping.

BACKSHORE: A shore placed under a concrete slab or beam after the formwork and original shores have been removed.

BAG (SACK): A quantity of Portland cement that weighs 94 pounds.

BAR: Metal rod used to reinforce concrete. Also called reinforcing steel, reinforcing bars, or rebar.

BAR CHAIR: Support for reinforcement bars during concrete placement.

BARREL (CEMENT): Weight measure of Portland cement (four bags or 376 pounds).

BASE SHEAR (EARTHQUAKE): Total horizontal seismic force exerted at the top of a foundation.

BASIC WIND SPEED: Probable fastest wind speed measured 33 feet above the ground in a flat open area.

BATCH: Quantity of concrete mixed at one time.

BATTER: Inclination from the vertical.

BATTERBOARDS: Elevated horizontal boards set at the corner of a building used to establish the locations of the corners and elevation of the building foundation wall.

BEAM: A horizontal load-bearing structural member.

BEAM BOTTOM: Soffit or bottom form for a concrete beam.

BEARING WALL: A wall that supports a load in addition to its own weight.

BENCH MARKS: accurately determined points of elevation that usually are marked by a monument.

BENDING MOMENT: Measure of bending effect due to a load acting on a member, which is usually measured in foot-pounds, inch-pounds, or inch-kips.

BENDING STRESS: The force per square inch of an area acting at a point along the length of the member, resulting from the bending moment applied at that point, usually expressed in pounds per square inch (psi).

BERM: a large, shaped bank of earth.

BID: Written proposal submitted by a bidder stating the prices for the work to be performed.

BIN-TYPE RETAINING WALL: A cellular wall used to resist sliding of a hillside across a roadway or property.

BLEEDING: Water on cement surface due to settlement of solids in the mix.

BLOCK GRADING: a rough grading method to establish basic ground form using every fifth contour line.

BLOCKING: Small wood pieces installed between studs, joists, or other members to prevent buckling of formwork.

BOARD FOOT: Unit of measure equivalent ,to a board foot square and 1 inch thick.

BOARDS: Lumber that is 2 or more inches wide and 11/2 inches (or less) thick.

BOND (ADHESION): The ability of an hardened cementitious coat of plaster to hold together with the substrate or to a previous cementitious coat by molecular interlocking action, or both.

BOND (CHEMICAL): The ability to hold together, by chemical process, whether by adhesion or cohesive bond.

BOND FAILURE: The failure of a cementitious coating to hold together or to remain adhered to the underlying substrate or another coat.

BOND: (Mechanical): The ability of a plaster coat to key into, embed with, or otherwise, to lock together with plaster undercoat or substrate.

BOW: Lumber distortion parallel to the grain.

BOX-OUT: Opening or pocket formed in concrete.

BRACE: Load-bearing member installed diagonally.

BUGGY: Two-wheel or motor-driven cart used to carry small quantities of concrete to forms.

BUG HOLE: Void in the surface of formed concrete caused by adhering air or water bubble not displaced during consolidation.

BUILDER'S LEVEL: Surveying instrument to control horizontal planes.

BUILDING PAPER: Heavy paper used to waterproof walls and roofs.

BUILDING PERMIT: Document issued by the building department certifying that plans have been approved for construction.

BUILDUP: Thickness of Shotcrete.

BUILT-UP MEMBER: Single structural wood member made from several pieces fastened together.

BULK CEMENT: Cement delivered in large quantity.

BULKHEAD: Partition built into wall forms to end each concrete pour.

CANT STRIP: See Chamfer.

CATCH BASIN: a drainage structure used to collect water from a specific area with a deep pit to catch sediment.

CATWALK: Narrow elevated walkway.

CELLULAR CONCRETE: Lightweight concrete made of Portland cement, lime/silica, or lime/pozzolan.

CEMENT: (Portland): A hydraulic cement that is made by fusing certain earth materials through pyro-processing to form hydraulic crystalline compounds, mostly calcium and aluminum silicates. These compounds are pulverized to a fine powder and a small amount of calcium sulfate is added to control set.

CEMENT MASON: A craftsman who smoothes and finishes surfaces of poured concrete walls, floors, and sidewalks.

CEMENT (MASONRY): A hydraulic cement for use in mortars for masonry and plaster; containing one or more of the following materials: Portland cement, granulated blast-furnace slag, pozzolan, hydrated lime, limestone, chalk, calcareous shell, tale, slag, clay, and in addition may contain other finely ground filler materials.

CENTERING: Temporary supports placed under arches, shells, and space structures that are removed or lowered as a unit to prevent destructive stresses on a structure caused by unequal support.

CHALK LINE (SNAP LINE): Spool-wound string encased in a chalk-filled container that is pulled taut across a surface, lifted and snapped directly downward so that it leaves a straight chalk mark.

CHAMFER: Beveled edge formed in concrete by a triangular strip of wood (chamfer strip) placed in a form corner.

CHANGE ORDER: Written order to the contractor, signed by the owner, authorizing him to add, delete, or revise work.

CHECK (LUMBER): Lengthwise separation of wood used in formwork that usually extends across or through annual growth rings.

CIRCUMFERENCE: the outside edge or perimeter of a circle. The formula for determining the length of a circumference is 2 multiplied by pi multiplied by the radius (expressed as "$2\pi r$"). (π = pi = 3.14).

CLEANOUT: Opening at the bottom of forms that allows access for removing refuse.

CLR: Abbreviation for clear when used to describe minimum distance between reinforcing bars or between the bars and concrete surface.

COEFFICIENT OF RUNOFF: a fixed ratio of runoff to rainfall used in the Rational Method of computing storm-water runoff.

COLUMN (POST): (1) A vertical load-bearing structural member. (2) A member with a ratio of height to least lateral dimension of 3 or greater. Used primarily to support axial loads.

COMBINED STRESS: Combination of axial and bending stresses acting on a member simultaneously.

COMMON NAIL: Steel wire nail.

COMPACTION: reduction in soil volume by pressure from rollers or tampers.

COMPRESSION: Force that tends to crush a structural member.

CONCENTRATED LOAD: Load centered at a given point.

CONCRETE: Mixture of Portland cement, fine aggregate, coarse aggregate, and water, with or without admixtures.

CONCRETE BLOCK: See Concrete Hollow Masonry Units.

CONCRETE, LIGHTWEIGHT: Low density concrete made by using lightweight aggregate. Usually weighs about, 90 pounds per cubic foot.

CONCRETE MIX DESIGNER: Technician who determines proper quantities of aggregates to meet specifications of a concrete mix.

CONCRETE MIXER OPERATOR (OR BATCH PLANT OPERATOR): Technician who tends mixing machines to mix sand, gravel, cement, and water to make concrete. Also cleans and maintains mixer.

CONCRETE, NORMAL WEIGHT: Concrete made with natural crushed rock and sand, usually having a unit weight of 135 to 165 pounds per cubic foot.

CONSTRUCTION DOCUMENTS: All the written, graphic, and pictorial documents describing the design, location, and physical characteristics of a building that are necessary for obtaining a building permit.

CONSTRUCTION JOINT: The surface where two successive placements of concrete meet; frequently there's a keyway or reinforcement across the joint.

CONTOUR: The physical features of the surface of a specific terrain or property. A line drawn on a plan that connects all points to equal elevation above or below a known or assumed reference point. See also Contour Line.

CONTOUR INTERVAL: The vertical distance between adjacent contour lines.

CONTOUR LINE: A line depicting the surface of the earth along a uniform height above sea level or arbitrary base.

CONTROL CHEMIST (BATCH PLANT): Technician who makes periodic chemical analyses of concrete mix to make sure it is uniform in content and meets specifications. He also keeps records and makes reports.

CRIBBING: See Bin-type retaining wall.

CROSS SECTION (BEAM): Section taken through a member perpendicular to its length.

CROWN: Upward bow in a horizontal structural member.

CROWNING: Installing a horizontal member with its crowned edge up.

CRSI: Concrete Reinforcing Steel Institute.

CULVERT: an underground conduit used to carry water by gravity flow.

CURING: (plaster): The act or process by which the cementitious surface coating continues hydration.

d: Abbreviation for "penny" used to designate nail size.

DAYLIGHT : term used to indicate the meeting at grade of a drain line and slope.

DEAD LOAD: Permanent weight of a building structure, including equipment.

DE-CENTER: To lower or remove centering or shoring.

DECKING: Sheathing material used for deck or slab soffit forms.

DEFLECTION (BEAM OR TRUSS): Amount of sag in a horizontal structural member. Usually expressed as a ratio of the amount of deflection to the span of the beam.

DEFORMATION: Changes in a member's shape, such as shortening, lengthening, twisting, buckling, or expanding. Also called strain.

DELAMINATION (PLYWOOD): Separation between plies, normally due to moisture.

DENSE SELECT STRUCTURAL: High-quality lumber used in formwork, relatively free of characteristics that impair its strength or stiffness.

DEPTH: Board dimension measured parallel to the direction of the principal load on the member.

DEPUTY INSPECTOR: Specially approved building inspector. Also called a special inspector.

DESIGN EARTHQUAKE: 90 percent probability of design not being exceeded in 50 years.

DESIGN LOAD: Total load that a structural member is designed to support.

DETENTION POND: A depression designed to detain storm runoff for a short period of time, usually until the storm passes.

DIAPHRAGM: Horizontal or nearly horizontal system designed to transmit lateral forces to the vertical structural members of a building.

DIAPHRAGM CHORD: Outer edges of a horizontal or vertical diaphragm.

DIAPHRAGM STRUT: Compression or tension member that transfers horizontal seismic loads to a diaphragm.

DIAPHRAGM (VERTICAL): See Shear wall.

DIMENSION LUMBER (STANDARD DRESSED LUMBER): Lumber that is 2 to 5 inches thick and up to 12 inches wide, including joists, rafters, studs, planks, posts, and small timbers.

DIVERSION DITCH: A man-made trench used to intercept the flow of rainwater down a slope.

DOWNSPOUT: Also see Leader

DRAIN PIPE: A pipe used to conduct rainwater or wastewater away from a building site.

DRY MIX: All the ingredients of a concrete mix without moisture.

DRYWELL: a large pit filled with gravel and reaching into lower sandy or gravelly soil to discharge water.

DUMPY LEVEL: Surveyor's instrument used to control horizontal planes.

DUST-ON: Casting dry cement on a concrete surface to absorb bleed water.

EARTHQUAKE MAGNITUDE: A system that measures the effect or intensity of an earthquake at a specified point in terms of a series of levels. This may be from I to XII in the Mercalli Scale, measured by the observed severity of damage, or from 1 to 8 in the Richter Magnitude Scale, based on the intensity of movement at the earthquake's epicenter by various seismographs.

ECOLOGY: The science or study of the interactions between plants, organisms and their environments.

EDGE NAILING (PLYWOOD SHEATHING): Series of nails along edge of each plywood panel.

ELASTIC LIMIT: Amount of stress that concrete can't recover from.

ELASTICITY: See Modulus of elasticity.

ELEPHANT TRUNK: Sectional metal tube for placing concrete in high forms.

ELEVATION: The elevation of a point is the vertical distance above or below another, the elevation of which is known or assumed. Elevations are generally based on mean sea level datum, which is the average of a series of observations of the levels of high and low tides of the ocean taken at convenient points over a long period of time.

END NAILING: Nails driven into the end of a board.

ENGINEERED WOOD: Specially-designed structural member or assembly that is usually built off-site.

ENTRAPPED AIR: Air present in concrete that is not added purposely, as by adding an air entraining admixture.

EQUILIBRIUM MOISTURE CONTENT: Moisture content of wood that is in balance with the relative humidity.

EROSION: The wearing away of land surface by detachment and transport of soils by water or wind.

EVAPOTRANSPIRATION: The moisture that escapes into the atmosphere by evaporation and transpiration from living plants. Evapotranspiration: The moisture that escapes into the atmosphere by evaporation and transpiration from living plants.

EXPOSURE (WIND): Description of the terrain surrounding a building and the highest wind velocity with regard to wind exposure.

FACE (PLYWOOD): Side of a plywood panel that is of higher veneer quality when front and back are of different veneer grades.

FACTOR OF SAFETY: Allowable unit stress based on judgment of a competent authority, the risk involved, consistent material quality, and loading condition control.

FALL LINE: The direction of drainage or a line perpendicular to the contours.

FALSE SET: Premature stiffening of concrete. Concrete can be made plastic again by vibrating.

FALSEWORK: Temporary structure erected to support work in progress, such as shores or vertical posts supporting formwork.

FASTEST MILE WIND SPEED: Highest sustained average wind speed on a mile-long sample of air passing a fixed point.

FAT MIX: Cement mixed with a high cement factor.

FAULT (EARTHQUAKE): Zone of weakness in the earth's crust allowing movement between adjacent crust blocks.

FBM: Feet board measure.

FIBERBOARD: Construction material made of wood or other plant fiber compressed into large sheets.

FINAL SET: Concrete that has hardened enough to resist penetration of a weighted test needle.

FINISH COAT: Final thin coat of Shotcrete or plaster preparatory to hand finishing.

FINISH GRADE ELEVATION: Usually refers to the surface level of the ground after completion of all work.

FLOOD PLAIN: Areas adjacent to streams or rivers subject to regular flooding.

FLOOR ELEVATION: The finish grade of existing or proposed buildings; usually given for the first floor, unless another floor meets grade.

FLOW LINE: A line representing the surface of water in a drainpipe or channel.

FLYING FORMS: Large mechanically handled sections of formwork; frequently includes supporting truss, beam, or scaffolding frames completely unitized.

FOG CURING: Storage of concrete test samples in a moist room, under a controlled temperature, maintained damp by a fine fog-like spray.

FOOTING: The widened base of a foundation used to spread the weight over the ground.

FORCE: Push or pull exerted by one object on another.

FORCE DIAGRAM: Graphic representation of forces as they interact within a structural system.

FORCE, EARTHQUAKE: Acceleration from rest to a velocity of a building resulting from sudden earth movement.

FORM LINE: An approximate contour (i.e., hachure, shading).

FORMWORK: Total system built to contain freshly-placed concrete, including sheathing, supporting members, hardware, and bracing.

FOUNDATION: Brick or concrete support wall that a house or building sits on.

FRENCH DRAIN: A linear drain consisting of a trench filled with loose stones, which discharges runoff back into the soil.

FROST-FREE FILL: Material such as clean sands and gravels for subgrade preparation of roads.

FULL-SIZE LUMBER (SAWN LUMBER): Undressed or rough lumber.

GABION: A large, heavy-gauge wire basket filled with stones used for retaining walls or erosion control. Gabions require no foundation, and can be easily installed in sensitive landscapes.

GANG NAILS: Light-gauge metal plates used as connectors for wood members.

GANGED FORMS: Prefabricated form panels joined to make a larger unit for efficiency in erecting, stripping, and reuse.

GIRDER: Major horizontal structural member that supports secondary beams, joists, or rafters.

GLUE-LAMINATED (GLU-LAM) TIMBERS: Structural members made of wood, plywood, or both, bonded together with adhesive.

GRADE: The given or proposed elevation at any spot. Grade also means the height of the ground level somewhere (synonymous with elevation); and indicates a slope or gradient.

GRADE-MARKED LUMBER: Lumber that has been inspected and stamped showing the specie and quality of the wood.

GRADIENT: The rate of slope between two points. Expressed by angle percentage of slope, or ratio of horizontal distance to vertical change in elevation.

GRADING: The movement of earth by cuts and fills to create landforms.

GREEN LUMBER: Freshly sawn, unseasoned, or undried wood.

GROUNDWATER LEVEL: The depth below finish grade at which the soil is saturated with water.

GUNITE: A dry mixture of cement and sand that is mixed with water at the nozzle and sprayed onto the contoured surface, having a reinforcement network in place, which when hardened forms the shell of the swimming pool.

GUSSET: Small piece of wood, plywood, or metal attached to the corners or intersections of a frame to add strength and stiffness.

GUTTER, ROOF: A rectangular, circular or trapezoidal sheet metal trough used to collect rainwater from a roof surface.

GUTTER, STREET: The trough along the edge of a roadway used to collect and conduct rainwater.

HACHURE: A technique to depict landforms in maps and drawings, using shading.

HORIZONTAL CURVE: A road curve in plan view.

HORIZONTAL SEISMIC FORCE: Reaction of a building or structure to the movement of the ground during an earthquake.

HYDROGRAPH WATER: A graph showing variation in stream-water depth or the volume of water flowing past a point in a stream over a period of time.

IDENTIFICATION INDEX (PLYWOOD): See Span rating.

IMPERVIOUS SOIL: Relatively waterproof soil.

INTERMEDIATE NAILING (SHEATHING): Series of nails within the interior of plywood panels.

INVERT ELEVATION: The flow line elevation of a pipe, culvert, channel, etc., given at all changes of alignment, and at each point of entry and departure of all structures.

INUNDATED AREA: Ground surface susceptible to flooding.

JACK: Mechanical device used to adjust the elevation of forms or form supports.

JACK SHORE: Telescoping adjustable single-post metal shore.

JACK TRUSS: a truss used to support another truss and eliminate a post.

JOURNEYMAN: Tradesman with the experience required to complete any task without supervision.

K.D.: Knocked down.

KEEL: Oil crayon used to mark the locations of framing members.

KERF: Notch or cut in a beam.

KILN: A long rotating drum in which crushed limestone is baked, then ground into a powder forming Portland cement.

KILN-DRIED LUMBER: Wood seasoned in a special chamber using artificial heat.

KILOMETER: One thousand (1000) meters or about 0.6 mile.

KIP: Unit of force representing a thousand pounds.

L-HEAD: Top of a shore that has a braced horizontal member projecting on one side, forming an inverted L-shaped assembly.

L-SHORE: Shore with an L-head.

LAGGING: Heavy sheathing used for underground work to temporarily support earthen walls.

LARGE SCALE: Detail plan scale enabling exact descriptions to be made (such as 1/8, 1/4, 3/16, etc., scale).

LATERAL BRACE (SUPPORT): Member installed at right angles to a chord or web members of trusses for alignment and support.

LATERAL LOAD (FORCE): Side-to-side force acting on a structure.

LAYER (PLYWOOD): Single veneer ply, or two or more plies, laminated with a parallel grain direction.

LEADER: See also Downspout

LEVEL: See also Dumpy Level

LOAD, DEAD: The weight of all permanent structural and nonstructural components of a building, such as walls, floors, roofs and fixed service equipment.

LOAD, FACTORED: Load multiplied by appropriate load factors, used to proportion members by strength design method.

LOAD, LIVE: Any load that isn't permanent, such as people and temporary construction loads.

LOAD, SERVICE: Live and dead loads.

LUMBER: Wood that has been sawed, planed, and cross-cut to length.

MACHINE STRESS-RATED LUMBER (MSR): Mechanically-graded lumber 2 inches or less thick and at least 2 inches wide used in formwork and shoring. Also called machine-evaluated lumber (MEL).

MANHOLE: An access hole and chamber in a drainage system to allow inspection, cleaning, and repair.

MBF: Thousand board feet.

MBM: Thousand (feet) board measure.

MECHANICALLY-LAMINATED: Laminated wood structural member held together with mechanical fasteners.

METER: A unit length equal to 39.37 inches, 1000 millimeters, or 100 centimeters.

MODULUS OF ELASTICITY: Ratio of normal stress to corresponding strain from tensile or compressive stresses below proportional limit of the material.

MODULUS OF RUPTURE: Maximum bending stress.

MOISTURE CONTENT (WOOD): Weight of water in wood divided by its dry weight.

MOMENT OF INERTIA (CROSS SECTION OF A BEAM): Property of a structural shape's measure of ability to resist changing shape, or to indicate member's strength.

NCMA: National Concrete Masonry Association.

NDS: National Design Specifications for Wood.

NER: National Evaluation Report.

NEUTRAL AXIS: In the cross section of a beam, the location within a board where there is neither tension nor compression stress.

NOMINAL SIZE (DIMENSION): Rough lumber size before finishing or surfacing.

NOMINAL SPAN: Horizontal distance between the edges of the supports of a beam or truss.

NOMINAL THICKNESS (PLYWOOD): Full designated thickness of plywood before sanding.

NON-BEARING WALL: Wall or partition that only carries its own weight.

NRMCA: National Ready Mixed Concrete Association.

OCCUPANCY: Purpose for which a building is to be used.

ON CENTER (OC, O.C.): Distance between the centers of adjacent repetitive structural members.

OPEN FLOW DRAINAGE: Removal of rainwater via surface drainage.

PARTICLEBOARD: Mat-formed panel made of wood particles or a combination of wood particles and wood fibers bonded together with synthetic resins and used in formwork.

PCA: Portland Cement Association.

PEDESTAL: Upright compression member with a ratio of unsupported height to average least lateral dimension of 3 or less.

PERCOLATION: Movement of water through pore spaces in the ground.

PERMEABILITY: The quality or state of being penetrated by water. Capacity of a soil to transmit water through it. Sand and gravels are highly permeable; clay and silt are not.

PIER: Masonry or concrete column used to support a beam.

PIEZOMETER: An instrument for measuring pressure or compressibility in soil compaction and surcharging.

PIER BLOCK: Preformed concrete footing that supports a post.

PILING: A structure composed of piles.

PLAIN CONCRETE: Structural concrete with no reinforcement, or with less reinforcement than minimum amount specified for reinforced concrete.

PLAIN REINFORCEMENT: Reinforcement that doesn't conform to definition of deformed reinforcement. Usually, smooth bars without bumps.

PLASTER: A combination of cement and sand that is mixed with water, with or without admixtures that when mixed thoroughly, placed properly, and finished accordingly, form a desired decorative aesthetic that is maintainable and near watertight finish applied over the shell of the swimming pool.

PLATEAU: A large, relatively level land area that is raised above the adjacent land; the flattened top of a hill or mountain.

PLUMB: True and level on a vertical plane; perfectly vertical or 90° to the horizontal. Term used to indicate that a wall, fence, post, etc., is straight (vertical).

PLUMB BOB: Metal weight suspended from a cord used to establish a vertical line.

PLY: Single veneer lamina in a glued plywood panel. Also, the number of thicknesses of veneer in a plywood panel or laminated member.

PLY/FORM: Plastic-coated plywood used for forming concrete.

PNEUMATICALLY-APPLIED CONCRETE: Gunite or Shotcrete

PNEUMATICALLY-DRIVEN FASTENERS: Air-driven nails, staples, or spikes.

POINT OF CURVATURE (P.C.): The point at which the curve departs from the tangent as one proceeds around the curve in the direction of change.

POINT OF INTERSECTION (P.I.): The point at which the two tangents intersect.

POINT OF TANGENCY (P.T.): The end of the curve and the beginning of the tangent to which the curve is connected.

POST (COLUMN): Vertical load-bearing structural member.

PRECAST CONCRETE: Structural concrete element that is cast and then placed in its final position in a structure.

PRESTRESSED CONCRETE: Structural concrete in which internal compressive stresses have been introduced by the steel tendons to reduce potential tensile stresses in concrete resulting from loads.

PRE-TENSIONING: Method of prestressing tendons in concrete before the concrete is placed.

PROFILE : The trace of the intersection of an imaginary vertical plane with the ground surface; normally used in road design.

RACKING: Twisting movement that can distort a framework.

RAINFALL INTENSITY: The amount of rain (measured in inches) falling on a specific area over a specific period of time.

RAINFALL RUNOFF: Rainwater that flows over an area or surface (not absorbed by the underlying earth).

RATIONAL METHOD: A method for computing approximate storm-water runoff volumes using a formula relating rainfall intensity, a coefficient of runoff, and watershed acreage.

REACTION: Load transmitted from a beam or truss to a support.

REGISTERED DESIGN PROFESSIONAL: Any architect or engineer registered or licensed to design a project in a given state according to the state's professional registration laws.

REINFORCED CONCRETE: Structural concrete that is reinforced with no less than the-minimum amounts of prestressing tendons or non-prestressed reinforcement specified in the building code.

RESIDUAL SOILS: Soils developed from the parent rock over which they now lie.

RETENTION POND: A permanent pond holding or retaining storm water.

RETROFIT: To add additional bracing, anchoring, or any improvement to a completed structure.

RUNOFF: The surface flow of water from an area or the total volume of surface flow during a specific time.

RUNOFF FACTOR: The coefficient that describes the percent of rainwater that is not absorbed by the soil, but flows over the earth's surface.

SCAB: Small piece of wood fastened to two formwork members to secure a butt joint.

SCAFFOLDING: Elevated platform erected to support workers, tools, and materials.

SCUPPER: A channel at the outer edge of a roof that conducts rainwater to a downspout or leader.

SEASONING (WOOD): Drying lumber by exposure to air and sun or by kiln.

SECTION OR CROSS SECTION: Sections are profile slices through the earth; cross sections are profiles taken at right angles to the center line of a project or road.

SECTION MODULUS: Property of the shape of a structural member that indicates its strength; the moment of inertia divided by the distance from the neutral axis to the extreme fiber of the section.

SEDIMENTATION : The deposition or accumulation of sediment.

SELECT STRUCTURAL LUMBER: High-quality lumber, free of characteristics that impair strength or stiffness.

SET: The reaction mechanism that takes place in which the physical, chemical, or mechanical properties develop as a concrete or plaster changes from a fresh slurry phase to a hardened solid phase in the presence of adequate moisture.

SHEAR LOAD: Side-to-side force(s) acting on a structure.

SHEAR STRESS: Stress that tends to keep two adjoining planes of a body from sliding on each other when two equal and opposite parallel forces act on them in opposite direction.

SHEAR WALL: A wall designed to resist horizontal loads.

SHEATHING: Structural covering applied to the outside surface of wall or roof frame. Also called sheeting.

SHEETING OR SHORING: Vertical piles driven around a deep excavation to prevent the sides from collapsing. Also refers to a thin layer of water moving across a surface.

SHOP DRAWING: Drawing, diagram, illustration, schedule, and other data prepared by a contractor, manufacturer, fabricator, supplier, or distributor to show some portion of the work.

SHORE: Temporary vertical or inclined member that supports formwork and fresh concrete until the structure has developed full strength.

SHOTCRETE: A wet mixture of cement, aggregate and delivered to the job site in a homogeneous state and typically sprayed onto the contoured surface , having a reinforcement in place, which when hardened forms the shell of the swimming pool.

SLOPE (OR GRADIENT) : The inclination of a surface expressed in percentage or as a proportion. Any geographic feature having a slope or gradient.

SOFFIT: Underside of a structural member of a building, such as a beam.

SOIL PERMEABILITY: The quality of a soil that enables water or air to move through it. Sand and gravel are permeable; clays and silts are not.

SOIL PROFILE: A vertical section showing the layers of soil at a site; usually three layers, called A, B, and C horizons, and varying from topsoil to bedrock.

SOIL TEXTURE: The relative proportions of sand, silt, and clay.

SOIL WATER-HOLDING CAPACITY: The capacity of a soil to hold water; useful in determining runoff rate.

SOLDIER: Vertical wales used to strengthen and align forms.

SPA, NON-PORTABLE: See "Swimming Pool."

SPAN: Horizontal distance between supports.

SPAN RATING (PLYWOOD): pair of numbers stamped on plywood sheathing to indicate its span capabilities over roofs and floors. Also called an identification index.

SPLASH BLOCK: A concrete version of the Splash pan.

SPLASH PAN: A sheet metal device at the outlet of a downspout or leader, used to prevent soil erosion.

SPECIFICATIONS: Technical descriptions of a project regarding materials, equipment, construction systems, standards, and workmanship.

SPIRAL REINFORCEMENT: Continuously wound reinforcement in the form of a cylindrical helix.

SPOT ELEVATIONS: Used to supplement the contour lines, to show variations from the normal gradient between the contour lines. Gradients between spot elevations are considered uniform unless vertical curves are indicated.

STANDARD PROCTOR TEST: A laboratory test used to determine soil compaction. The test uses a metal cylinder with a weighted ram to compact soil samples. Most compacted fills are specified to meet 95 percent of the Standard Proctor Test, which is sometimes referred to as the Standard AASHO compaction.

STORM SEWER: A drain used for conveying rainwater, but not sewage or industrial wastes, to a point of disposal.

STRAP: (1) A metal bar used in concrete reinforcement; (2) A metal supporting strap used to hold one end of a beam or joist and connect it to another structural member.

STRENGTH, DESIGN: Nominal strength multiplied by a strength reduction factor f .

STRESS: Intensity of force per unit area.

STRIKE: To lower or remove formwork or centering.

STRINGER: A beam that supports floor or deck sheathing.

STRIPPING: Disassembling forming and shoring, usually for reuse.

STRIPS: Boards less than 6 inches wide.

STRUCTURAL CONCRETE: All concrete used for structural purposes, including plain and reinforced concrete.

STRUT: Horizontal or inclined compression member.

STUD: Vertical wood or metal member that supports the sheathing in wall forming.

SUBCONTRACTOR: Individual, firm, or corporation having a direct contract with a general contractor or subcontractor for the performance of part of a project.

SUBGRADE: The upper surface of the native soil on which is placed the road, foundations, topsoil, or other final materials.

SUBSOIL: The B horizon of soils, or the soil below which roots normally grow.

SUPER-ELEVATION: A horizontal curve with the outside radius higher than the inside to counter the pull of gravity while traveling at high speeds.

SURCHARGE: A temporary loading of soil to induce settlement. Following settlement, the extra soil will be removed as the permanent structure replaces it.

SWALE: A wide, shallow, slightly sloping ditch that collects and transports runoff as open flow drainage.

SWAMP LAND: See Inundated area.

SWIMMING POOL: Any structure intended for swimming or recreational bathing that contains water over 24 inches (610 mm) deep. This includes in-ground, above ground and on-ground swimming pools, hot tub and spas.

SWIMMING POOL, INDOOR: A swimming pool which is totally contained within a structure and surrounded on all four sides by walls of said structure.

SWIMMING POOL, OUTDOOR: Any swimming pool which is not an indoor pool.

T&G: Tongue-and-groove.

T-HEAD: Top of a shore with a braced horizontal member projecting on two sides, forming a T-shaped assembly.

TELL-TALE: Any device designed to indicate the movement of formwork.

TEMPLATE: Thin plate or board frame used as a guide in positioning or spacing form parts, reinforcement, anchors, etc.

TENSILE STRENGTH: Resistance to tensile forces.

TENSILE STRESS: Pulling force over a unit area expressed in psi, or tensile force divided by the tensile area.

TENSION: Force exerted on a structural member that tends to pull it apart or elongate it.

TIE: Loop of reinforcing bar or wire enclosing lengthwise reinforcement. A continuous wound bar or wire in the form of a circle, rectangle, or other polygon shape.

TIE WIRE: Metal wires used to hold opposing forms in position.

TIMBERS: Lumber 5 or more inches in the least dimension, including beams, stringers, posts, sills, girders, and purlins.

TOPOGRAPHIC MAP: A drawing or chart that shows the physical features of the land, especially altitudes or contours.

TRENCH DRAIN: Sometimes called strip drain. A linear drain with a concrete channel beneath used to collect water along a linear space.

TRIBUTARY AREA (DOMAIN): Roof, floor or wall area that contributes to the load on a structural member.

TRUSS: Shop-fabricated frame used as a roof support.

ULTIMATE STRESS: The maximum stress that a material can stand before it breaks apart.

UNIFORM LOAD: Load that is equally distributed over a given length of a beam, and is usually expressed as pounds per lineal foot (plf).

UNIT STRESS: Amount of stress on 1 square inch of a material.

UPLIFT: Force(s) acting to lift a structure caused by earthquake or wind.

WALE (WALER): Horizontal member used to align and brace studs on concrete forms.

WANE: Lumber defect located near the edge or corner of a board caused by the lack of bark.

WATER COURSE: A natural channel for rainwater flow (e.g., a gully, stream or river).

WATER-HOLDING CAPACITY: The ability of soil to hold water against gravity. Clay soils have a high capacity; sandy soils low.

WATERSHED: The total catchment area above a given point on a stream that contributes water to the stream or river. Watersheds are separated by a summit or ridge.

WATER TABLE: The upper limit of the soil that is wholly saturated with water.

WEIR: An adjustable dam across a stream to control the water level or measure flow.

WET LAND: See Inundated area.

WORK: Construction required under the contract documents.

WORKING STRESS: Unit stress that experience has shown to be safe for a material.

WRI: Wire Reinforcing Institute.

X-BRACE: Paired set of tension sway braces made of steel rods, angles, or other members used to resist sideways overturning from lateral forces on a structure. Also see Sway brace.

YARD LUMBER: Wood members used for common construction.

YIELD STRENGTH: Specified minimum yield strength or yield point of reinforcement, in psi.

YOKE: Tie or clamping device placed around column forms or over the top of wall or footing forms to keep them from spreading due to lateral concrete pressure.

ZERO-PERCENT INCREASE: An approach to drainage that restricts the amount of storm water that can leave the project to the amount of runoff prior to construction.

CHAPTER 12

References for Drainage and Retaining Walls

"1997 Uniform Building Code," International Building Officials, Whittier, CA 1997

"2001 California Building Code," California Building Standards Commission, 2002

"Architectural Sheet metal Manual," Sheet Metal and Air-conditioning Contractors National Association, Inc., 1979

"Concrete and Masonry" Headquarters Dept of the army, Supt of documents, Washington DC 1970

"County Engineer Standards," Dept. of County Engineer County of Los Angeles. Sanitation Division, 1971

"Design Manual," Department of the Navy, Navy Facilities Engineering Command, Alexandria, VA 1971

"General Grading Requirements," City of Los Angeles, Building and Safety, 1983

"Handbook of Drainage and Construction Products," Armco Drainage & Metal Products, Inc., Middleton, Ohio, 1955

"Minimum property Standards for One and Two Family Living Units," Federal Housing Administration, 1958

"Plan Review Manual." International Conference of Building Officials, Inc., Whittier, CA 1971

"Standard Plans," Dept. of Public Works, Bureau of Engineering, City of Los Angeles, 1947

"Street Design: Bureau of Engineering Manual." Dept. of Public Works, Bureau of Engineering, City of Los Angeles.

Blendermann, Louis, "Design of Plumbing and Drainage Systems," The Industrial Press, NY, 1959.

Callender, John Hancock, Time Saver Standards for Architectural Design," McGraw-Hill Book Co., Inc. NY 1954.

Gaylord, Edwin H. and Gaylord, Charles N., "Structural Engineering Handbook," McGraw-Hill Book Co., Inc. NY 1968.

Havers, John and Stubbs, Frank W., "Handbook of Heavy Construction," McGraw-Hill Book Co., Inc., NY 1971.

Kidder, Frank E. and Parker, Harry, "Architects and Builders Handbook, John Wiley & Sons, Inc. NY 1947.

Manas, Vincent T., "National Plumbing Code Handbook," McGraw-Hill Book Co., Inc.,. NY 1957.

Ramsey, Charles George and Sleeper, Harold Reeve, "Architectural Graphic Standards," John Wiley & Sons, London, 1961.

Seely, Elwyn E., "Data Book for Civil Engineers," John Wiley & Sons, Inc. London, 1960.

Retaining Wall Details

Details provided by Contributing Editor

Hamid Azizi, P.E.
President

EXECUTIVE ENGINEERING
19510 Ventura Blvd., Suite 110
Tarzana, CA 91356-4234
Phone: (818) 708-1962
Facsimile: (818) 708-1958
E-mail: EE101@aol.com

Hamid Azizi is a Registered Professional Engineer in civil engineering, successfully serving the industry since 1983. Mr. Azizi is a graduate of the University of Nevada, Reno (UNR). His diverse expertise includes the analysis and design of low-rise residential and commercial structures, and field investigation and engineering evaluations of existing structures. He is an active member and past president of the Consulting Structural Engineers Society (CSES).

Mr. Azizi has a standard pool plan registered with the City and County of Los Angeles.

PAGE	BACKFILL	SURCHARGE	KEY	FOOTING
A-3 through A-5	Double Wall			
A-6 through A-19	Garden Wall			
A-20 through A-22	Level	Yes	No	Out of Hill
A-23 and A-24	Level	Yes	Yes	Out of Hill
A-25 through A-33	Level		No	Into Hill
A-34 through A-55	Level		No	Out of Hill
A-56 through A-61	Level		Yes	Into Hill
A-62 through A-72	Level		Yes	Out of Hill
A-73 and A-74	Sloped		No	Into Hill
A-75 through A-83	Sloped		No	Out of Hill
A-84 through A-89	Sloped		Yes	Into Hill
A-90 through A-102	Sloped		Yes	Out of Hill

ILLUSTRATION ONLY. NOT FOR CONSTRUCTION.

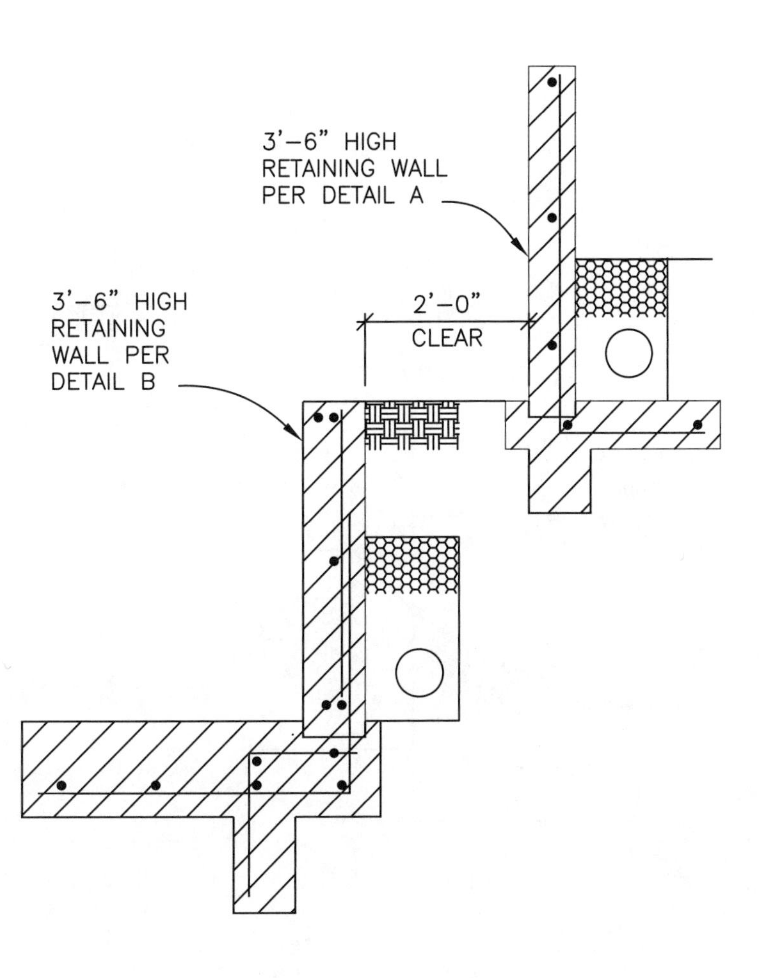

DOUBLE WALL

ILLUSTRATION ONLY. NOT FOR CONSTRUCTION.

EXECUTIVE ENGINEERING
19510 Ventura Blvd., Suite 110
Tarzana, California 91356-4234
Phone: (818) 708-1962 / Facsimile: (818) 708-1958
E-mail: EE110@aol.com

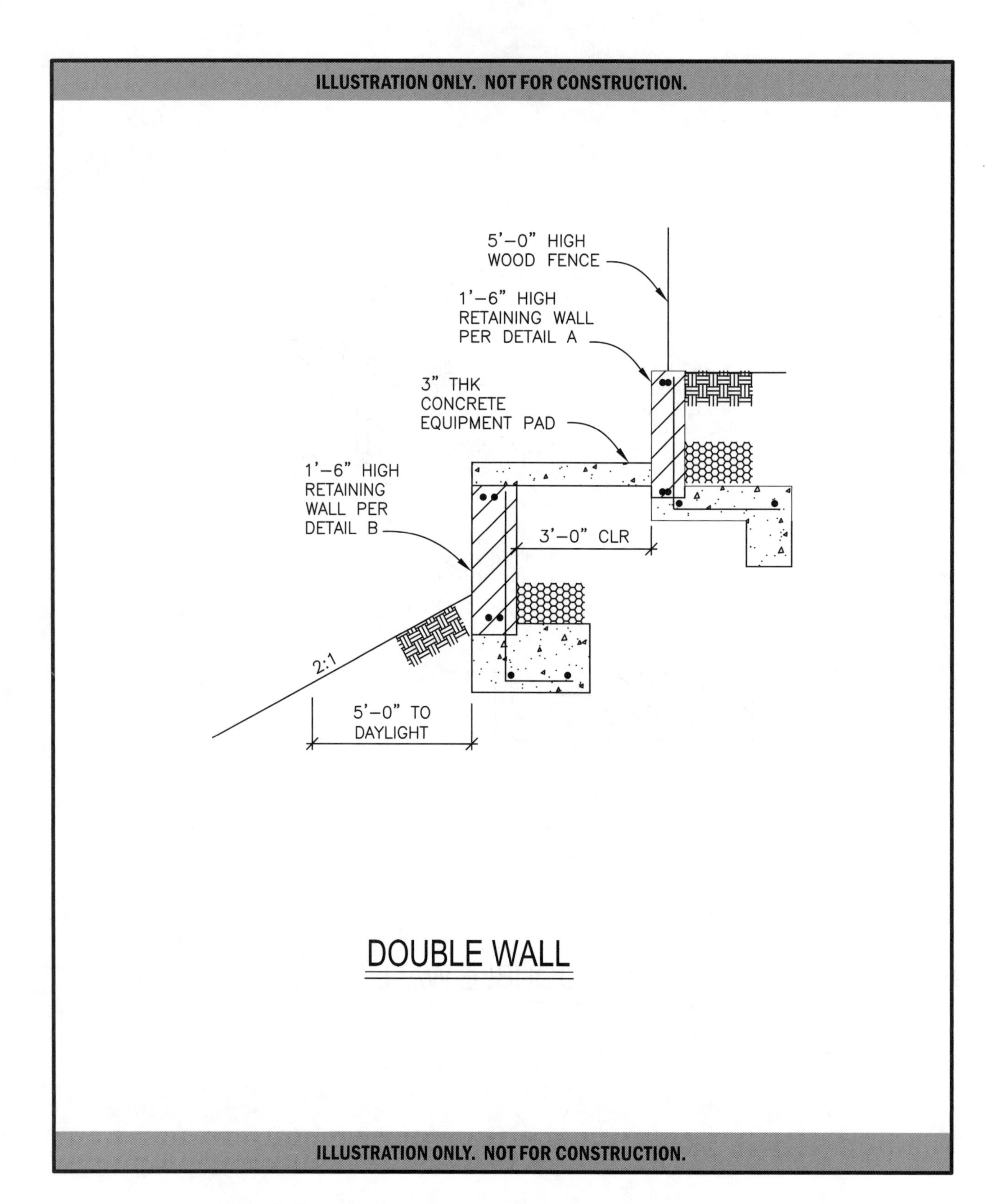

EXECUTIVE ENGINEERING
19510 Ventura Blvd., Suite 110
Tarzana, California 91356-4234
Phone: (818) 708-1962 / Facsimile: (818) 708-1958
E-mail: EE110@aol.com

ILLUSTRATION ONLY. NOT FOR CONSTRUCTION.

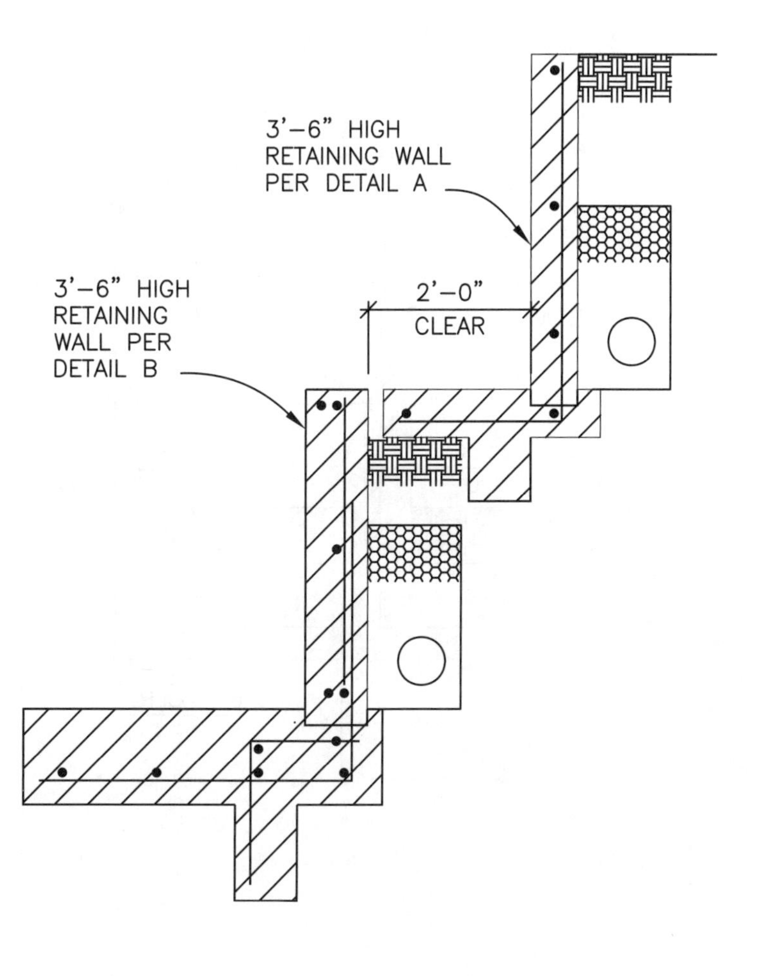

DOUBLE WALL

ILLUSTRATION ONLY. NOT FOR CONSTRUCTION.

EXECUTIVE ENGINEERING
19510 Ventura Blvd., Suite 110
Tarzana, California 91356-4234
Phone: (818) 708-1962 / Facsimile: (818) 708-1958
E-mail: EE110@aol.com

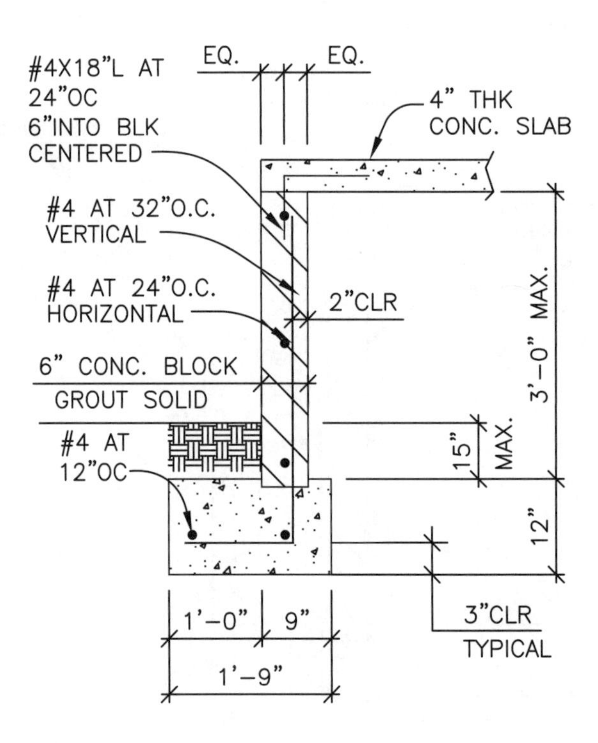

GARDEN WALL

ILLUSTRATION ONLY. NOT FOR CONSTRUCTION.

EXECUTIVE ENGINEERING
19510 Ventura Blvd., Suite 110
Tarzana, California 91356-4234
Phone: (818) 708-1962 / Facsimile: (818) 708-1958
E-mail: EE110@aol.com

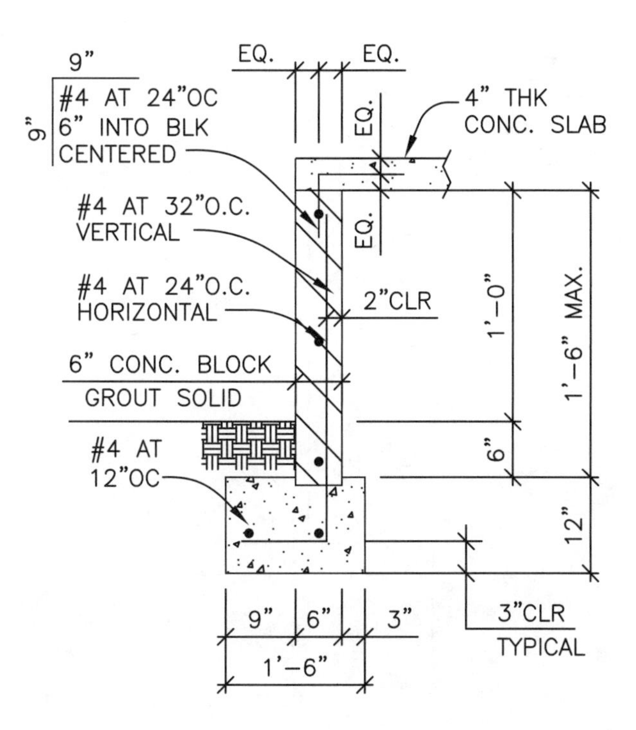

GARDEN WALL

ILLUSTRATION ONLY. NOT FOR CONSTRUCTION.

EXECUTIVE ENGINEERING
19510 Ventura Blvd., Suite 110
Tarzana, California 91356-4234
Phone: (818) 708-1962 / Facsimile: (818) 708-1958
E-mail: EE110@aol.com

ILLUSTRATION ONLY. NOT FOR CONSTRUCTION.

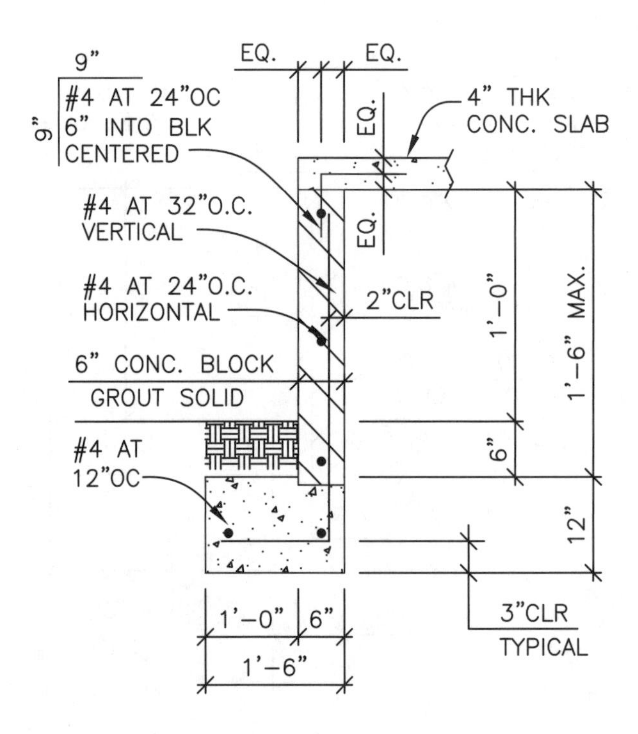

GARDEN WALL

ILLUSTRATION ONLY. NOT FOR CONSTRUCTION.

EXECUTIVE ENGINEERING
19510 Ventura Blvd., Suite 110
Tarzana, California 91356-4234
Phone: (818) 708-1962 / Facsimile: (818) 708-1958
E-mail: EE110@aol.com

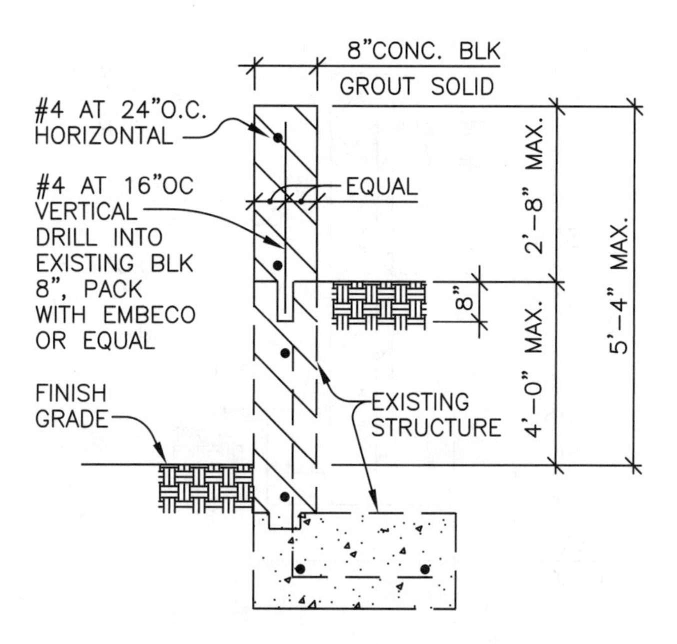

GARDEN WALL

EXECUTIVE ENGINEERING
19510 Ventura Blvd., Suite 110
Tarzana, California 91356-4234
Phone: (818) 708-1962 / Facsimile: (818) 708-1958
E-mail: EE110@aol.com

ILLUSTRATION ONLY. NOT FOR CONSTRUCTION.

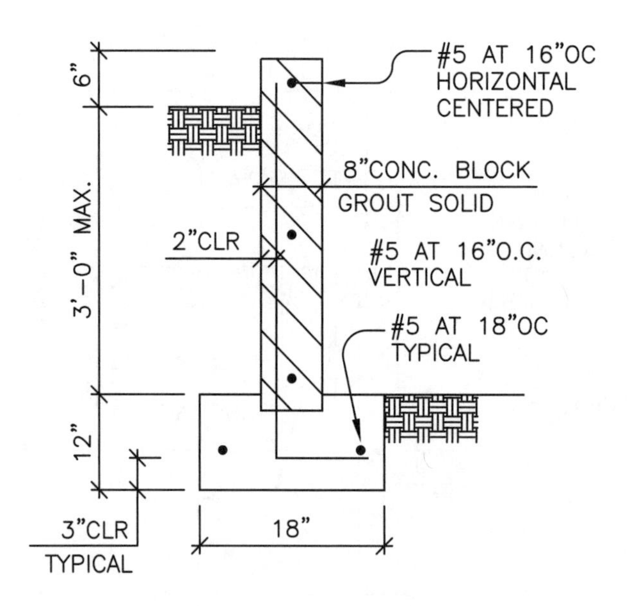

GARDEN WALL

ILLUSTRATION ONLY. NOT FOR CONSTRUCTION.

EXECUTIVE ENGINEERING
19510 Ventura Blvd., Suite 110
Tarzana, California 91356-4234
Phone: (818) 708-1962 / Facsimile: (818) 708-1958
E-mail: EE110@aol.com

ILLUSTRATION ONLY. NOT FOR CONSTRUCTION.

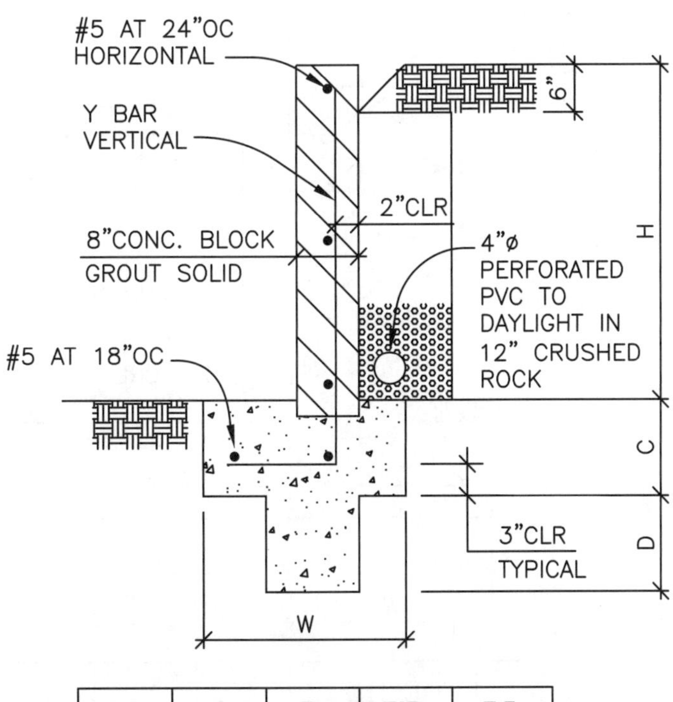

H	C	D	W	Y
4'–0"	12"	12"	2'–4"	#5@24"

GARDEN WALL

ILLUSTRATION ONLY. NOT FOR CONSTRUCTION.

EXECUTIVE ENGINEERING
19510 Ventura Blvd., Suite 110
Tarzana, California 91356-4234
Phone: (818) 708-1962 / Facsimile: (818) 708-1958
E-mail: EE110@aol.com

ILLUSTRATION ONLY. NOT FOR CONSTRUCTION.

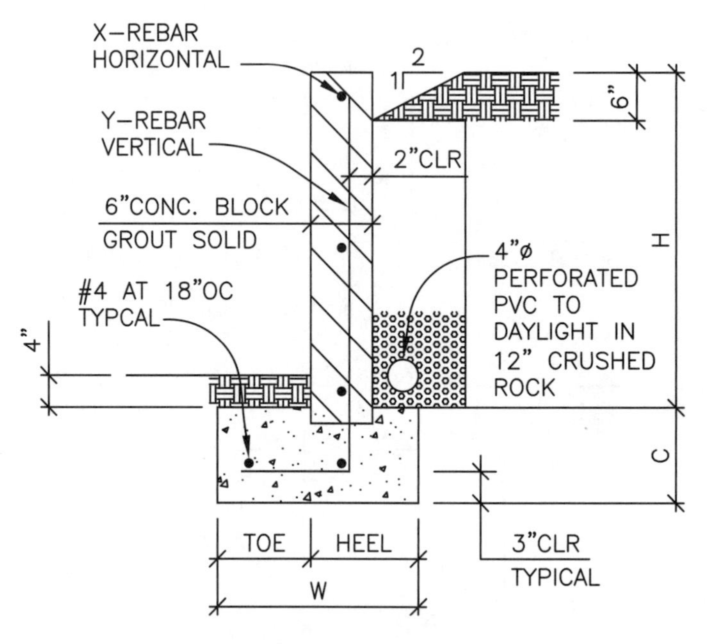

H	C	TOE	HEEL	W	X	Y
3'–6"	12"	1'–6"	1'–0"	2'–6"	#4@24"	#4@24"

GARDEN WALL

ILLUSTRATION ONLY. NOT FOR CONSTRUCTION.

EXECUTIVE ENGINEERING
19510 Ventura Blvd., Suite 110
Tarzana, California 91356-4234
Phone: (818) 708-1962 / Facsimile: (818) 708-1958
E-mail: EE110@aol.com

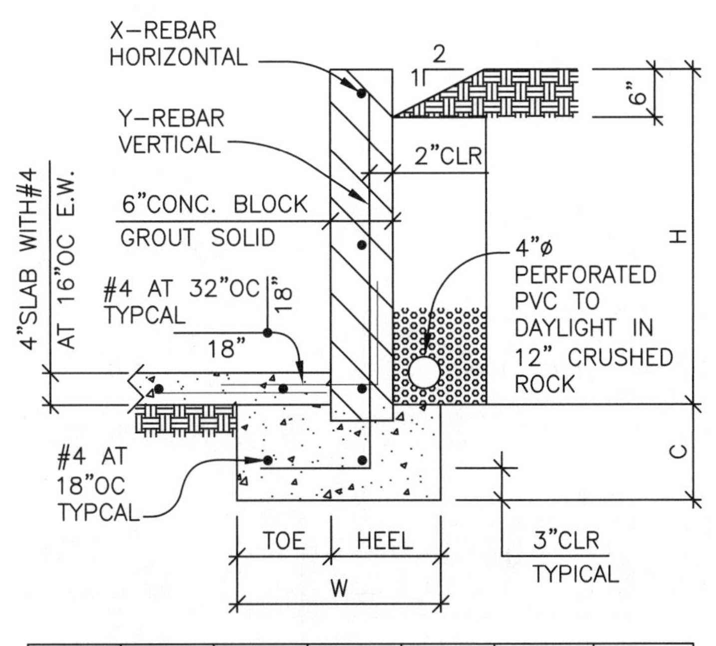

H	C	TOE	HEEL	W	X	Y
3'–6"	12"	1'–6"	1'–0"	2'–6"	#4@24"	#4@24"

GARDEN WALL

ILLUSTRATION ONLY. NOT FOR CONSTRUCTION.

EXECUTIVE ENGINEERING
19510 Ventura Blvd., Suite 110
Tarzana, California 91356-4234
Phone: (818) 708-1962 / Facsimile: (818) 708-1958
E-mail: EE110@aol.com

ILLUSTRATION ONLY. NOT FOR CONSTRUCTION.

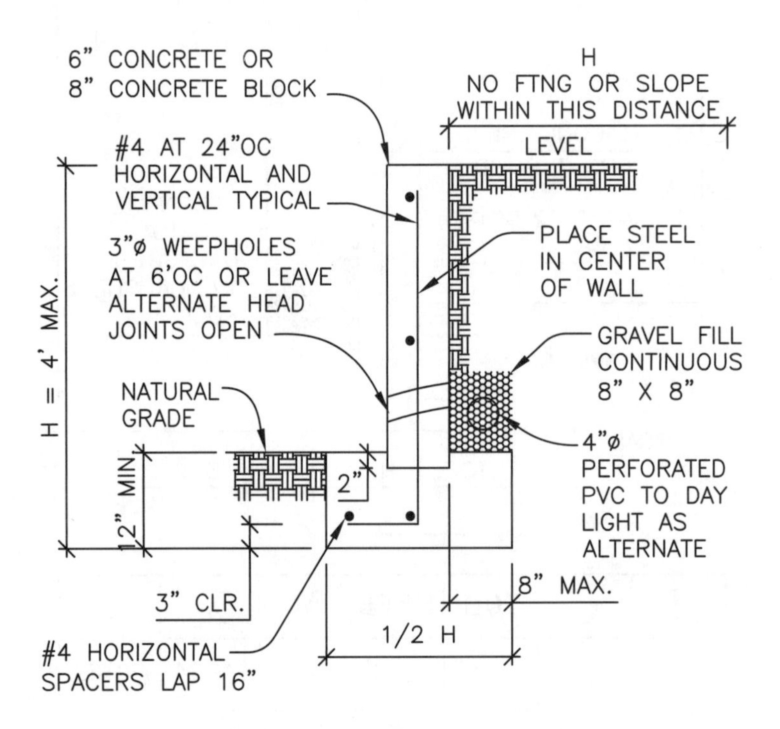

GARDEN WALL

ILLUSTRATION ONLY. NOT FOR CONSTRUCTION.

EXECUTIVE ENGINEERING
19510 Ventura Blvd., Suite 110
Tarzana, California 91356-4234
Phone: (818) 708-1962 / Facsimile: (818) 708-1958
E-mail: EE110@aol.com

ILLUSTRATION ONLY. NOT FOR CONSTRUCTION.

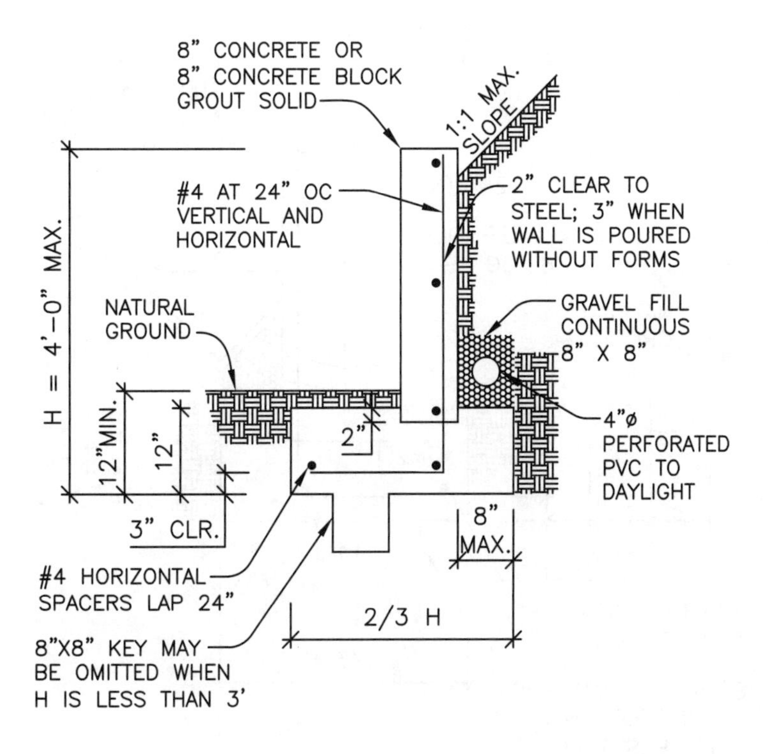

GARDEN WALL

ILLUSTRATION ONLY. NOT FOR CONSTRUCTION.

EXECUTIVE ENGINEERING
19510 Ventura Blvd., Suite 110
Tarzana, California 91356-4234
Phone: (818) 708-1962 / Facsimile: (818) 708-1958
E-mail: EE110@aol.com

ILLUSTRATION ONLY. NOT FOR CONSTRUCTION.

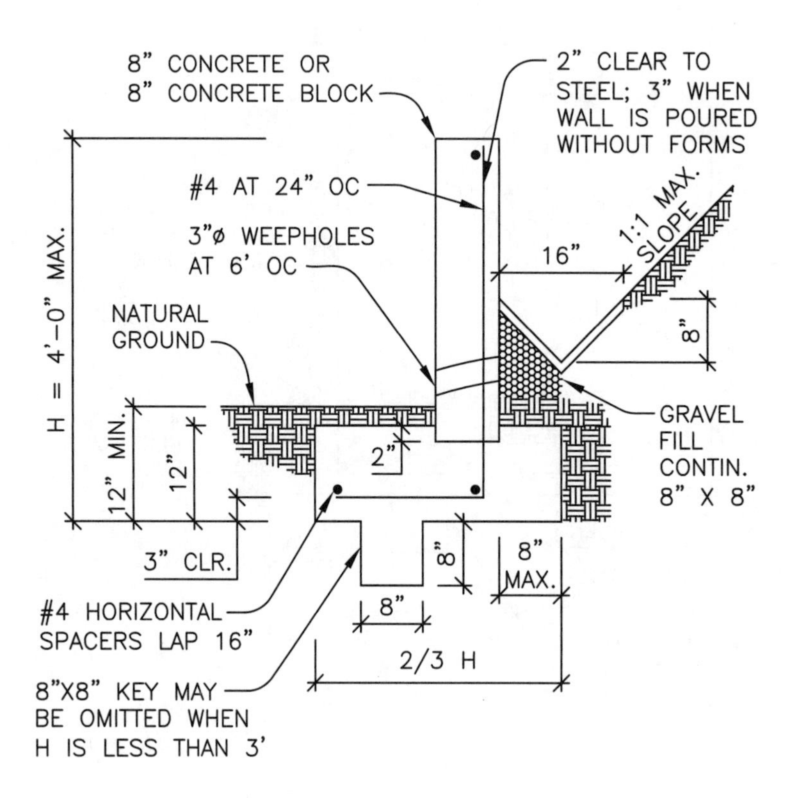

GARDEN WALL

ILLUSTRATION ONLY. NOT FOR CONSTRUCTION.

EXECUTIVE ENGINEERING
19510 Ventura Blvd., Suite 110
Tarzana, California 91356-4234
Phone: (818) 708-1962 / Facsimile: (818) 708-1958
E-mail: EE110@aol.com

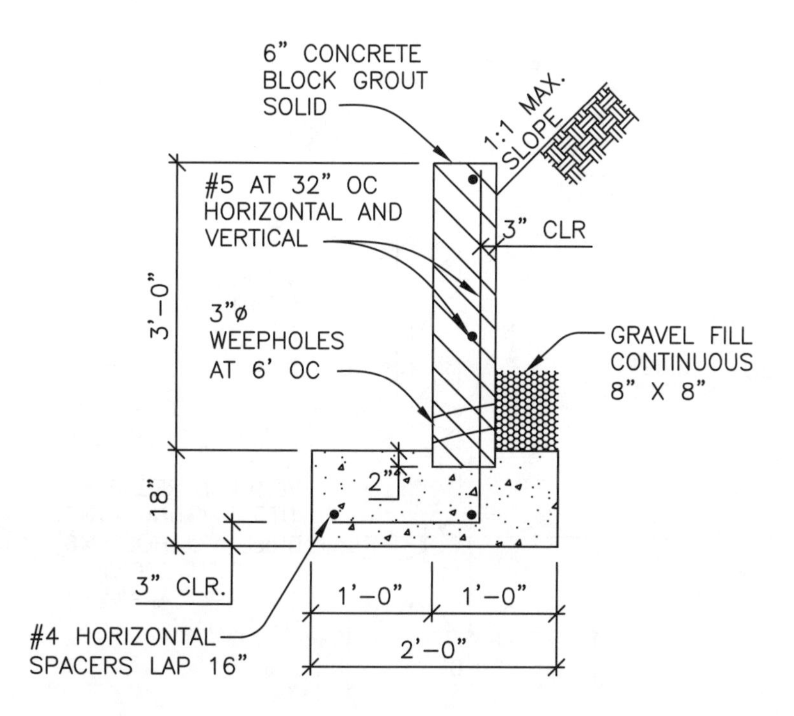

GARDEN WALL

EXECUTIVE ENGINEERING
19510 Ventura Blvd., Suite 110
Tarzana, California 91356-4234
Phone: (818) 708-1962 / Facsimile: (818) 708-1958
E-mail: EE110@aol.com

ILLUSTRATION ONLY. NOT FOR CONSTRUCTION.

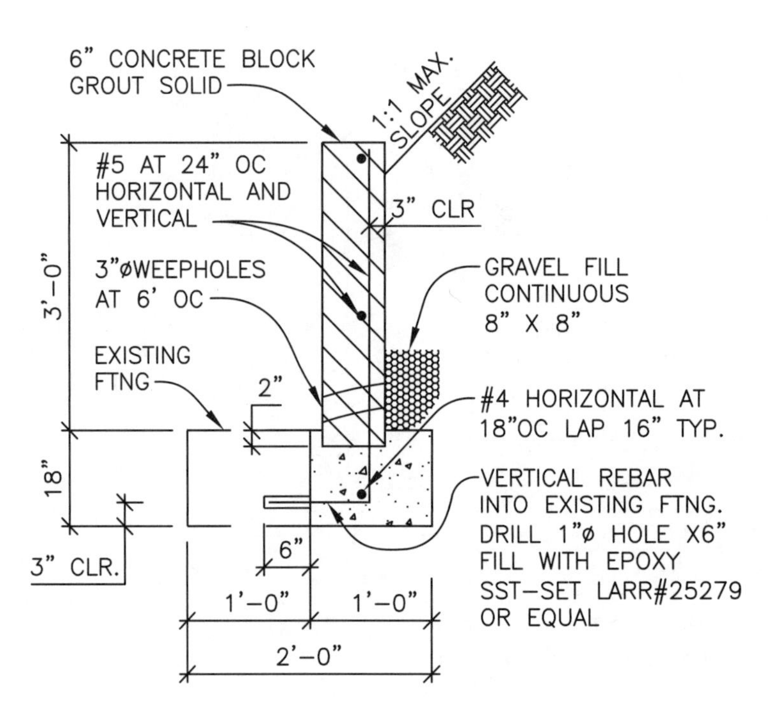

GARDEN WALL

ILLUSTRATION ONLY. NOT FOR CONSTRUCTION.

EXECUTIVE ENGINEERING
19510 Ventura Blvd., Suite 110
Tarzana, California 91356-4234
Phone: (818) 708-1962 / Facsimile: (818) 708-1958
E-mail: EE110@aol.com

ILLUSTRATION ONLY. NOT FOR CONSTRUCTION.

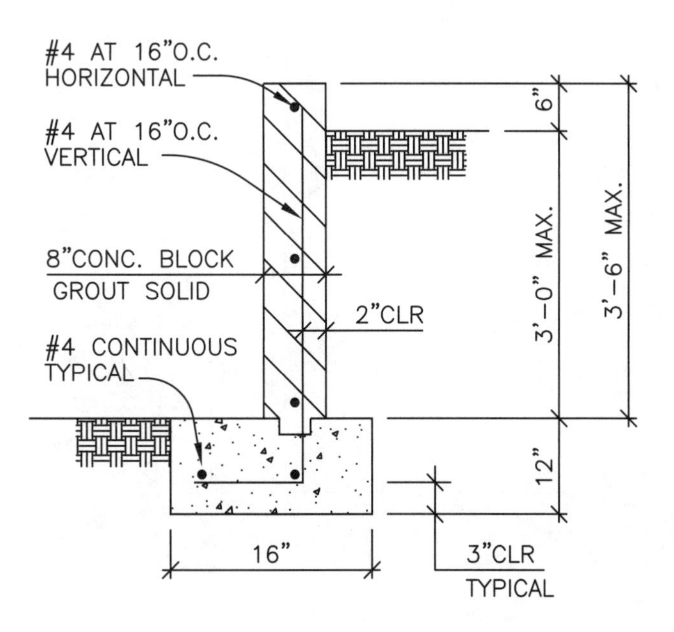

GARDEN WALL

ILLUSTRATION ONLY. NOT FOR CONSTRUCTION.

EXECUTIVE ENGINEERING
19510 Ventura Blvd., Suite 110
Tarzana, California 91356-4234
Phone: (818) 708-1962 / Facsimile: (818) 708-1958
E-mail: EE110@aol.com

ILLUSTRATION ONLY. NOT FOR CONSTRUCTION.

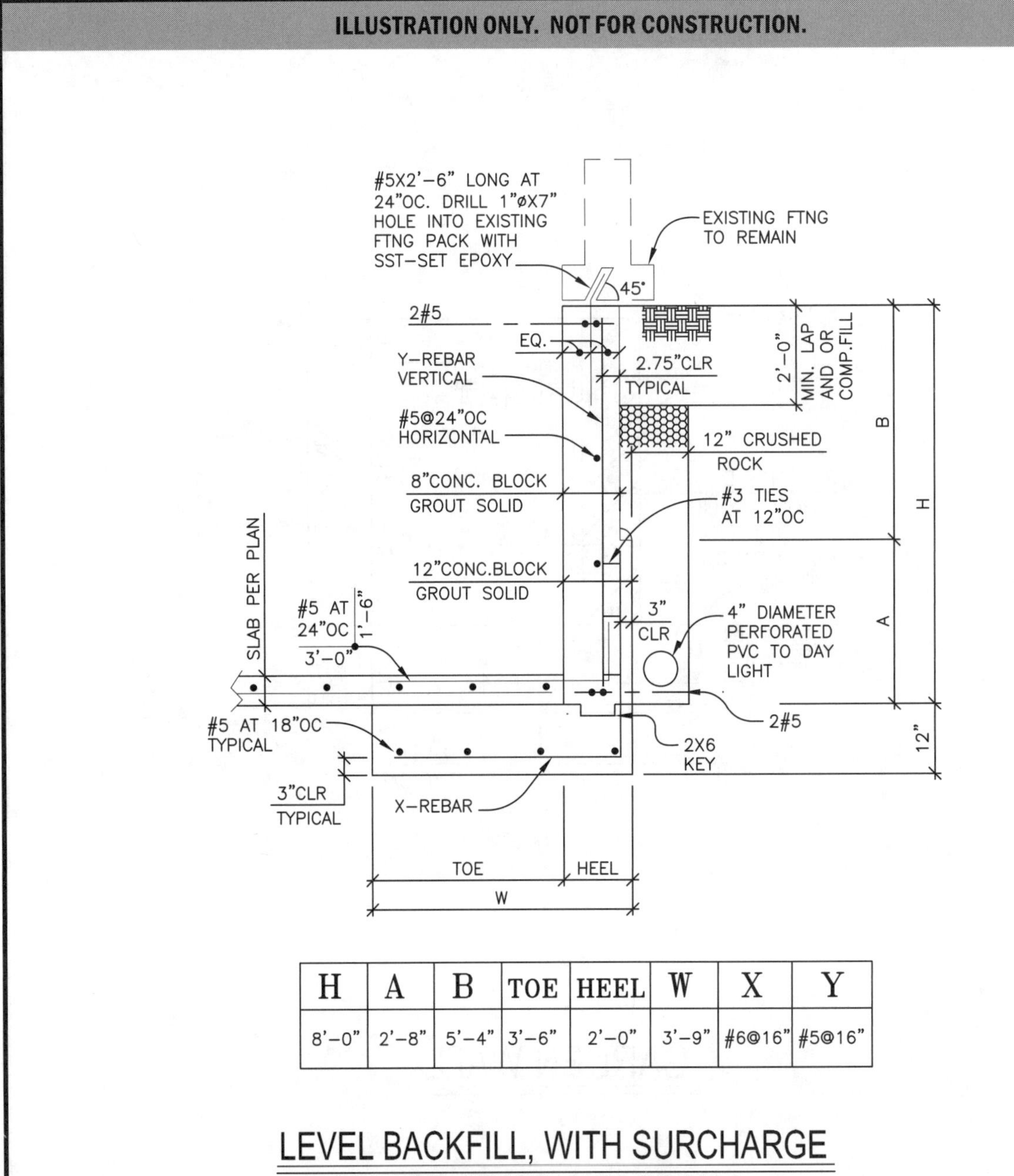

H	A	B	TOE	HEEL	W	X	Y
8'-0"	2'-8"	5'-4"	3'-6"	2'-0"	3'-9"	#6@16"	#5@16"

LEVEL BACKFILL, WITH SURCHARGE

NO KEY, FOOTING OUT OF HILL

ILLUSTRATION ONLY. NOT FOR CONSTRUCTION.

EXECUTIVE ENGINEERING
19510 Ventura Blvd., Suite 110
Tarzana, California 91356-4234
Phone: (818) 708-1962 / Facsimile: (818) 708-1958
E-mail: EE110@aol.com

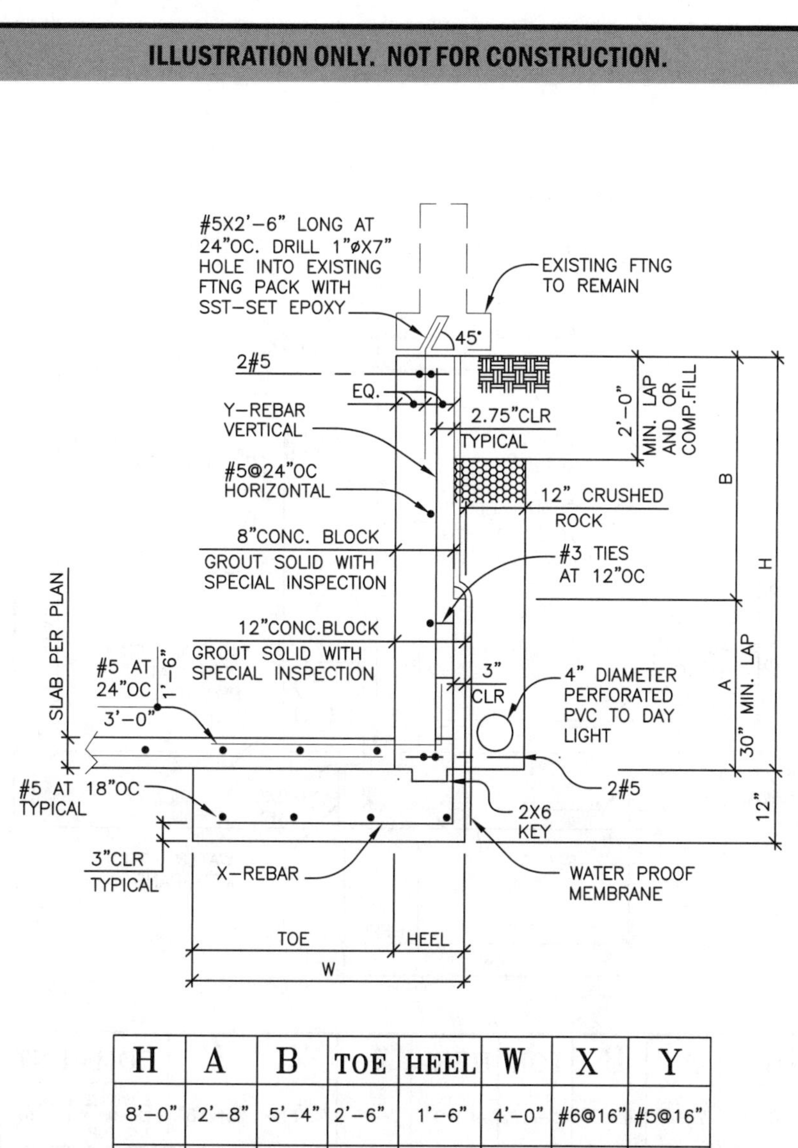

H	A	B	TOE	HEEL	W	X	Y
8'-0"	2'-8"	5'-4"	2'-6"	1'-6"	4'-0"	#6@16"	#5@16"
7'-0"	2'-0"	5'-0"	2'-6"	1'-0"	3'-6"	#6@16"	#5@16"

LEVEL BACKFILL, WITH SURCHARGE
NO KEY, FOOTING OUT OF HILL

EXECUTIVE ENGINEERING
19510 Ventura Blvd., Suite 110
Tarzana, California 91356-4234
Phone: (818) 708-1962 / Facsimile: (818) 708-1958
E-mail: EE110@aol.com

ILLUSTRATION ONLY. NOT FOR CONSTRUCTION.

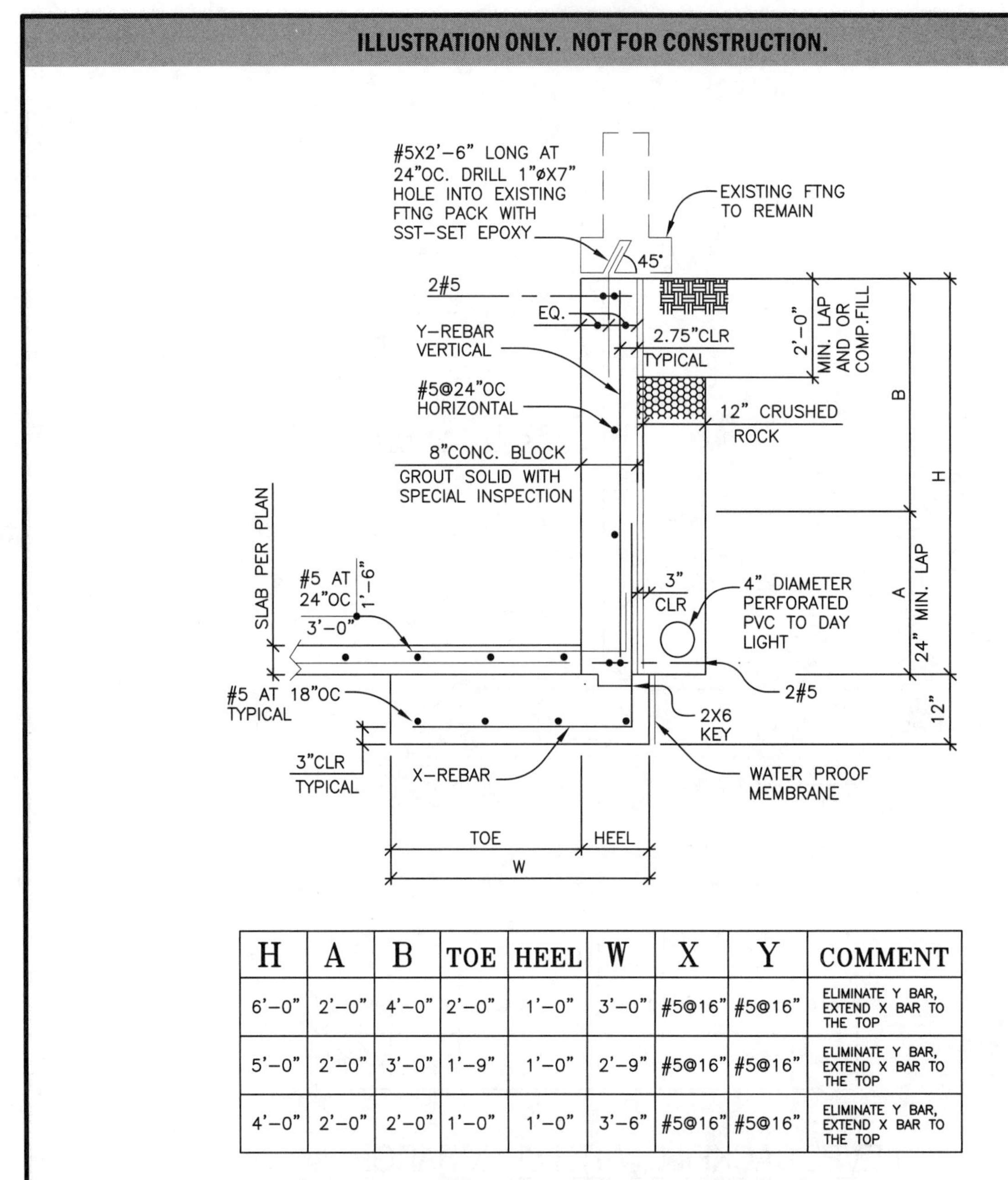

H	A	B	TOE	HEEL	W	X	Y	COMMENT
6'-0"	2'-0"	4'-0"	2'-0"	1'-0"	3'-0"	#5@16"	#5@16"	ELIMINATE Y BAR, EXTEND X BAR TO THE TOP
5'-0"	2'-0"	3'-0"	1'-9"	1'-0"	2'-9"	#5@16"	#5@16"	ELIMINATE Y BAR, EXTEND X BAR TO THE TOP
4'-0"	2'-0"	2'-0"	1'-0"	1'-0"	3'-6"	#5@16"	#5@16"	ELIMINATE Y BAR, EXTEND X BAR TO THE TOP

LEVEL BACKFILL, WITH SURCHARGE

NO KEY, FOOTING OUT OF HILL

ILLUSTRATION ONLY. NOT FOR CONSTRUCTION.

EXECUTIVE ENGINEERING
19510 Ventura Blvd., Suite 110
Tarzana, California 91356-4234
Phone: (818) 708-1962 / Facsimile: (818) 708-1958
E-mail: EE110@aol.com

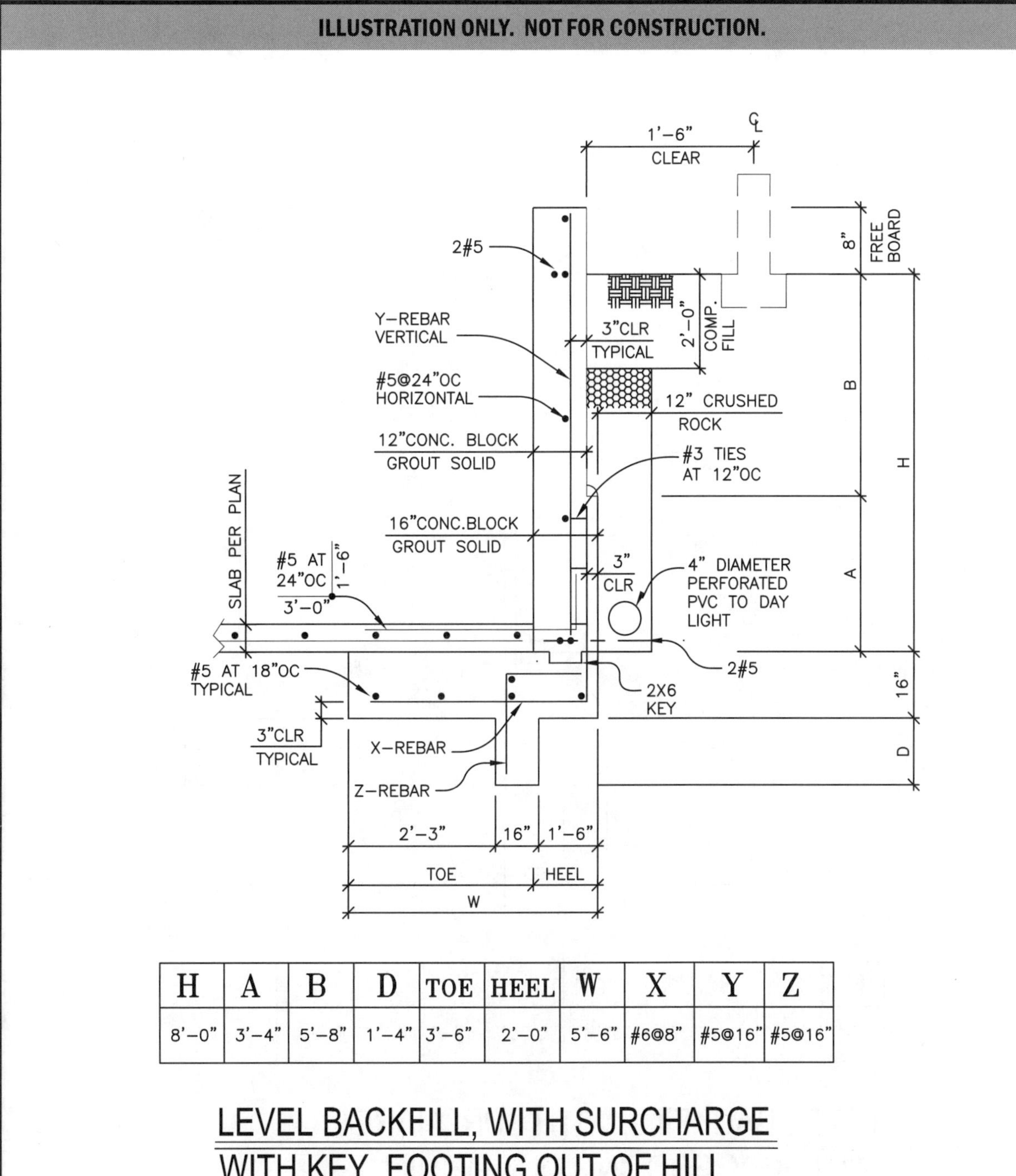

H	A	B	D	TOE	HEEL	W	X	Y	Z
8'-0"	3'-4"	5'-8"	1'-4"	3'-6"	2'-0"	5'-6"	#6@8"	#5@16"	#5@16"

LEVEL BACKFILL, WITH SURCHARGE WITH KEY, FOOTING OUT OF HILL

ILLUSTRATION ONLY. NOT FOR CONSTRUCTION.

EXECUTIVE ENGINEERING
19510 Ventura Blvd., Suite 110
Tarzana, California 91356-4234
Phone: (818) 708-1962 / Facsimile: (818) 708-1958
E-mail: EE110@aol.com

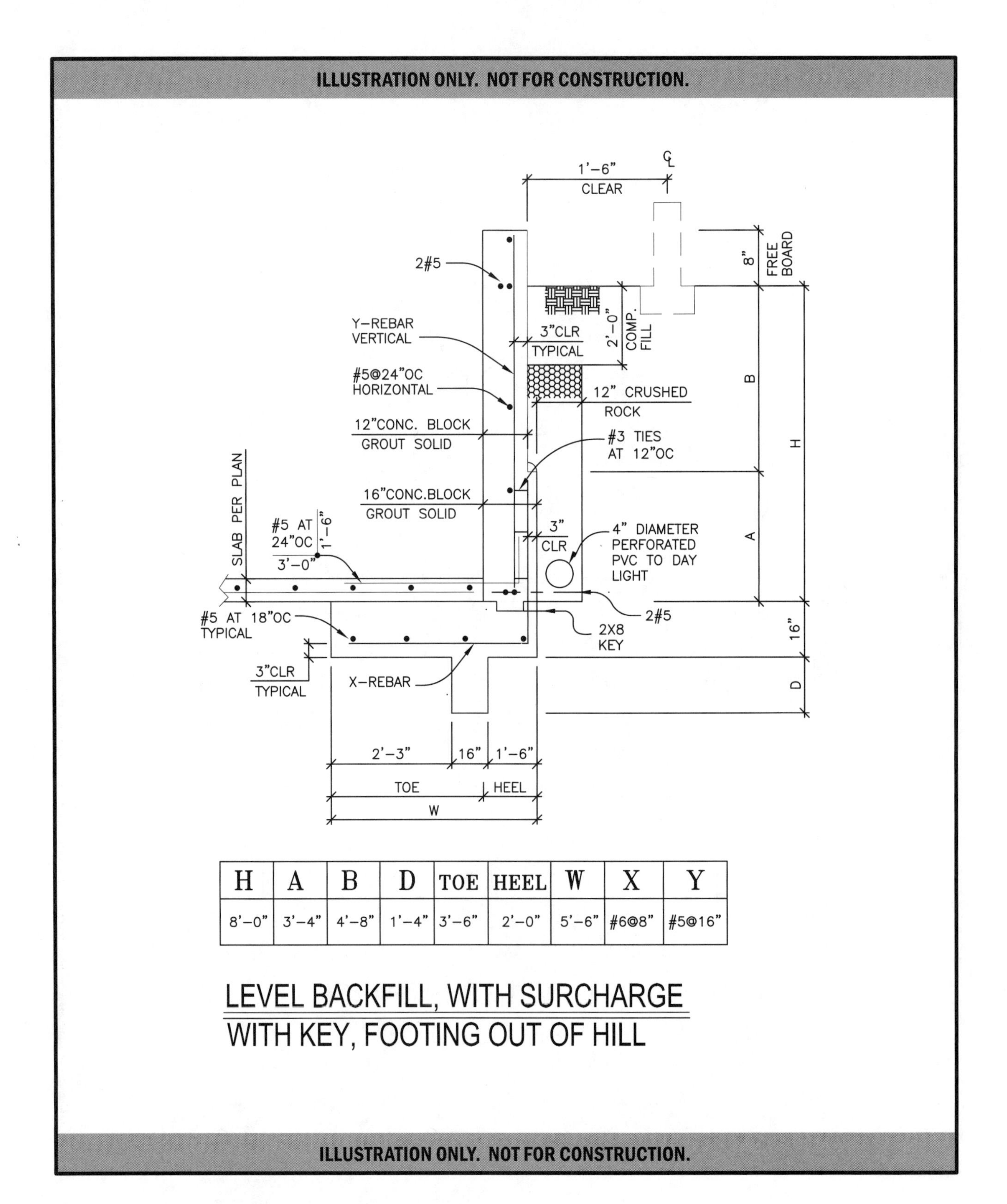

H	A	B	D	TOE	HEEL	W	X	Y
8'–0"	3'–4"	4'–8"	1'–4"	3'–6"	2'–0"	5'–6"	#6@8"	#5@16"

LEVEL BACKFILL, WITH SURCHARGE
WITH KEY, FOOTING OUT OF HILL

ILLUSTRATION ONLY. NOT FOR CONSTRUCTION.

EXECUTIVE ENGINEERING
19510 Ventura Blvd., Suite 110
Tarzana, California 91356-4234
Phone: (818) 708-1962 / Facsimile: (818) 708-1958
E-mail: EE110@aol.com

ILLUSTRATION ONLY. NOT FOR CONSTRUCTION.

2#5 TYPICAL

Y–REBAR VERTICAL

#5 AT 24"OC HORIZONTAL TYPICAL

8"CONC. BLOCK GROUT SOLID

12"CONC. BLK GROUT SOLID

24" MAX. COMPACTED FILL WHERE OCUURS

8" FREE BOARD

2'–0"

COMPACTED FILL

2–3/4" CLR

12" CRUSHED ROCK

2#5 WITH #3 TIES AT 32"OC

4" DIAMETER PERFORATED PVC TO DAY LIGHT

3" CLR

#5 AT 18"OC TYPICAL

A

H

B

LAP

C

X–REBAR

3" CLR TYPICAL

TOE

HEEL

W

LEVEL BACKFILL, NO KEY, FOOTING INTO HILL

ILLUSTRATION ONLY. NOT FOR CONSTRUCTION.

EXECUTIVE ENGINEERING
19510 Ventura Blvd., Suite 110
Tarzana, California 91356-4234
Phone: (818) 708-1962 / Facsimile: (818) 708-1958
E-mail: EE110@aol.com

ILLUSTRATION ONLY. NOT FOR CONSTRUCTION.

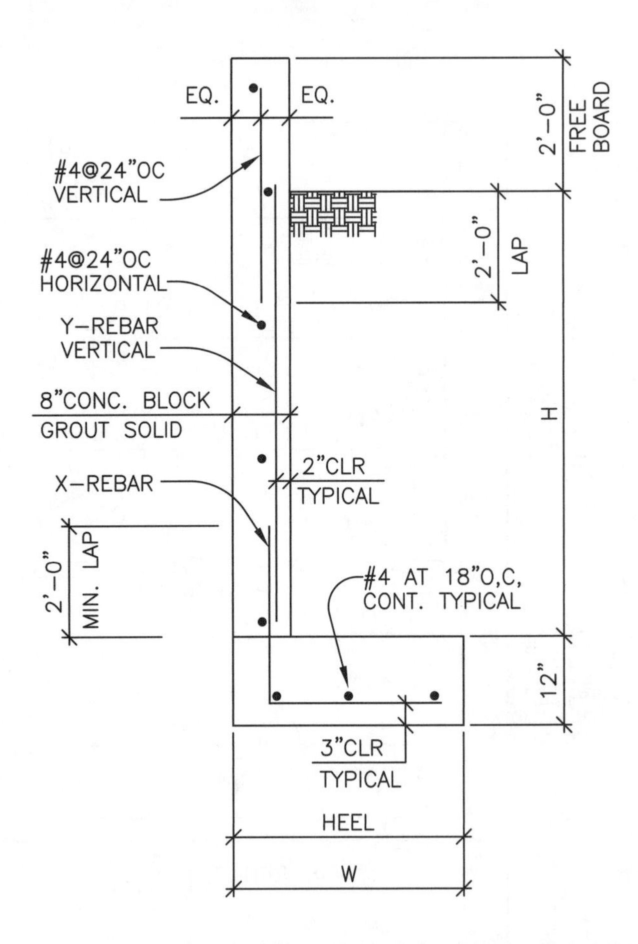

H	TOE	HEEL	W	X	Y	COMMENT
3'-0"	—	1'-9"	1'-9"	#4@24"	#4@24"	ELIMINATE Y-REBAR EXTEND X-REBAR TO TOP OF THE WALL
4'-0"	—	2'-6"	2'-6"	#4@24"	#4@24"	ELIMINATE Y-REBAR EXTEND X-REBAR TO TOP OF THE WALL

LEVEL BACKFILL,
NO KEY, FOOTING INTO HILL

ILLUSTRATION ONLY. NOT FOR CONSTRUCTION.

EXECUTIVE ENGINEERING
19510 Ventura Blvd., Suite 110
Tarzana, California 91356-4234
Phone: (818) 708-1962 / Facsimile: (818) 708-1958
E-mail: EE110@aol.com

ILLUSTRATION ONLY. NOT FOR CONSTRUCTION.

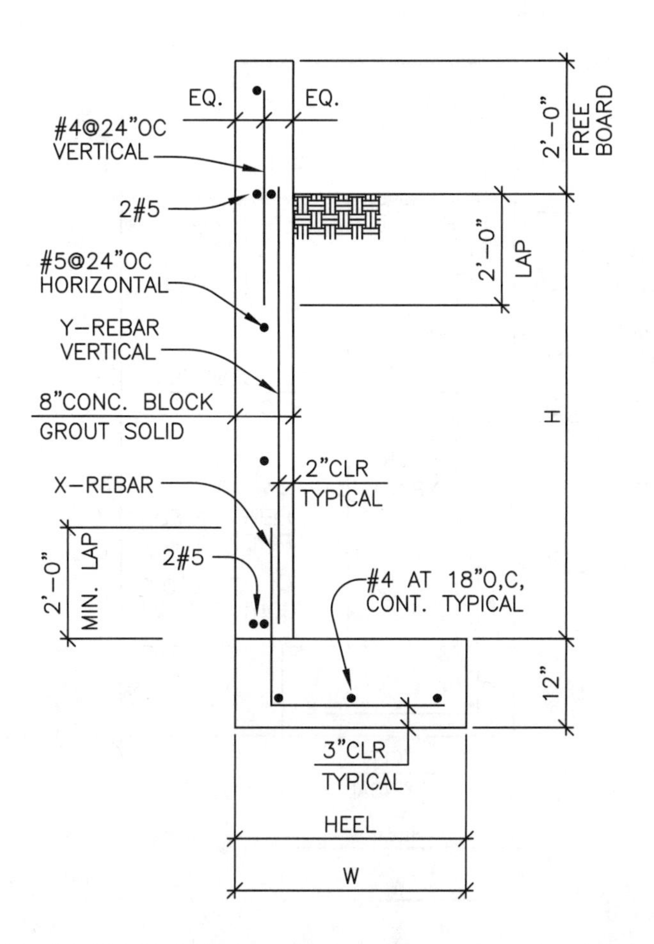

H	TOE	HEEL	W	X	Y	COMMENT
5'-0"	—	3'-6"	3'-6"	#4@16"	#4@16"	ELIMINATE Y-REBAR EXTEND X-REBAR TO TOP OF THE WALL
6'-0"	—	4'-6"	4'-6"	#5@8"	#5@8"	ELIMINATE Y-REBAR EXTEND X-REBAR TO TOP OF THE WALL

LEVEL BACKFILL, NO KEY, FOOTING INTO HILL

ILLUSTRATION ONLY. NOT FOR CONSTRUCTION.

EXECUTIVE ENGINEERING
19510 Ventura Blvd., Suite 110
Tarzana, California 91356-4234
Phone: (818) 708-1962 / Facsimile: (818) 708-1958
E-mail: EE110@aol.com

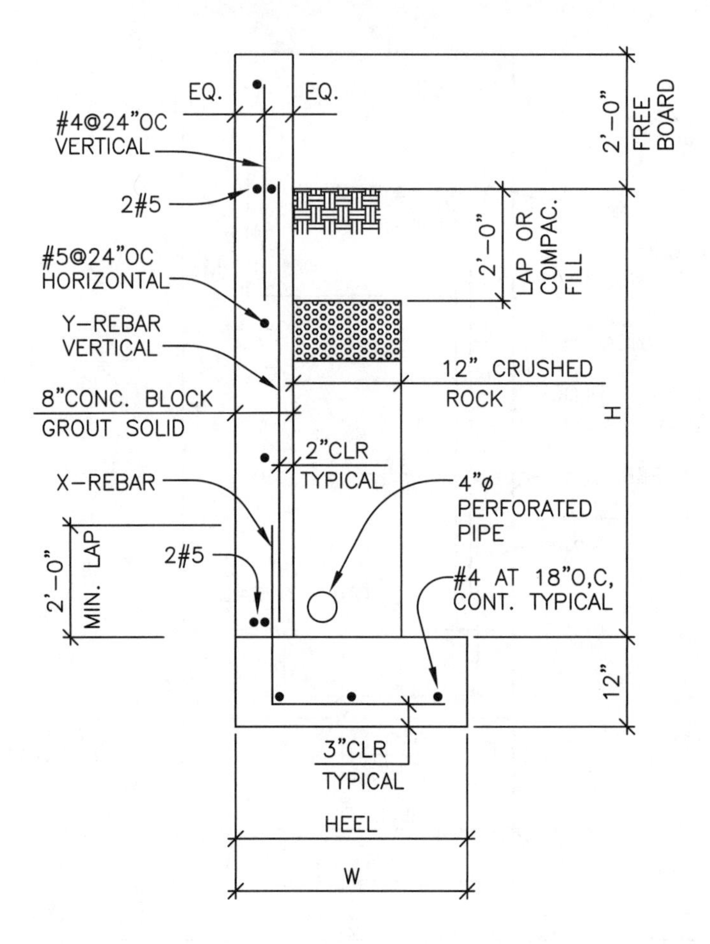

H	TOE	HEEL	W	X	Y	COMMENT
5'–0"	—	3'–6"	3'–6"	#4@16"	#4@16"	ELIMINATE Y–REBAR EXTEND X–REBAR TO TOP OF THE WALL
6'–0"	—	4'–6"	4'–6"	#5@8"	#5@8"	ELIMINATE Y–REBAR EXTEND X–REBAR TO TOP OF THE WALL

LEVEL BACKFILL, NO KEY, FOOTING INTO HILL

ILLUSTRATION ONLY. NOT FOR CONSTRUCTION.

EXECUTIVE ENGINEERING
19510 Ventura Blvd., Suite 110
Tarzana, California 91356-4234
Phone: (818) 708-1962 / Facsimile: (818) 708-1958
E-mail: EE110@aol.com

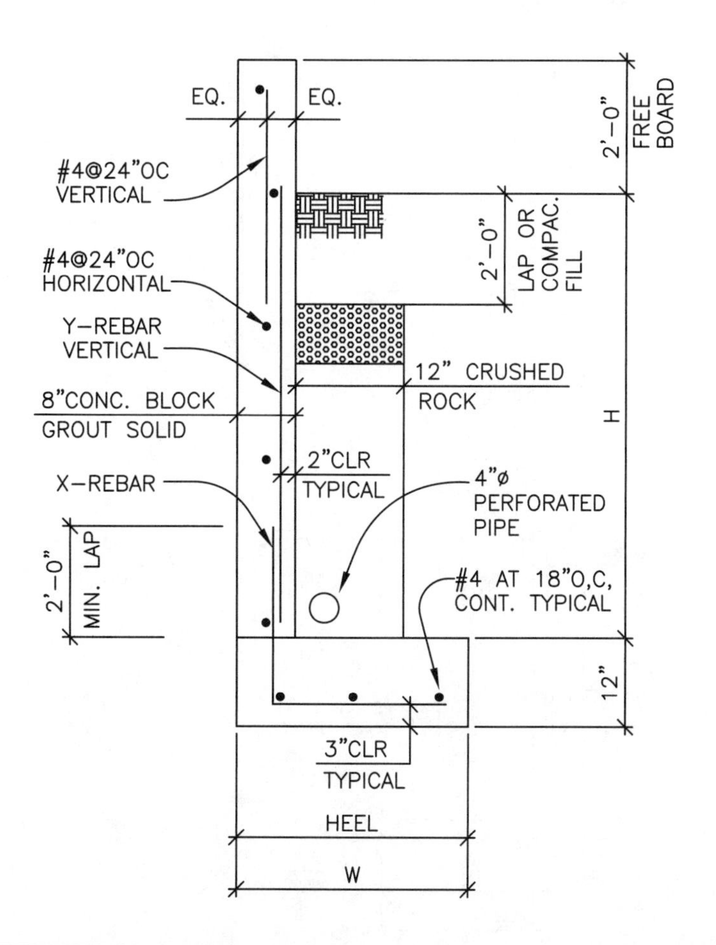

H	TOE	HEEL	W	X	Y	COMMENT
3'–0"	—	1'–9"	1'–9"	#4@24"	#4@24"	ELIMINATE Y–REBAR EXTEND X–REBAR TO TOP OF THE WALL
4'–0"	—	2'–6"	2'–6"	#4@24"	#4@24"	ELIMINATE Y–REBAR EXTEND X–REBAR TO TOP OF THE WALL

LEVEL BACKFILL,
NO KEY, FOOTING INTO HILL

ILLUSTRATION ONLY. NOT FOR CONSTRUCTION.

EXECUTIVE ENGINEERING
19510 Ventura Blvd., Suite 110
Tarzana, California 91356-4234
Phone: (818) 708-1962 / Facsimile: (818) 708-1958
E-mail: EE110@aol.com

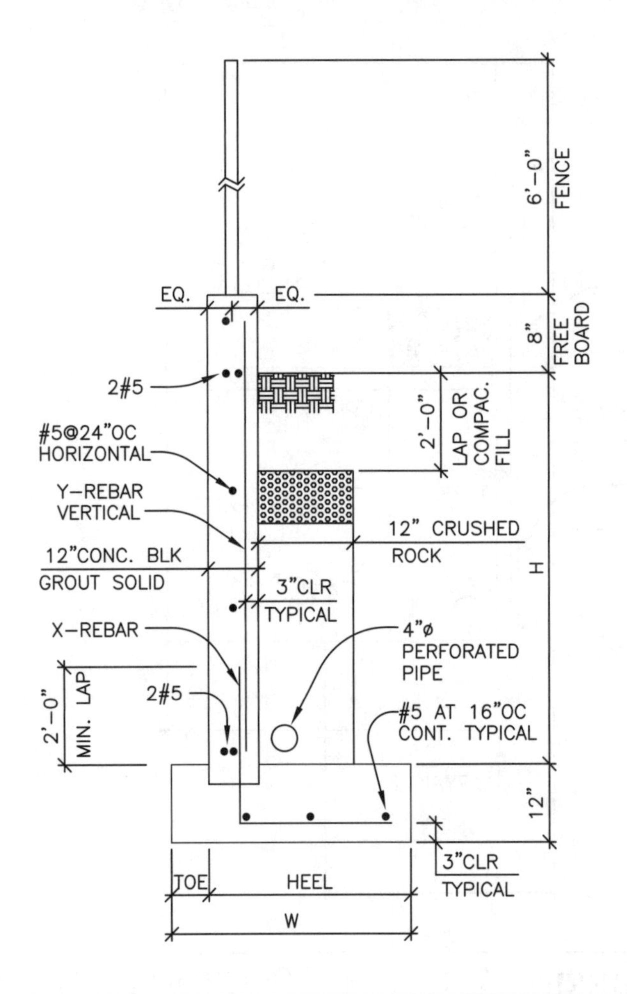

H	TOE	HEEL	W	X	Y	COMMENT	SPECS
6'-0"	6"	4'-6"	5'-0"	#5@16"	#5@16"	ELIMINATE Y-REBAR EXTEND X-REBAR TO TOP OF THE WALL	F'c=2,500 PSI Fy=40,000 PSI

LEVEL BACKFILL, NO KEY, FOOTING INTO HILL

ILLUSTRATION ONLY. NOT FOR CONSTRUCTION.

EXECUTIVE ENGINEERING
19510 Ventura Blvd., Suite 110
Tarzana, California 91356-4234
Phone: (818) 708-1962 / Facsimile: (818) 708-1958
E-mail: EE110@aol.com

ILLUSTRATION ONLY. NOT FOR CONSTRUCTION.

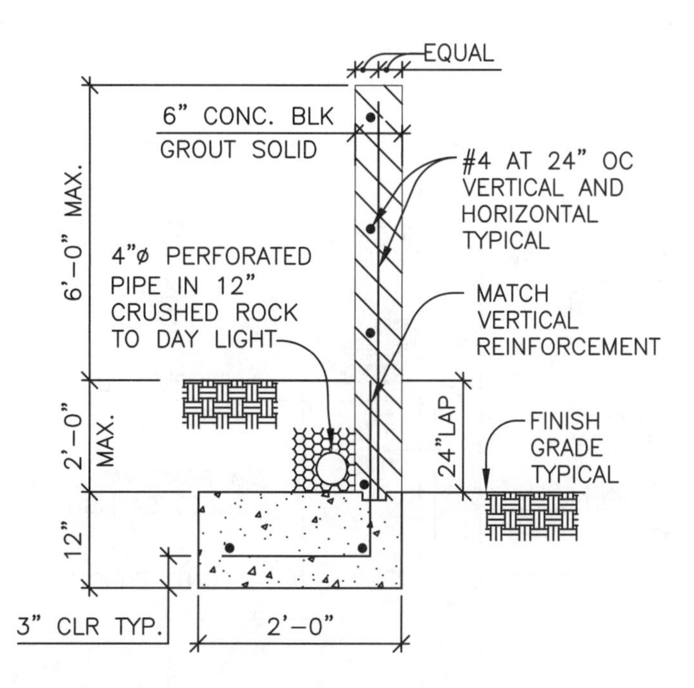

LEVEL BACKFILL,
NO KEY, FOOTING INTO HILL

ILLUSTRATION ONLY. NOT FOR CONSTRUCTION.

EXECUTIVE ENGINEERING
19510 Ventura Blvd., Suite 110
Tarzana, California 91356-4234
Phone: (818) 708-1962 / Facsimile: (818) 708-1958
E-mail: EE110@aol.com

ILLUSTRATION ONLY. NOT FOR CONSTRUCTION.

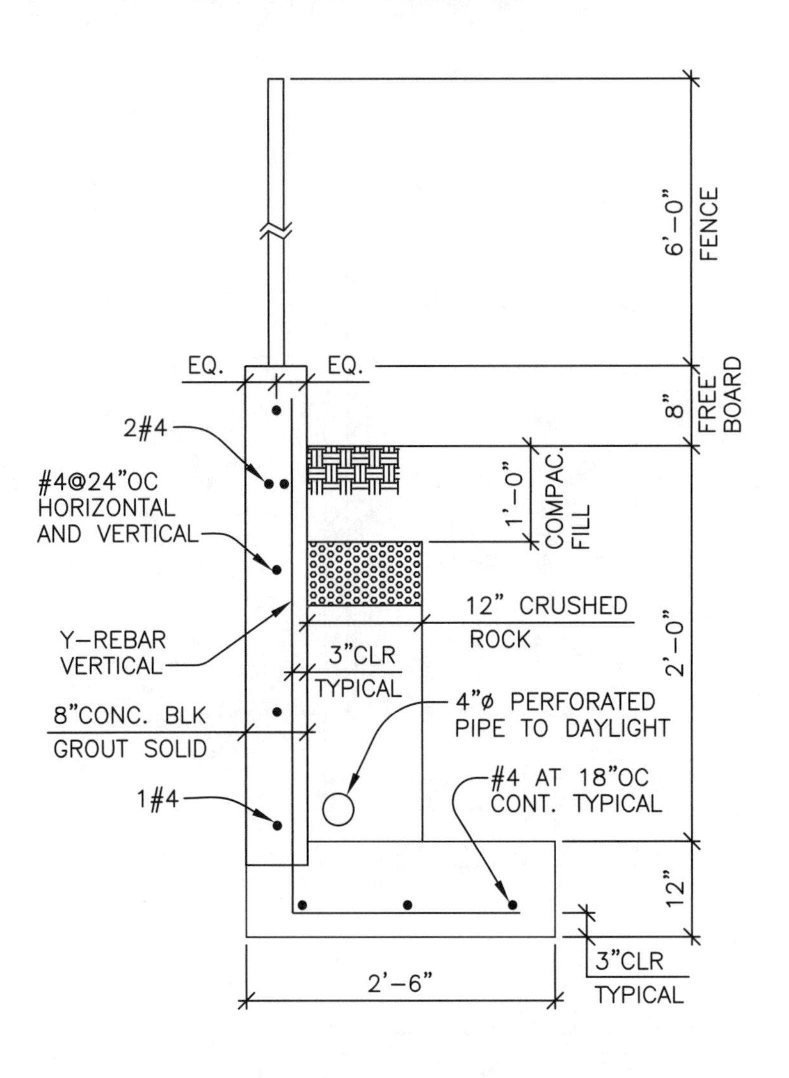

LEVEL BACKFILL, NO KEY, FOOTING INTO HILL

ILLUSTRATION ONLY. NOT FOR CONSTRUCTION.

EXECUTIVE ENGINEERING
19510 Ventura Blvd., Suite 110
Tarzana, California 91356-4234
Phone: (818) 708-1962 / Facsimile: (818) 708-1958
E-mail: EE110@aol.com

ILLUSTRATION ONLY. NOT FOR CONSTRUCTION.

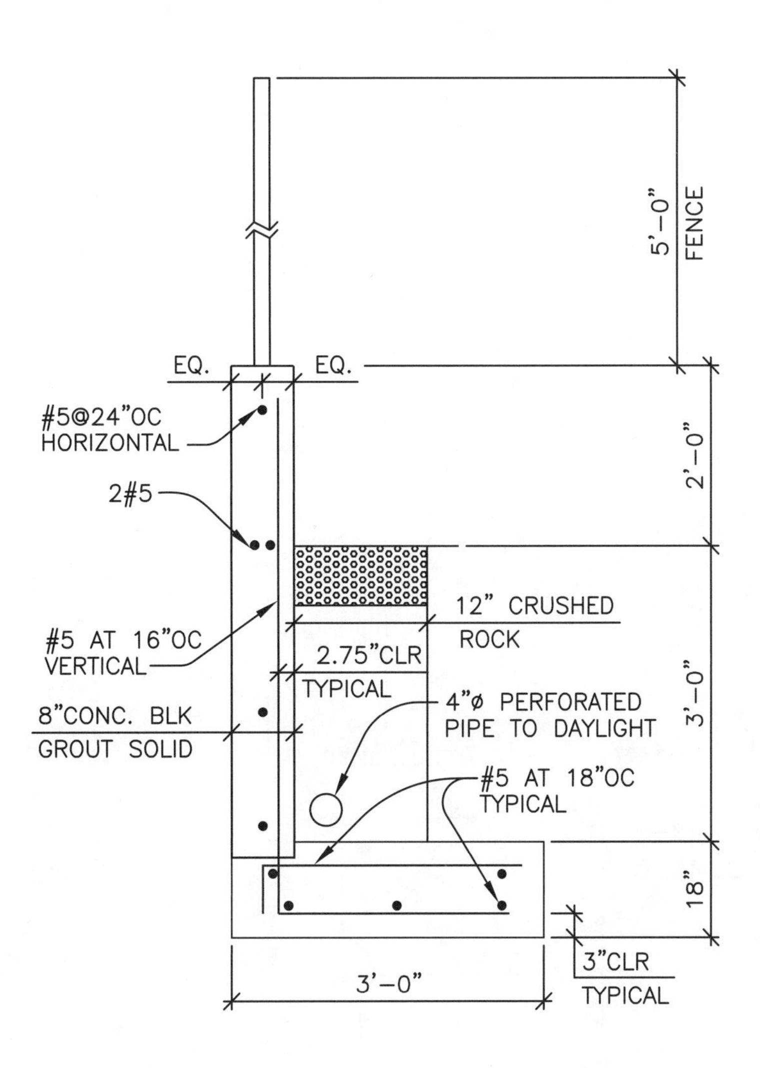

LEVEL BACKFILL, NO KEY, FOOTING INTO HILL

ILLUSTRATION ONLY. NOT FOR CONSTRUCTION.

EXECUTIVE ENGINEERING
19510 Ventura Blvd., Suite 110
Tarzana, California 91356-4234
Phone: (818) 708-1962 / Facsimile: (818) 708-1958
E-mail: EE110@aol.com

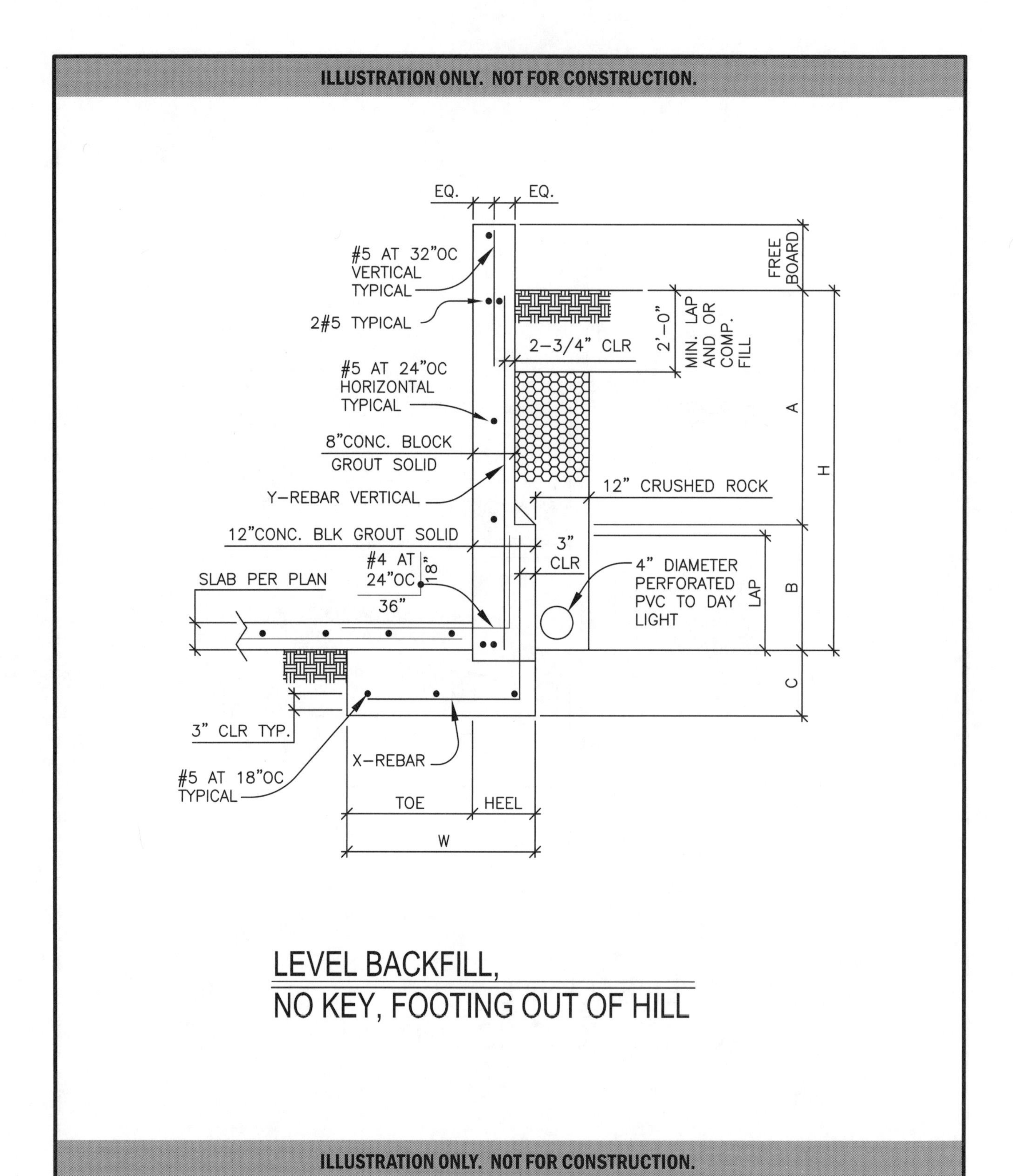

EXECUTIVE ENGINEERING
19510 Ventura Blvd., Suite 110
Tarzana, California 91356-4234
Phone: (818) 708-1962 / Facsimile: (818) 708-1958
E-mail: EE110@aol.com

EQ. EQ.

Z–REBAR VERTICAL

LAP

2#5 TYPICAL

#5 AT 24"OC HORIZONTAL TYPICAL

12"CONC. BLOCK GROUT SOLID

Y–REBAR VERTICAL

16"CONC. BLK GROUT SOLID

#4 AT 24"OC

18"

36"

SLAB PER PLAN

3" CLR TYP.

#5 AT 18"OC TYPICAL

X–REBAR

TOE HEEL

W

2–3/4" CLR

2'–0"

COMP. FILL

12" CRUSHED ROCK

3" CLR

4" DIAMETER PERFORATED PVC TO DAY LIGHT

LAP

F

FREE BOARD

A

H

B

C

LEVEL BACKFILL, NO KEY, FOOTING OUT OF HILL

ILLUSTRATION ONLY. NOT FOR CONSTRUCTION.

EXECUTIVE ENGINEERING
19510 Ventura Blvd., Suite 110
Tarzana, California 91356-4234
Phone: (818) 708-1962 / Facsimile: (818) 708-1958
E-mail: EE110@aol.com

EQ. EQ.

#5 AT 24"OC
VERTICAL
TYPICAL

2#5 TYPICAL

#5 AT 24"OC
HORIZONTAL
TYPICAL

8"CONC. BLOCK
GROUT SOLID

Y–REBAR VERTICAL

#4 AT
24"OC
18"
36"

SLAB PER PLAN

3" CLR TYP.

#5 AT 18"OC
TYPICAL

X–REBAR

TOE HEEL

W

FREE
BOARD

2'–0"

MIN. LAP
AND OR
COMP.
FILL

12" CRUSHED ROCK

2–3/4"
CLR TYPICAL

H

4" DIAMETER
PERFORATED
PVC TO DAY
LIGHT

LAP

WATER PROOF
MEMBRANE
WHERE OCCURS

C

LEVEL BACKFILL,
NO KEY, FOOTING OUT OF HILL

ILLUSTRATION ONLY. NOT FOR CONSTRUCTION.

EXECUTIVE ENGINEERING
19510 Ventura Blvd., Suite 110
Tarzana, California 91356-4234
Phone: (818) 708-1962 / Facsimile: (818) 708-1958
E-mail: EE110@aol.com

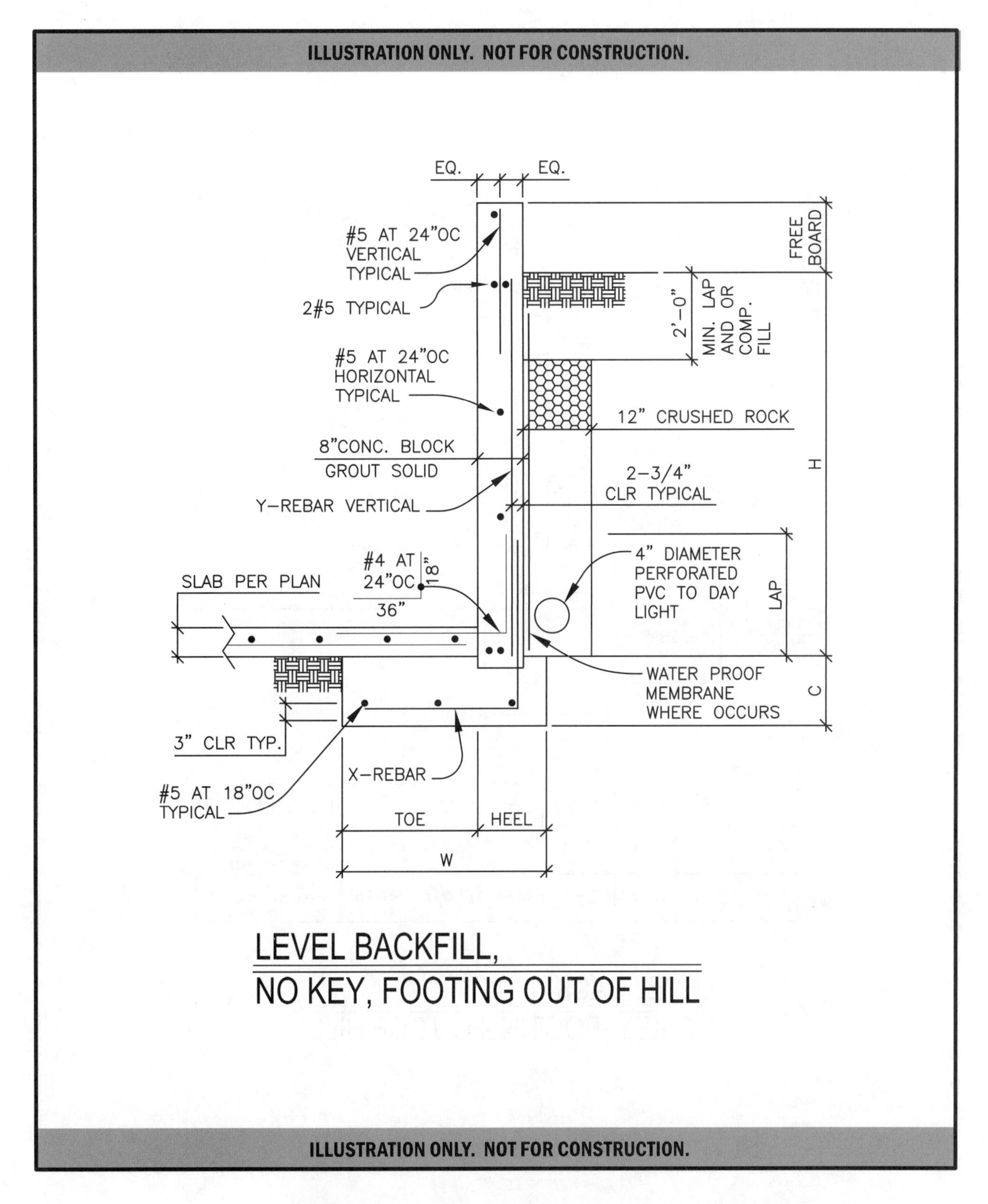

EXECUTIVE ENGINEERING
19510 Ventura Blvd., Suite 110
Tarzana, California 91356-4234
Phone: (818) 708-1962 / Facsimile: (818) 708-1958
E-mail: EE110@aol.com

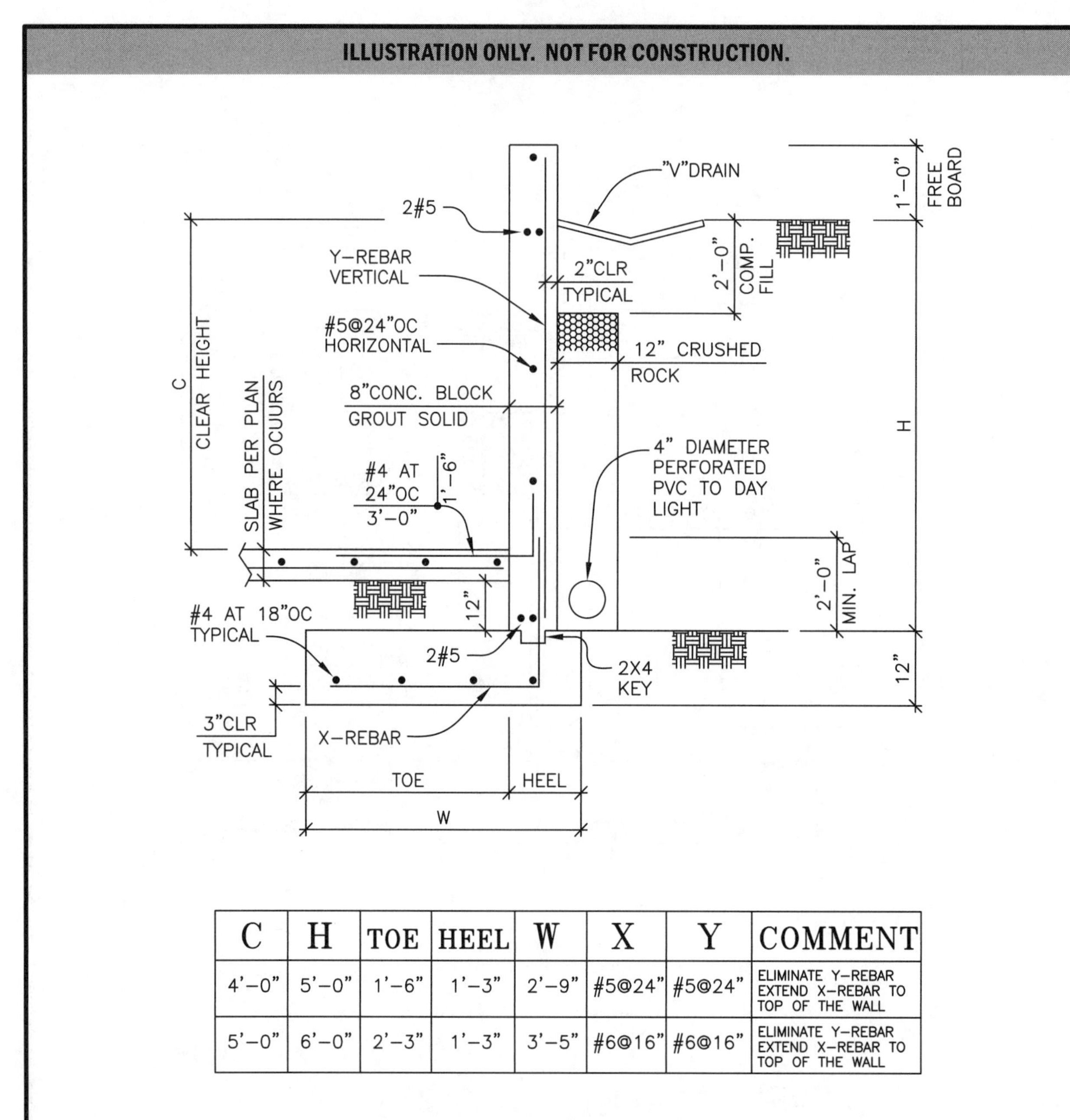

C	H	TOE	HEEL	W	X	Y	COMMENT
4'–0"	5'–0"	1'–6"	1'–3"	2'–9"	#5@24"	#5@24"	ELIMINATE Y–REBAR EXTEND X–REBAR TO TOP OF THE WALL
5'–0"	6'–0"	2'–3"	1'–3"	3'–5"	#6@16"	#6@16"	ELIMINATE Y–REBAR EXTEND X–REBAR TO TOP OF THE WALL

LEVEL BACKFILL, NO KEY, FOOTING OUT OF HILL

ILLUSTRATION ONLY. NOT FOR CONSTRUCTION.

EXECUTIVE ENGINEERING
19510 Ventura Blvd., Suite 110
Tarzana, California 91356-4234
Phone: (818) 708-1962 / Facsimile: (818) 708-1958
E-mail: EE110@aol.com

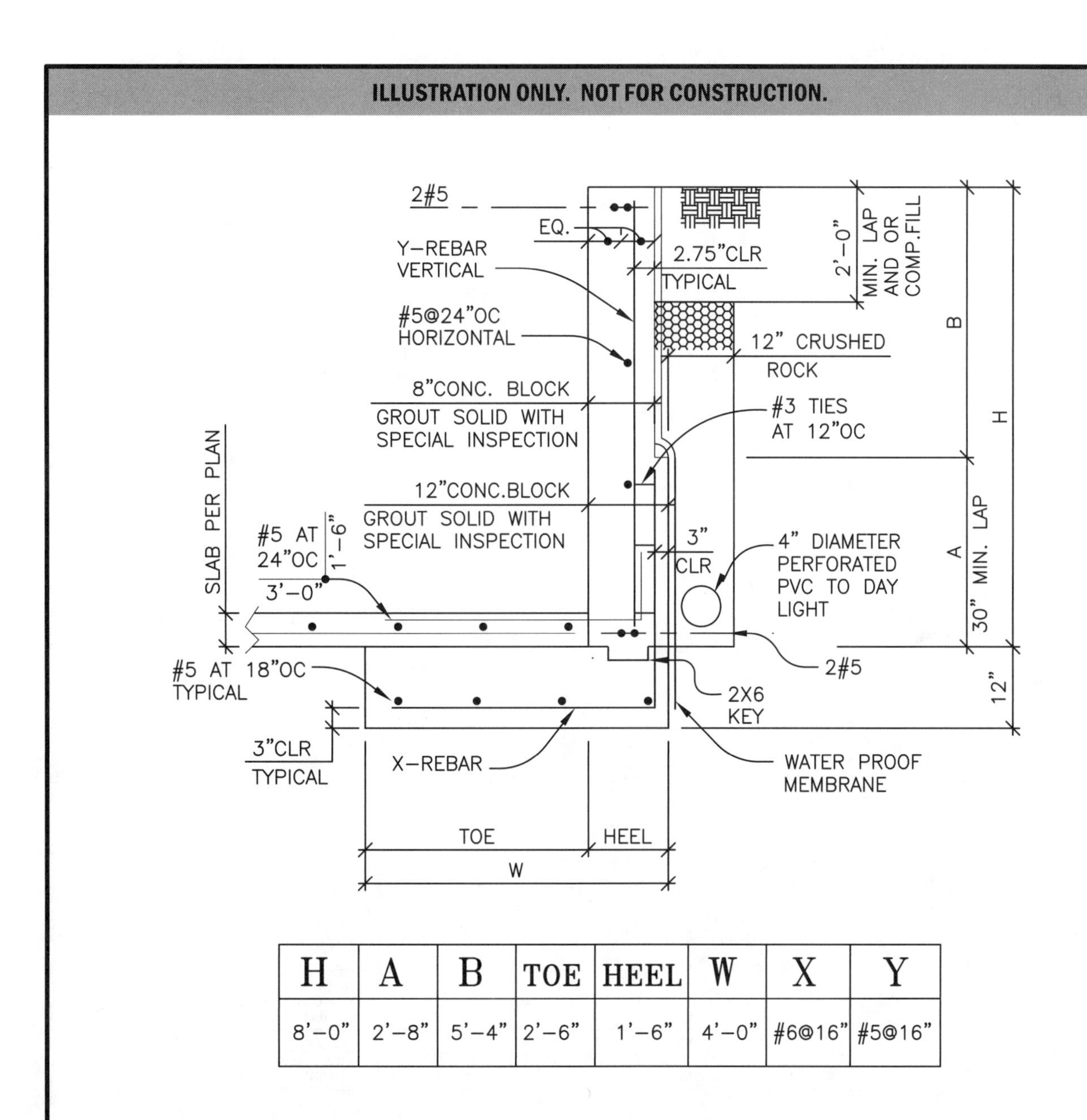

H	A	B	TOE	HEEL	W	X	Y
8'–0"	2'–8"	5'–4"	2'–6"	1'–6"	4'–0"	#6@16"	#5@16"

LEVEL BACKFILL, NO KEY, FOOTING OUT OF HILL

ILLUSTRATION ONLY. NOT FOR CONSTRUCTION.

EXECUTIVE ENGINEERING
19510 Ventura Blvd., Suite 110
Tarzana, California 91356-4234
Phone: (818) 708-1962 / Facsimile: (818) 708-1958
E-mail: EE110@aol.com

ILLUSTRATION ONLY. NOT FOR CONSTRUCTION.

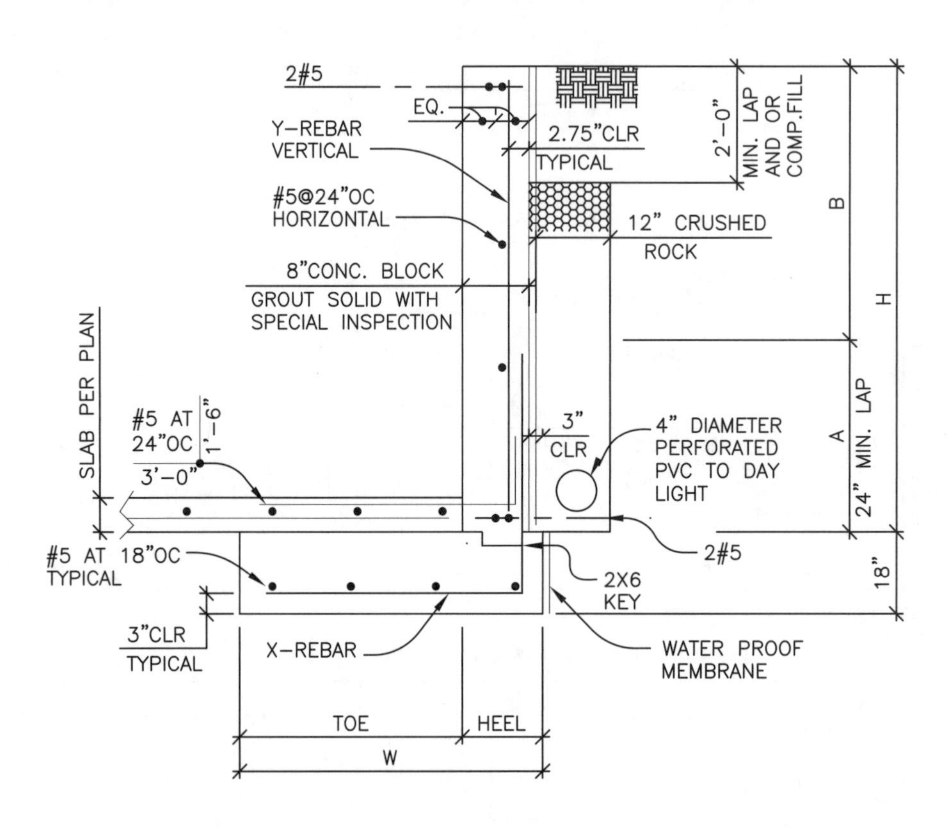

H	A	B	TOE	HEEL	W	X	Y	COMMENT
6'–0"	2'–0"	4'–0"	2'–0"	1'–0"	3'–0"	#6@16"	#6@16"	ELIMINATE Y BAR, EXTEND X BAR TO THE TOP
5'–0"	2'–0"	3'–0"	1'–9"	1'–0"	2'–9"	#6@16"	#5@16"	ELIMINATE Y BAR, EXTEND X BAR TO THE TOP
4'–0"	2'–0"	2'–0"	1'–0"	1'–0"	3'–6"	#5@16"	#5@16"	ELIMINATE Y BAR, EXTEND X BAR TO THE TOP

LEVEL BACKFILL,
NO KEY, FOOTING OUT OF HILL

ILLUSTRATION ONLY. NOT FOR CONSTRUCTION.

EXECUTIVE ENGINEERING
19510 Ventura Blvd., Suite 110
Tarzana, California 91356-4234
Phone: (818) 708-1962 / Facsimile: (818) 708-1958
E-mail: EE110@aol.com

ILLUSTRATION ONLY. NOT FOR CONSTRUCTION.

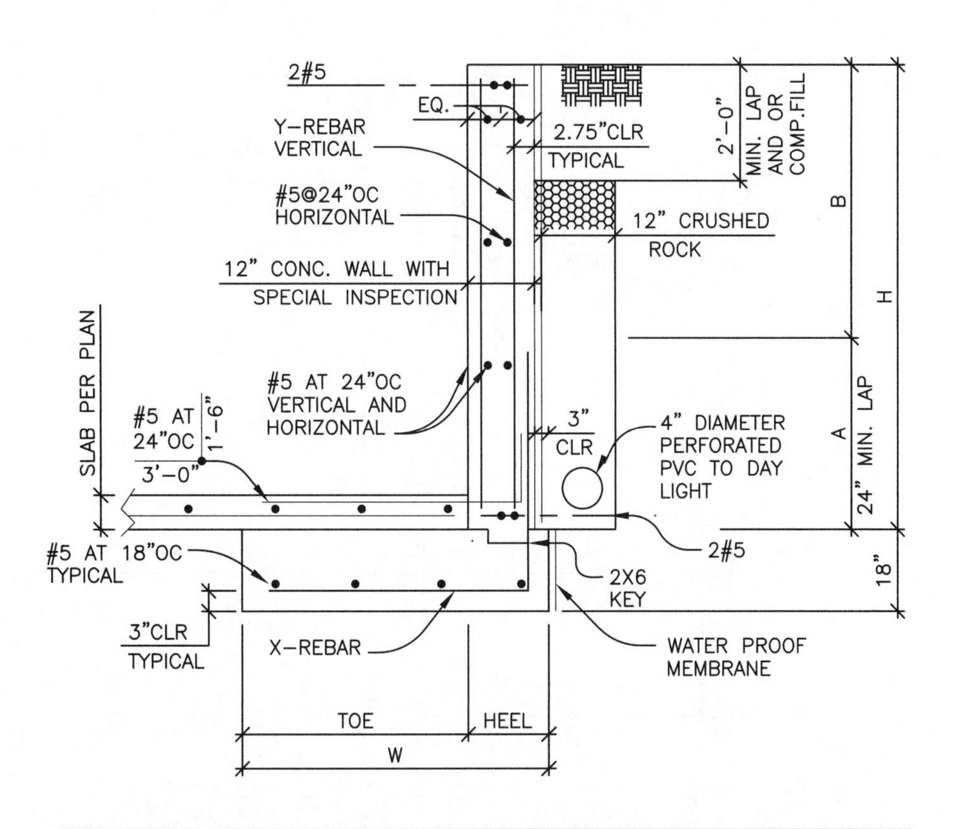

H	A	B	TOE	HEEL	W	X	Y	COMMENT
6'-0"	2'-0"	4'-0"	2'-0"	1'-0"	3'-0"	#6@16"	#6@16"	ELIMINATE Y BAR, EXTEND X BAR TO THE TOP
5'-0"	2'-0"	3'-0"	1'-9"	1'-0"	2'-9"	#6@16"	#5@16"	ELIMINATE Y BAR, EXTEND X BAR TO THE TOP
4'-0"	2'-0"	2'-0"	1'-0"	1'-0"	3'-6"	#5@16"	#5@16"	ELIMINATE Y BAR, EXTEND X BAR TO THE TOP

LEVEL BACKFILL, NO KEY, FOOTING OUT OF HILL

ILLUSTRATION ONLY. NOT FOR CONSTRUCTION.

EXECUTIVE ENGINEERING
19510 Ventura Blvd., Suite 110
Tarzana, California 91356-4234
Phone: (818) 708-1962 / Facsimile: (818) 708-1958
E-mail: EE110@aol.com

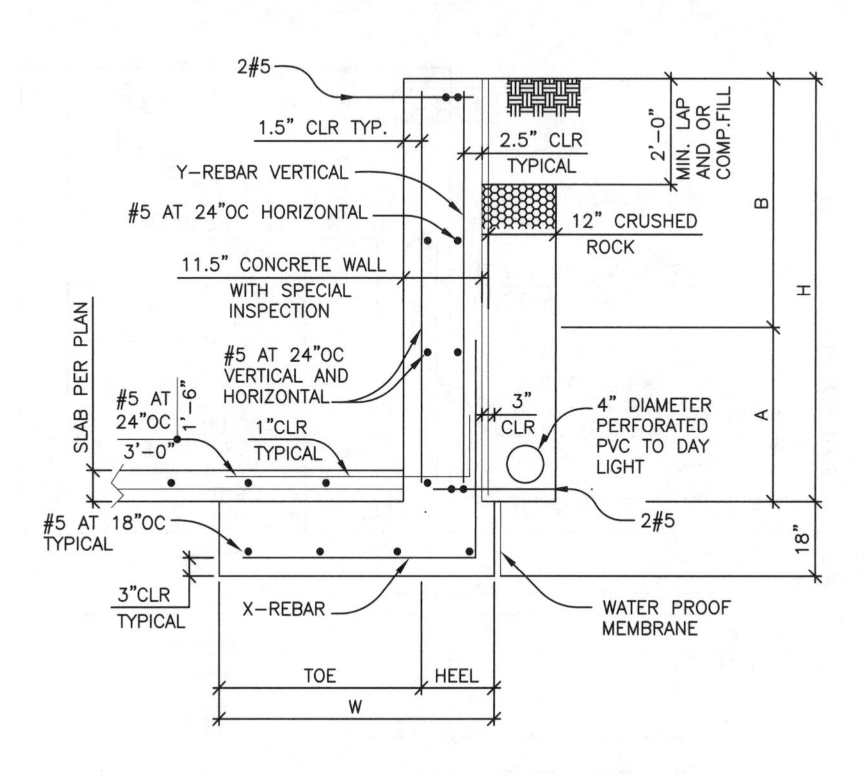

H	A	B	TOE	HEEL	W	X	Y	COMMENT
8'-0"	2'-8"	5'-4"	3'-0"	1'-0"	4'-0"	#6@16"	#6@16"	
7'-0"	2'-8"	4'-4"	2'-6"	1'-0"	3'-6"	#6@16"	#6@16"	
6'-0"	2'-0"	4'-0"	1'-6"	1'-0"	2'-6"	#5@16"	#5@16"	ELIMINATE Y BAR, EXTEND X BAR TO THE TOP
5'-0"	2'-0"	3'-0"	1'-6"	1'-0"	2'-6"	#5@16"	#5@16"	ELIMINATE Y BAR, EXTEND X BAR TO THE TOP
4'-0"	2'-0"	2'-0"	1'-0"	1'-0"	2'-0"	#5@16"	#5@16"	ELIMINATE Y BAR, EXTEND X BAR TO THE TOP

LEVEL BACKFILL,
NO KEY, FOOTING OUT OF HILL

ILLUSTRATION ONLY. NOT FOR CONSTRUCTION.

EXECUTIVE ENGINEERING
19510 Ventura Blvd., Suite 110
Tarzana, California 91356-4234
Phone: (818) 708-1962 / Facsimile: (818) 708-1958
E-mail: EE110@aol.com

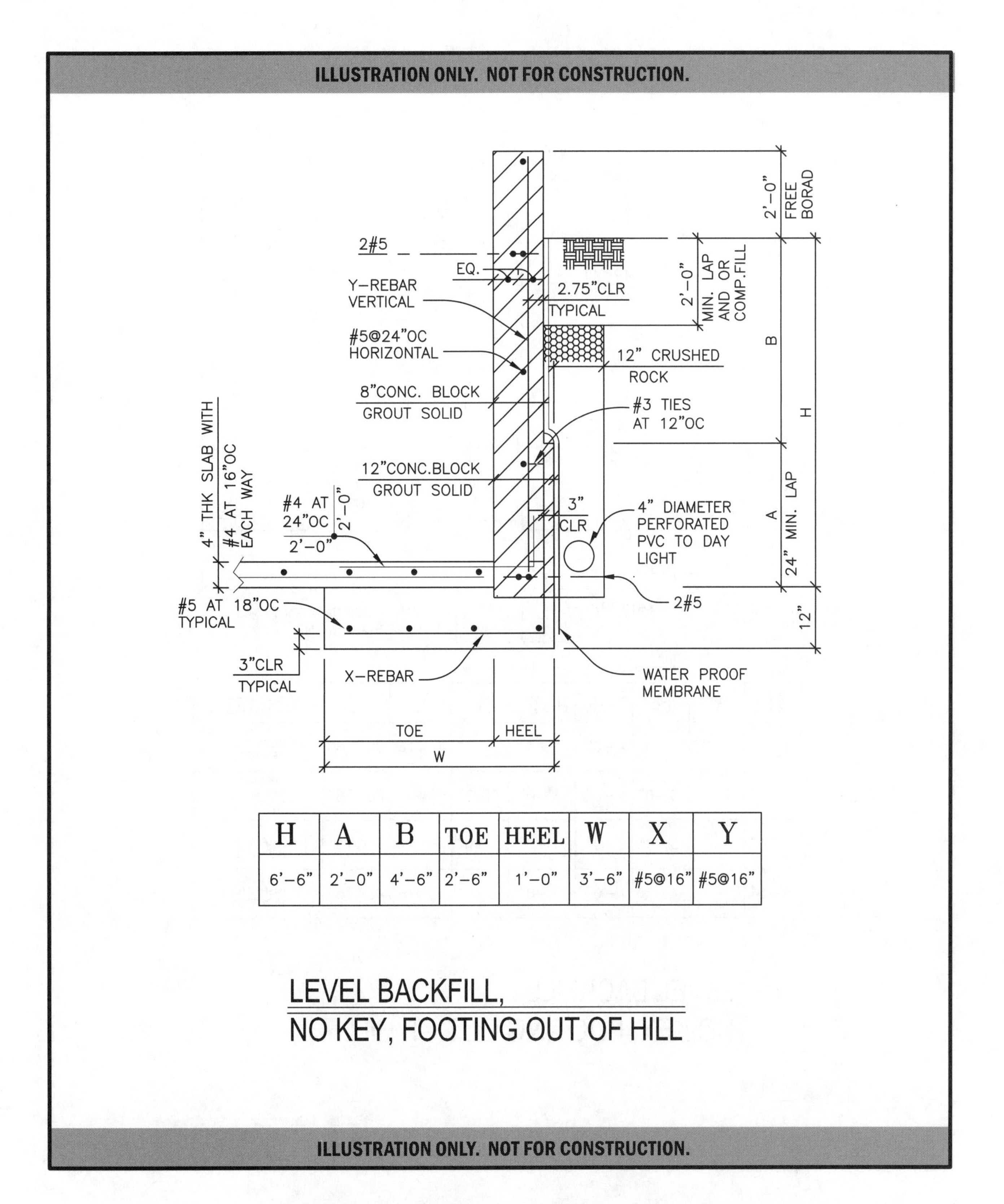

H	A	B	TOE	HEEL	W	X	Y
6'–6"	2'–0"	4'–6"	2'–6"	1'–0"	3'–6"	#5@16"	#5@16"

LEVEL BACKFILL,
NO KEY, FOOTING OUT OF HILL

ILLUSTRATION ONLY. NOT FOR CONSTRUCTION.

EXECUTIVE ENGINEERING
19510 Ventura Blvd., Suite 110
Tarzana, California 91356-4234
Phone: (818) 708-1962 / Facsimile: (818) 708-1958
E-mail: EE110@aol.com

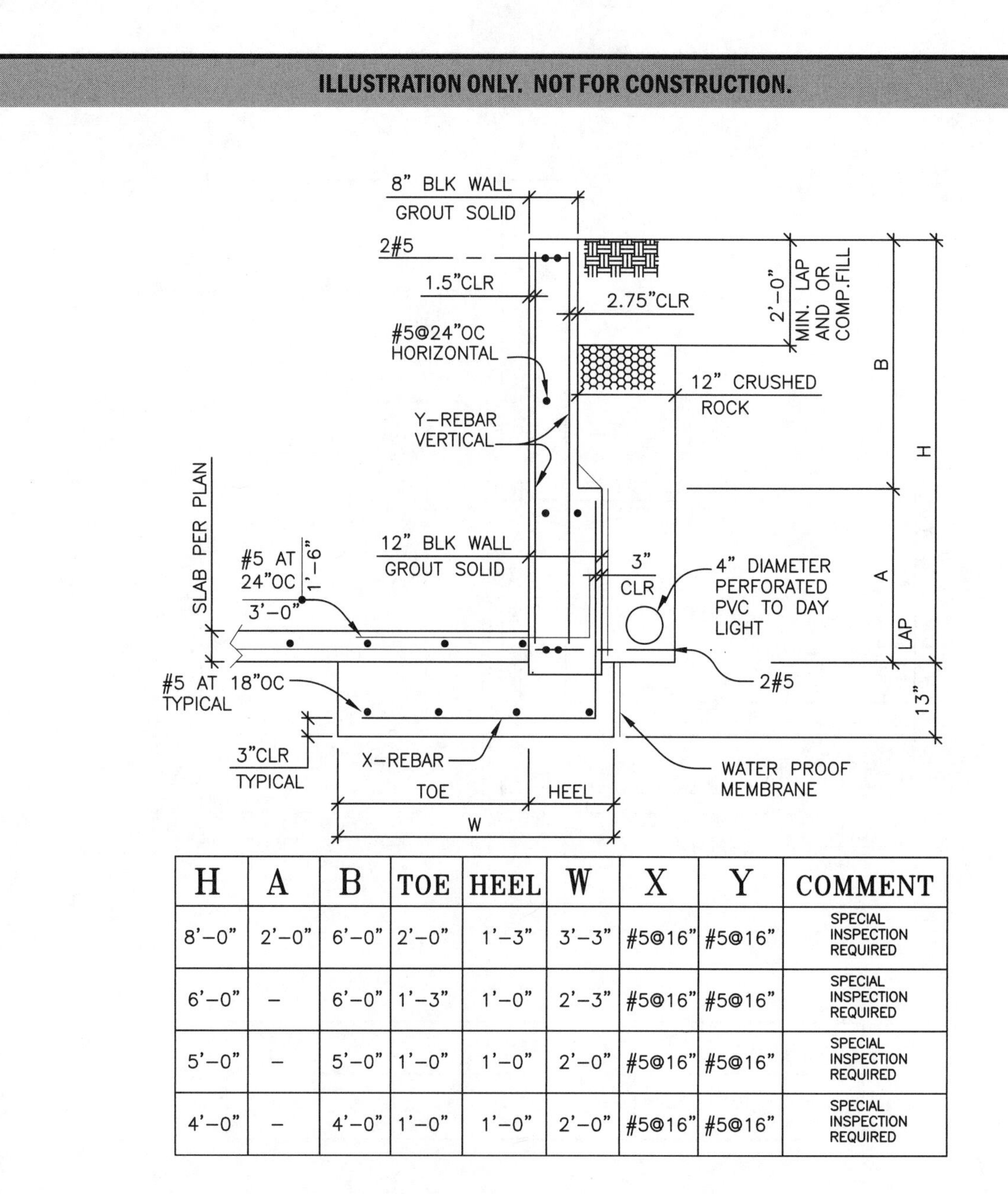

H	A	B	TOE	HEEL	W	X	Y	COMMENT
8'-0"	2'-0"	6'-0"	2'-0"	1'-3"	3'-3"	#5@16"	#5@16"	SPECIAL INSPECTION REQUIRED
6'-0"	–	6'-0"	1'-3"	1'-0"	2'-3"	#5@16"	#5@16"	SPECIAL INSPECTION REQUIRED
5'-0"	–	5'-0"	1'-0"	1'-0"	2'-0"	#5@16"	#5@16"	SPECIAL INSPECTION REQUIRED
4'-0"	–	4'-0"	1'-0"	1'-0"	2'-0"	#5@16"	#5@16"	SPECIAL INSPECTION REQUIRED

LEVEL BACKFILL,
NO KEY, FOOTING OUT OF HILL

ILLUSTRATION ONLY. NOT FOR CONSTRUCTION.

EXECUTIVE ENGINEERING
19510 Ventura Blvd., Suite 110
Tarzana, California 91356-4234
Phone: (818) 708-1962 / Facsimile: (818) 708-1958
E-mail: EE110@aol.com

ILLUSTRATION ONLY. NOT FOR CONSTRUCTION.

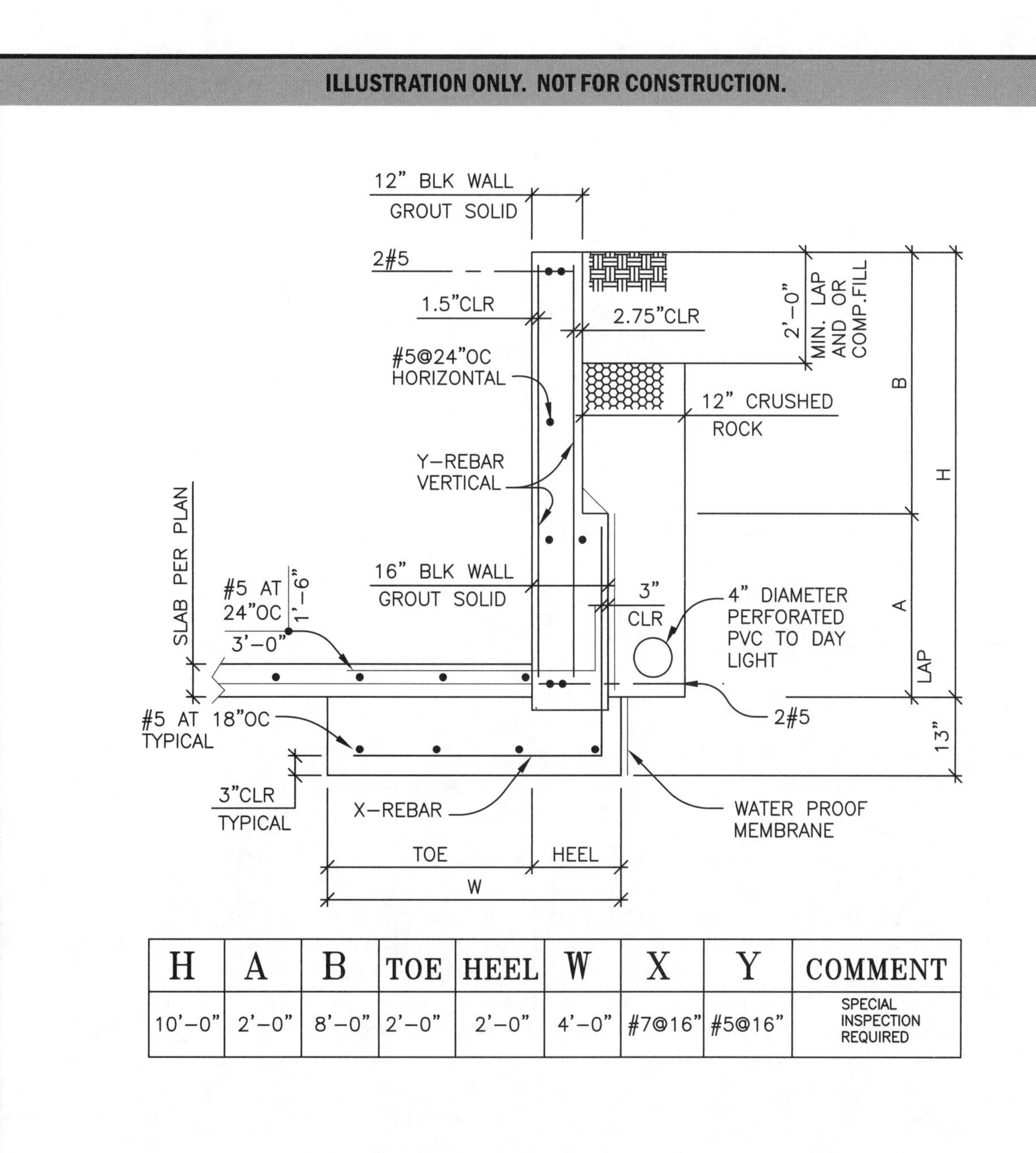

H	A	B	TOE	HEEL	W	X	Y	COMMENT
10'-0"	2'-0"	8'-0"	2'-0"	2'-0"	4'-0"	#7@16"	#5@16"	SPECIAL INSPECTION REQUIRED

LEVEL BACKFILL, NO KEY, FOOTING OUT OF HILL

ILLUSTRATION ONLY. NOT FOR CONSTRUCTION.

EXECUTIVE ENGINEERING
19510 Ventura Blvd., Suite 110
Tarzana, California 91356-4234
Phone: (818) 708-1962 / Facsimile: (818) 708-1958
E-mail: EE110@aol.com

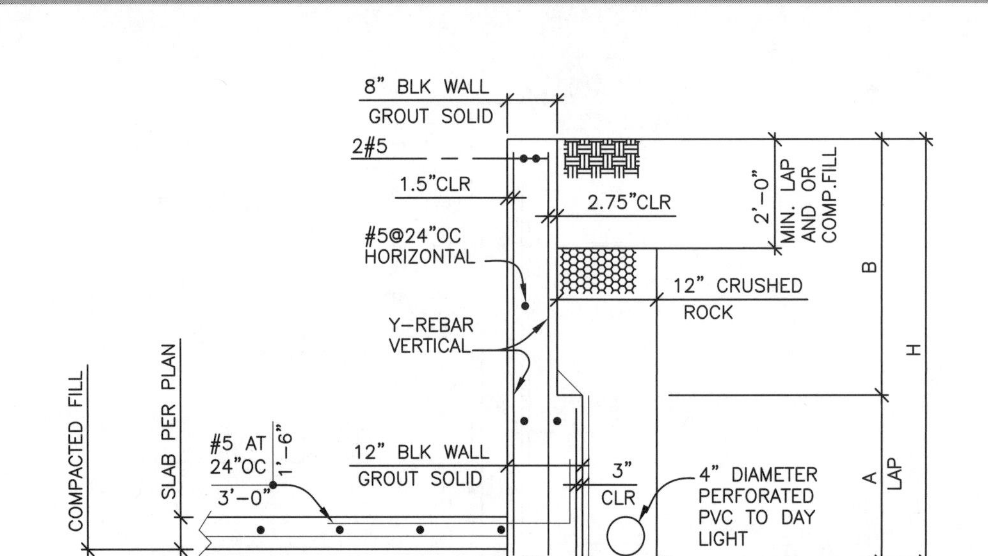

H	A	B	D	TOE	HEEL	W	X	Y	COMMENT
10'–0"	10'–0"	–	24"	3'–9"	1'–9"	5'–6"	#7@16"	#7@16"	SPECIAL INSPECTION REQUIRED
8'–0"	8'–0"	–	24"	2'–0"	1'–9"	3'–9"	#6@16"	#6@16"	SPECIAL INSPECTION REQUIRED
6'–0"	–	6'–0"	18"	1'–6"	1'–6"	3'–0"	#5@16"	#5@16"	SPECIAL INSPECTION REQUIRED
4'–0"	–	4'–0"	16"	1'–0"	1'–0"	2'–0"	#5@16"	#5@16"	SPECIAL INSPECTION REQUIRED

LEVEL BACKFILL,
NO KEY, FOOTING OUT OF HILL

ILLUSTRATION ONLY. NOT FOR CONSTRUCTION.

EXECUTIVE ENGINEERING
19510 Ventura Blvd., Suite 110
Tarzana, California 91356-4234
Phone: (818) 708-1962 / Facsimile: (818) 708-1958
E-mail: EE110@aol.com

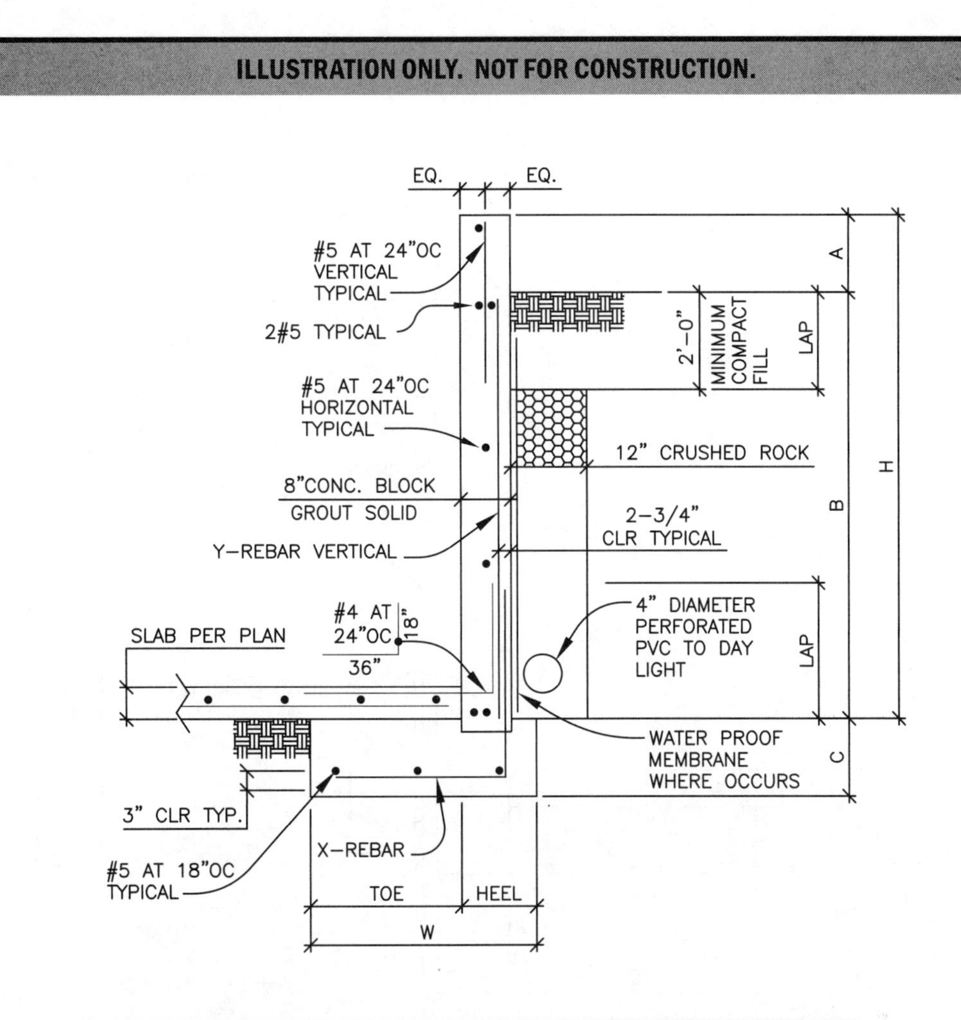

H	A	B	C	TOE	HEEL	W	X	Y*	LAP
8'-0"	4'-0"	4'-0"	12"	1'-3"	1'-0"	2'-3"	#5@16"	#5@16"	24"
8'-0"	2'-6"	5'-6"	12"	1'-6"	1'-0"	2'-6"	#5@16"	#5@16"	24"

COMMENT: ELIMINATE Y-REBAR EXTEND
X-REBAR TO TOP OF THE WALL WHERE APPLIES

LEVEL BACKFILL, NO KEY, FOOTING OUT OF HILL

ILLUSTRATION ONLY. NOT FOR CONSTRUCTION.

EXECUTIVE ENGINEERING
19510 Ventura Blvd., Suite 110
Tarzana, California 91356-4234
Phone: (818) 708-1962 / Facsimile: (818) 708-1958
E-mail: EE110@aol.com

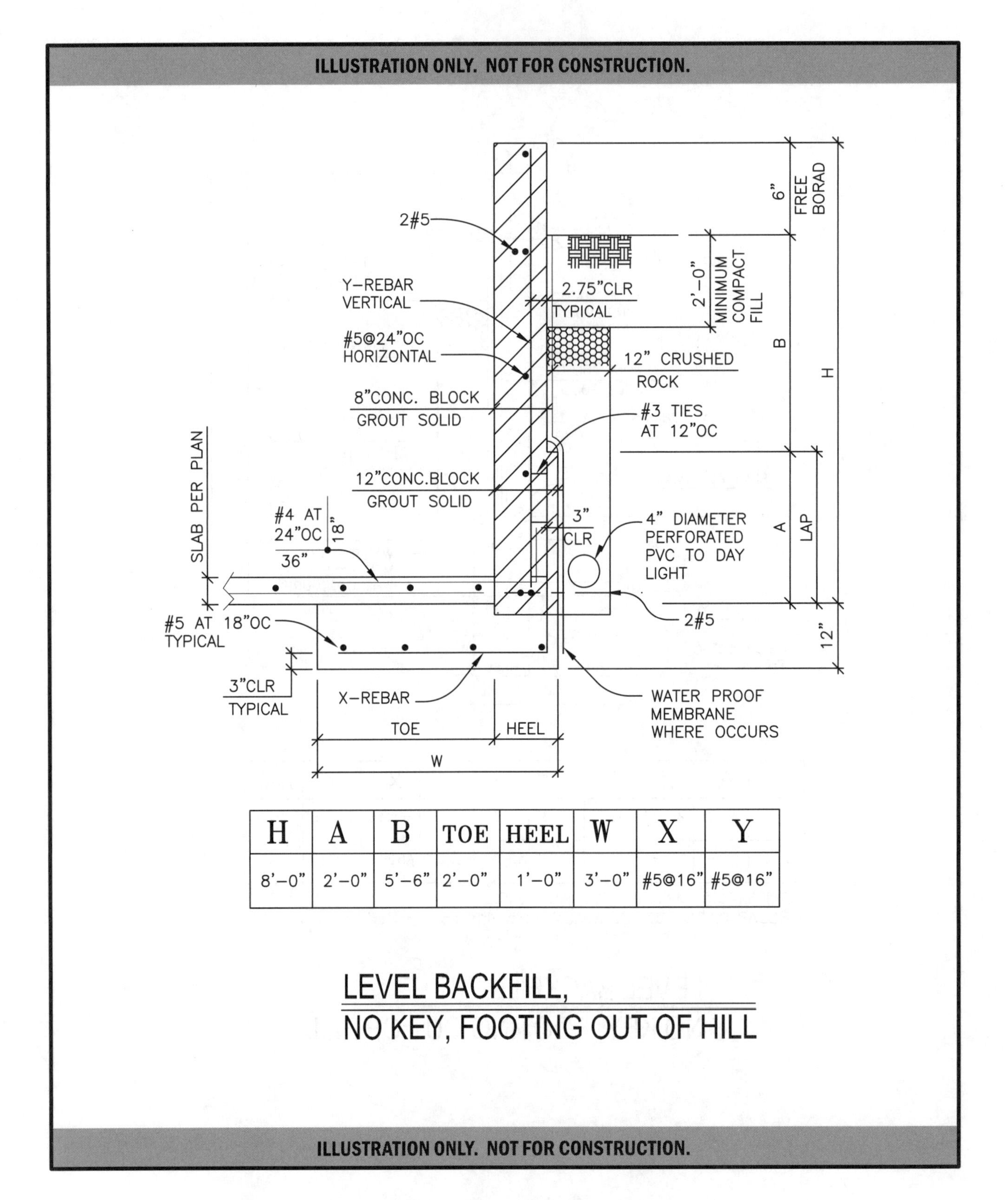

H	A	B	TOE	HEEL	W	X	Y
8'–0"	2'–0"	5'–6"	2'–0"	1'–0"	3'–0"	#5@16"	#5@16"

LEVEL BACKFILL,
NO KEY, FOOTING OUT OF HILL

EXECUTIVE ENGINEERING
19510 Ventura Blvd., Suite 110
Tarzana, California 91356-4234
Phone: (818) 708-1962 / Facsimile: (818) 708-1958
E-mail: EE110@aol.com

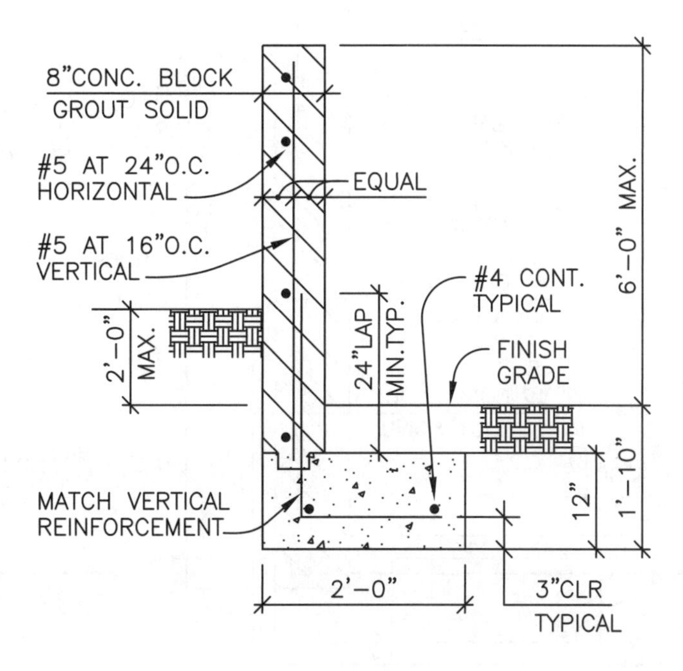

LEVEL BACKFILL,
NO KEY, FOOTING OUT OF HILL

ILLUSTRATION ONLY. NOT FOR CONSTRUCTION.

EXECUTIVE ENGINEERING
19510 Ventura Blvd., Suite 110
Tarzana, California 91356-4234
Phone: (818) 708-1962 / Facsimile: (818) 708-1958
E-mail: EE110@aol.com

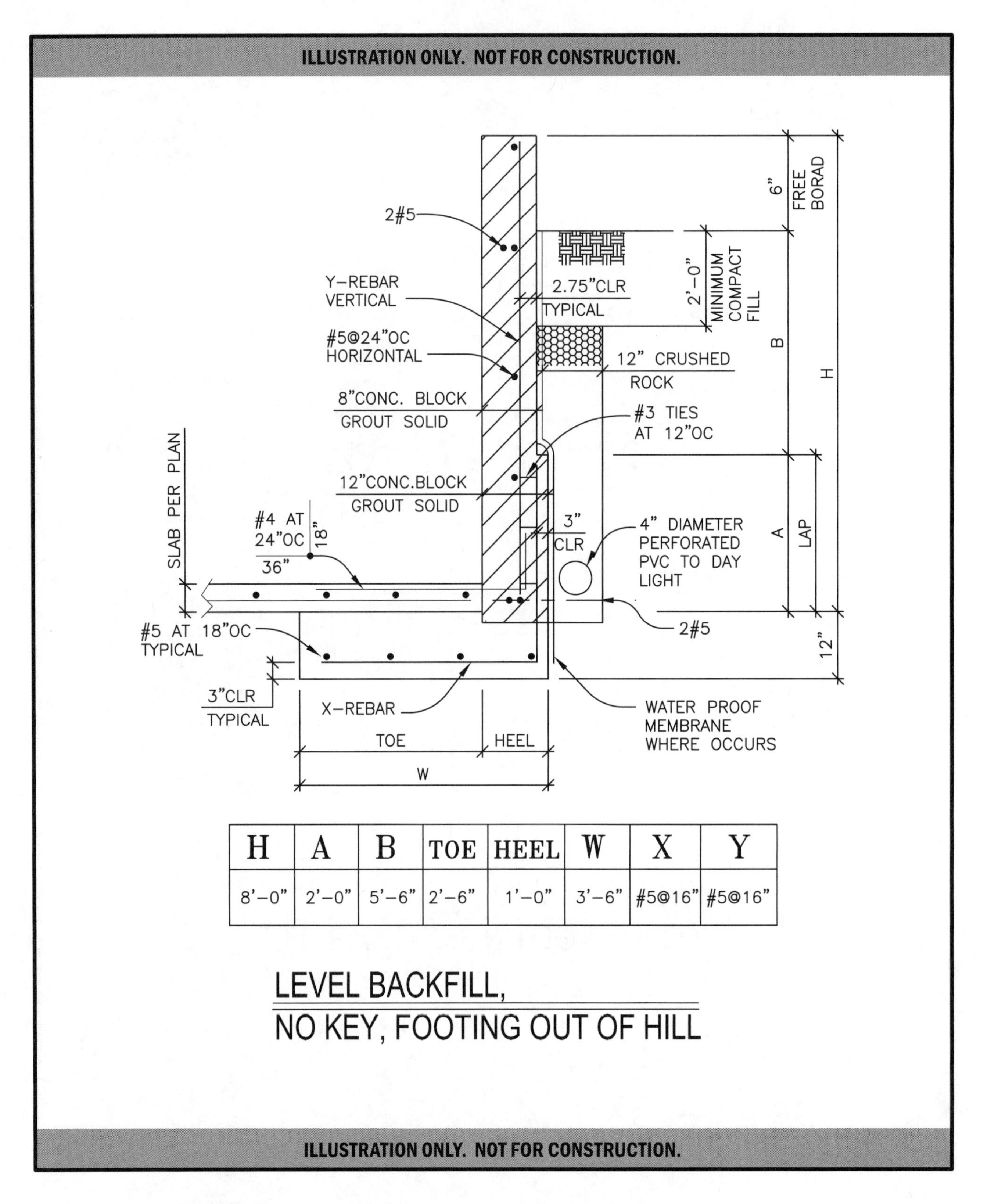

H	A	B	TOE	HEEL	W	X	Y
8'–0"	2'–0"	5'–6"	2'–6"	1'–0"	3'–6"	#5@16"	#5@16"

LEVEL BACKFILL, NO KEY, FOOTING OUT OF HILL

ILLUSTRATION ONLY. NOT FOR CONSTRUCTION.

EXECUTIVE ENGINEERING
19510 Ventura Blvd., Suite 110
Tarzana, California 91356-4234
Phone: (818) 708-1962 / Facsimile: (818) 708-1958
E-mail: EE110@aol.com

ILLUSTRATION ONLY. NOT FOR CONSTRUCTION.

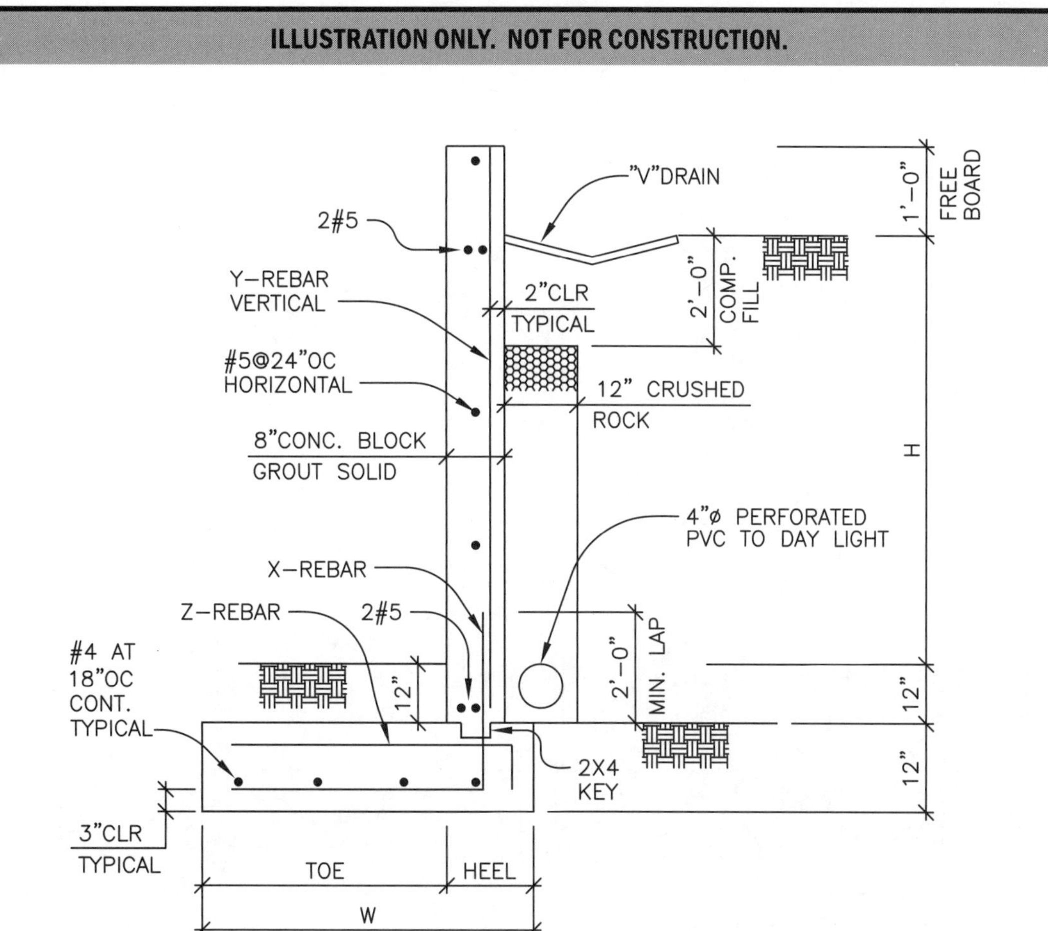

H	TOE	HEEL	W	X	Y	Z	COMMENT
4'–0"	1'–6"	1'–3"	2'–9"	#5@24"	#5@24"	#5@16"	ELIMINATE Y-REBAR EXTEND X-REBAR TO TOP OF THE WALL
5'–0"	2'–3"	1'–3"	3'–5"	#6@16"	#6@16"	#5@16"	ELIMINATE Y-REBAR EXTEND X-REBAR TO TOP OF THE WALL

LEVEL BACKFILL, NO KEY, FOOTING OUT OF HILL

ILLUSTRATION ONLY. NOT FOR CONSTRUCTION.

EXECUTIVE ENGINEERING
19510 Ventura Blvd., Suite 110
Tarzana, California 91356-4234
Phone: (818) 708-1962 / Facsimile: (818) 708-1958
E-mail: EE110@aol.com

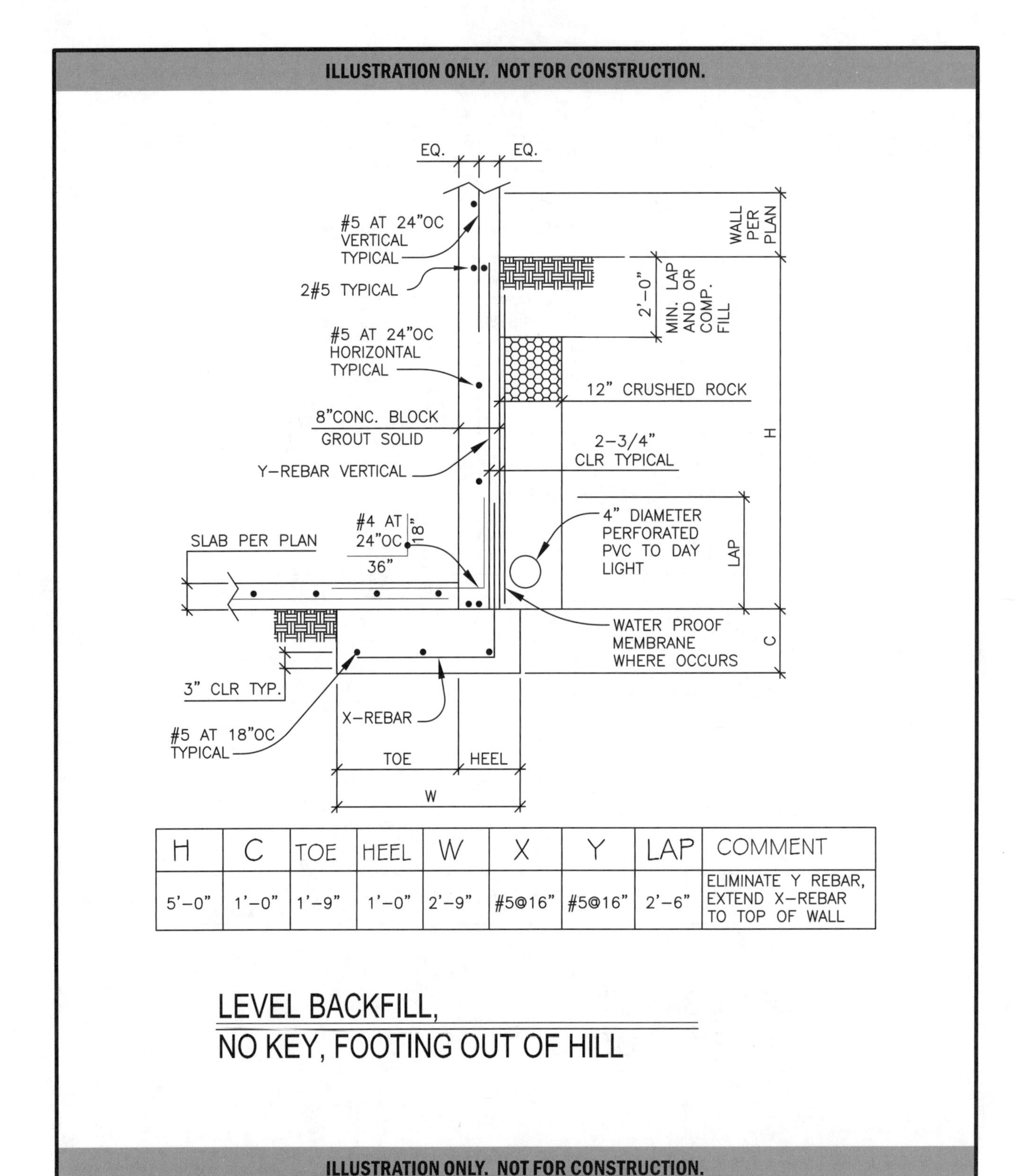

H	C	TOE	HEEL	W	X	Y	LAP	COMMENT
5'-0"	1'-0"	1'-9"	1'-0"	2'-9"	#5@16"	#5@16"	2'-6"	ELIMINATE Y REBAR, EXTEND X-REBAR TO TOP OF WALL

LEVEL BACKFILL, NO KEY, FOOTING OUT OF HILL

EXECUTIVE ENGINEERING
19510 Ventura Blvd., Suite 110
Tarzana, California 91356-4234
Phone: (818) 708-1962 / Facsimile: (818) 708-1958
E-mail: EE110@aol.com

ILLUSTRATION ONLY. NOT FOR CONSTRUCTION.

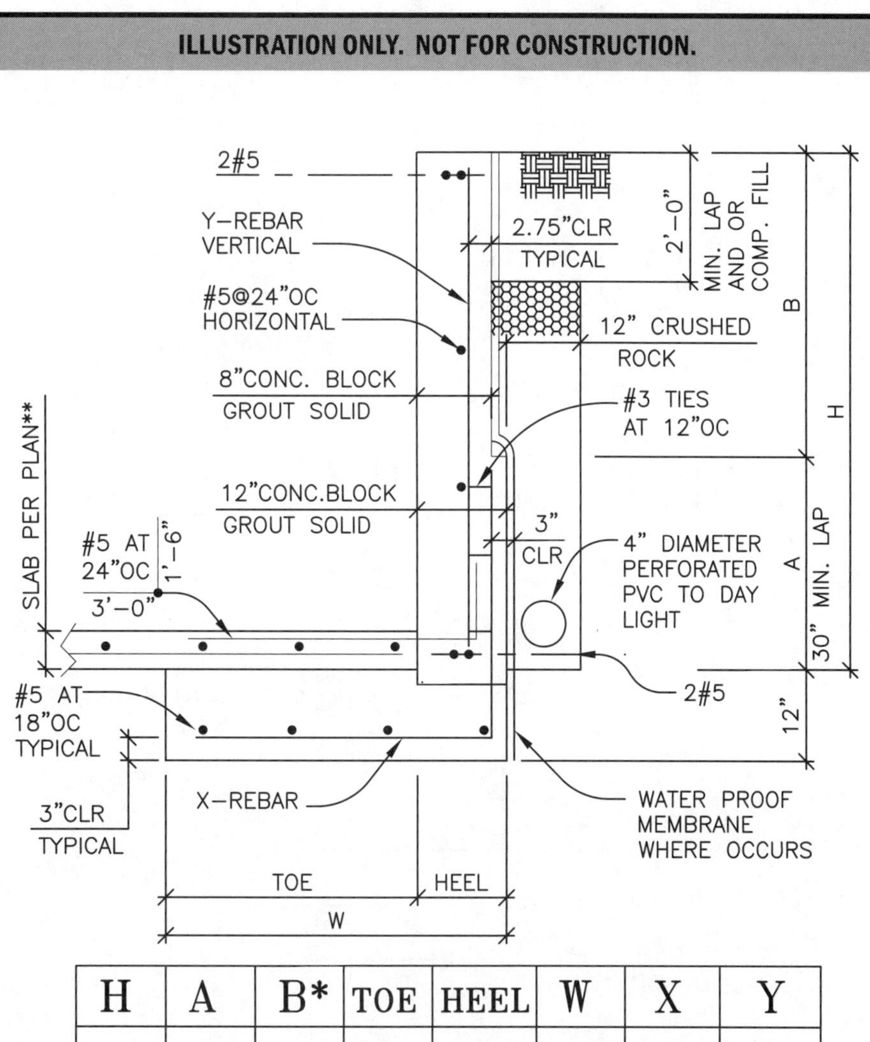

H	A	B*	TOE	HEEL	W	X	Y
6'–0"	3'–0"	3'–0"	4'–6"	1'–0"	5'–6"	#5@8"	#5@8"

*12" BLOCK WALL MAY BE EXTENDED TO THE TOP

** MAY USE 4" THICK SLAB WITH #4 AT 16" ON CENTER EACH WAY OVER 2" GRAVEL

LEVEL BACKFILL, NO KEY, FOOTING OUT OF HILL

ILLUSTRATION ONLY. NOT FOR CONSTRUCTION.

EXECUTIVE ENGINEERING
19510 Ventura Blvd., Suite 110
Tarzana, California 91356-4234
Phone: (818) 708-1962 / Facsimile: (818) 708-1958
E-mail: EE110@aol.com

ILLUSTRATION ONLY. NOT FOR CONSTRUCTION.

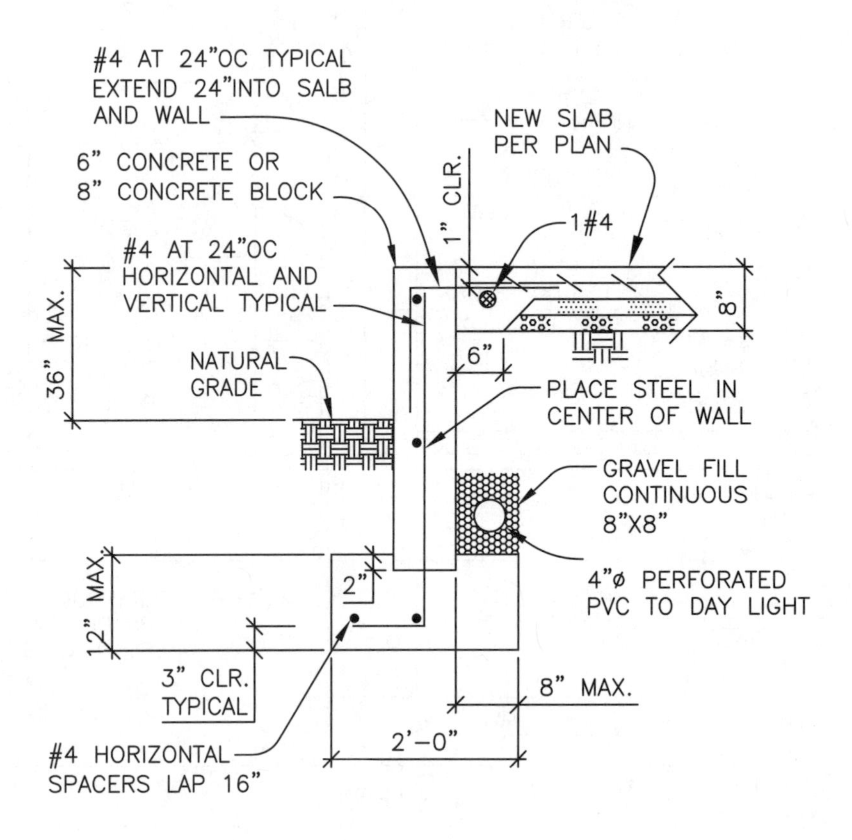

LEVEL BACKFILL,
NO KEY, FOOTING OUT OF HILL

ILLUSTRATION ONLY. NOT FOR CONSTRUCTION.

EXECUTIVE ENGINEERING
19510 Ventura Blvd., Suite 110
Tarzana, California 91356-4234
Phone: (818) 708-1962 / Facsimile: (818) 708-1958
E-mail: EE110@aol.com

ILLUSTRATION ONLY. NOT FOR CONSTRUCTION.

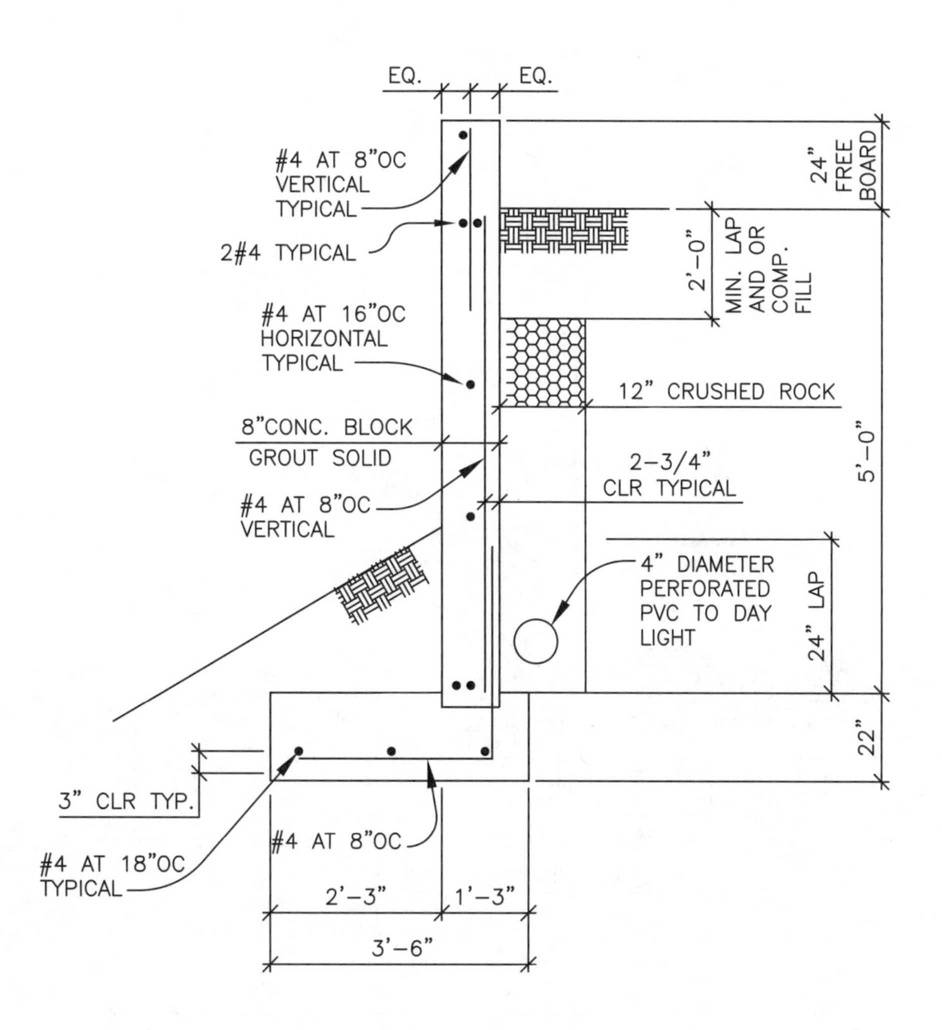

LEVEL BACKFILL,
NO KEY, FOOTING OUT OF HILL

ILLUSTRATION ONLY. NOT FOR CONSTRUCTION.

EXECUTIVE ENGINEERING
19510 Ventura Blvd., Suite 110
Tarzana, California 91356-4234
Phone: (818) 708-1962 / Facsimile: (818) 708-1958
E-mail: EE110@aol.com

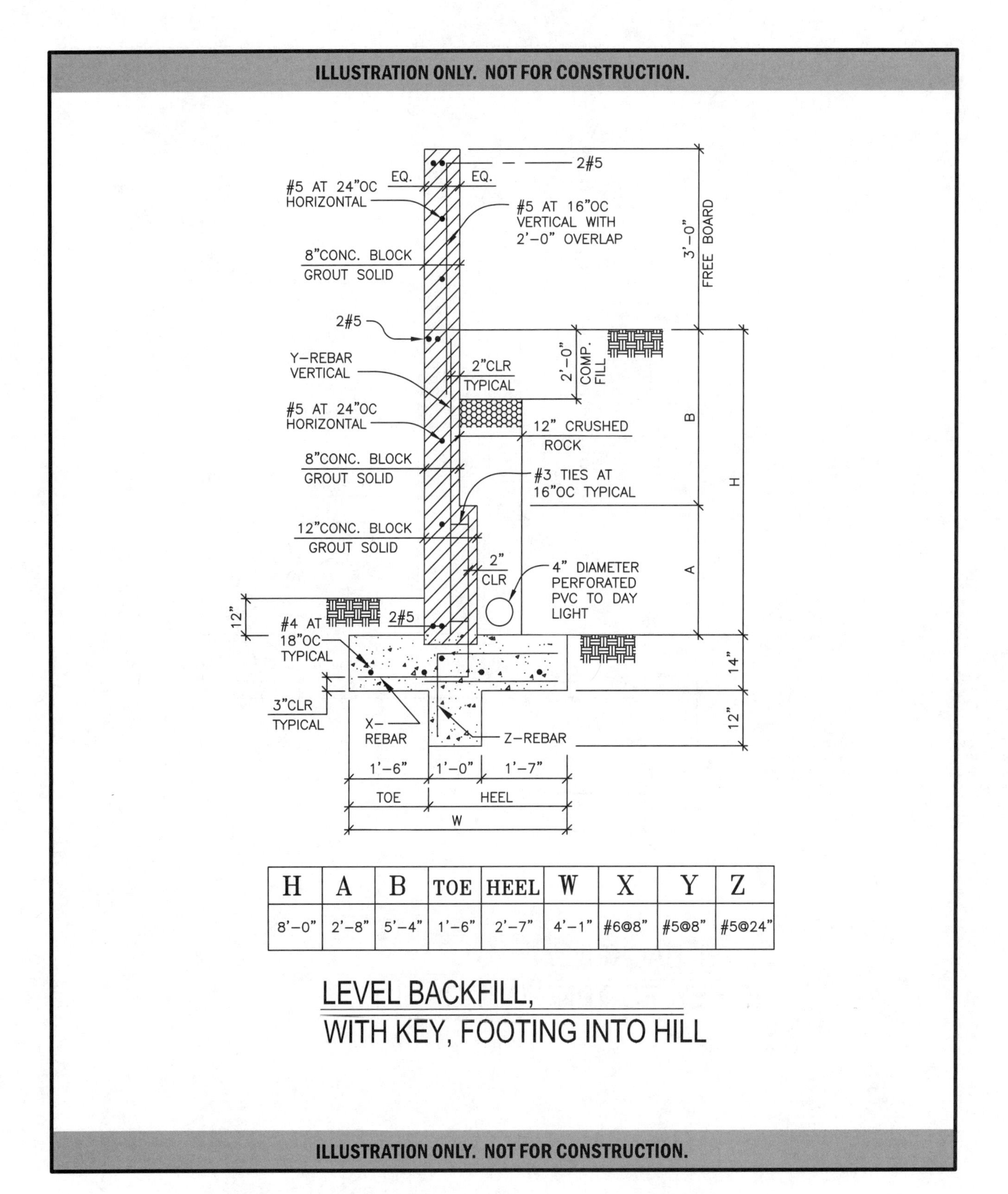

H	A	B	TOE	HEEL	W	X	Y	Z
8'-0"	2'-8"	5'-4"	1'-6"	2'-7"	4'-1"	#6@8"	#5@8"	#5@24"

LEVEL BACKFILL, WITH KEY, FOOTING INTO HILL

EXECUTIVE ENGINEERING
19510 Ventura Blvd., Suite 110
Tarzana, California 91356-4234
Phone: (818) 708-1962 / Facsimile: (818) 708-1958
E-mail: EE110@aol.com

ILLUSTRATION ONLY. NOT FOR CONSTRUCTION.

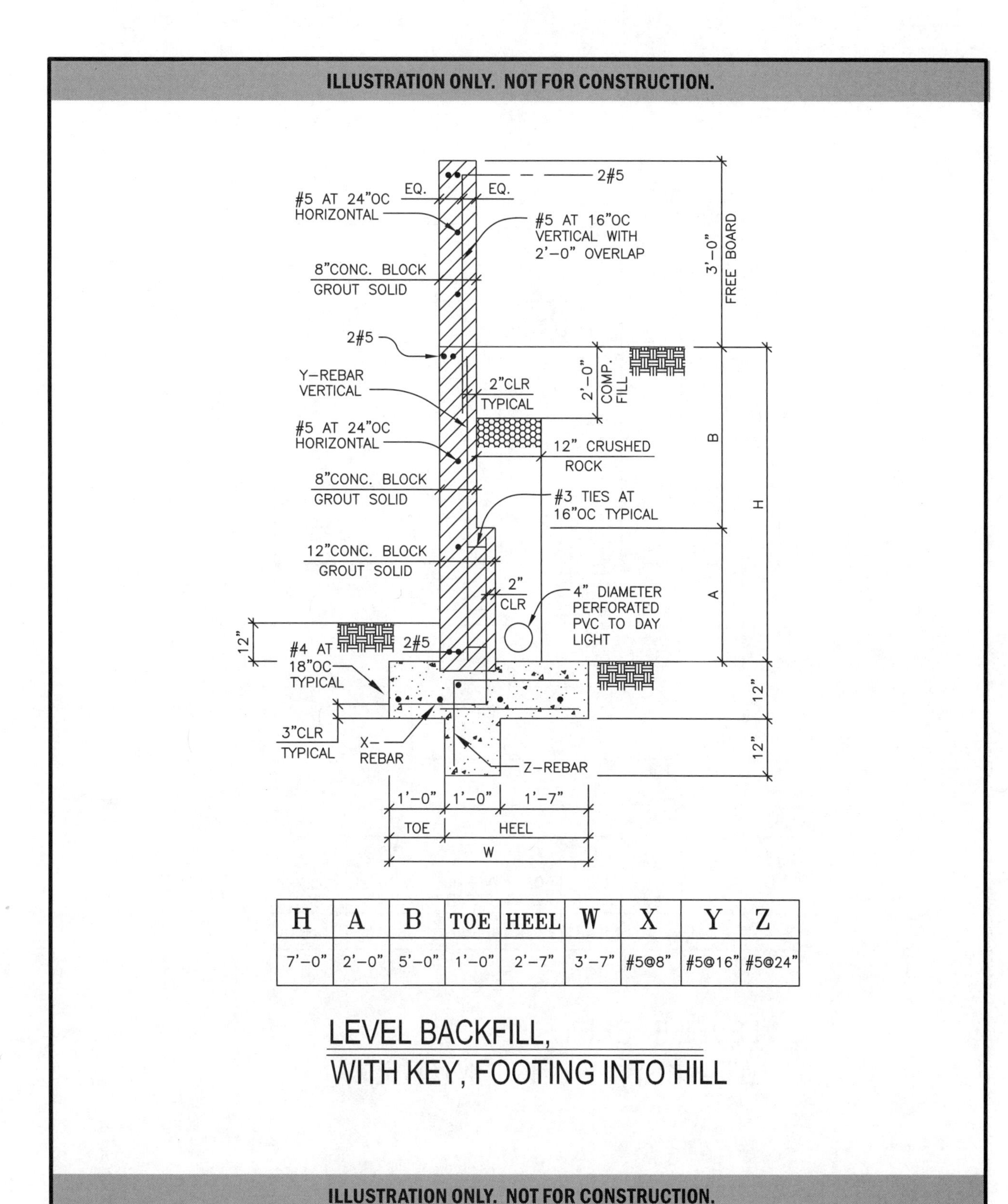

H	A	B	TOE	HEEL	W	X	Y	Z
7'-0"	2'-0"	5'-0"	1'-0"	2'-7"	3'-7"	#5@8"	#5@16"	#5@24"

LEVEL BACKFILL, WITH KEY, FOOTING INTO HILL

ILLUSTRATION ONLY. NOT FOR CONSTRUCTION.

EXECUTIVE ENGINEERING
19510 Ventura Blvd., Suite 110
Tarzana, California 91356-4234
Phone: (818) 708-1962 / Facsimile: (818) 708-1958
E-mail: EE110@aol.com

ILLUSTRATION ONLY. NOT FOR CONSTRUCTION.

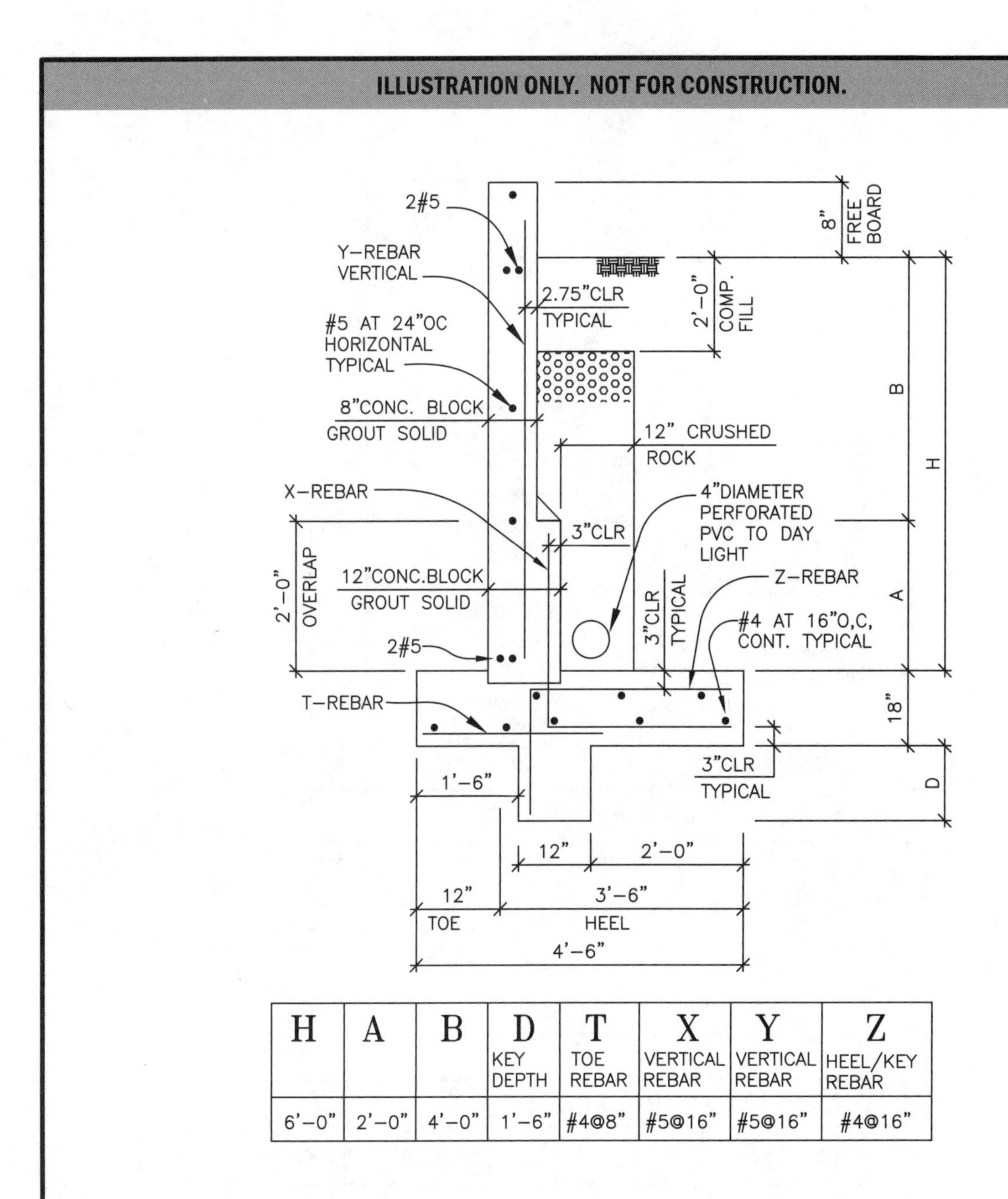

H	A	B	D KEY DEPTH	T TOE REBAR	X VERTICAL REBAR	Y VERTICAL REBAR	Z HEEL/KEY REBAR
6'-0"	2'-0"	4'-0"	1'-6"	#4@8"	#5@16"	#5@16"	#4@16"

LEVEL BACKFILL,
WITH KEY, FOOTING INTO HILL

ILLUSTRATION ONLY. NOT FOR CONSTRUCTION.

EXECUTIVE ENGINEERING
19510 Ventura Blvd., Suite 110
Tarzana, California 91356-4234
Phone: (818) 708-1962 / Facsimile: (818) 708-1958
E-mail: EE110@aol.com

ILLUSTRATION ONLY. NOT FOR CONSTRUCTION.

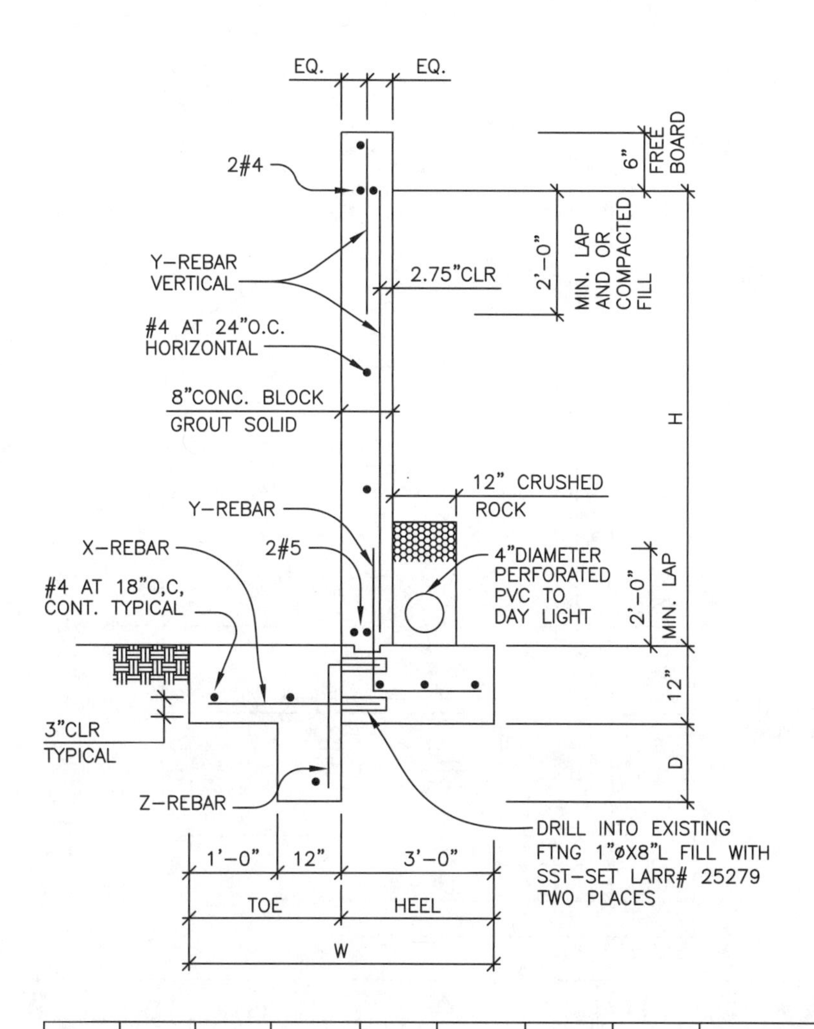

H	D	TOE	HEEL	W	X	Y	Z	COMMENT
6'-0"	1'-0"	2'-0"	3'-0"	5'-0"	#6@18"	#4@8"	#6@18"	ELIMINATE Y-REBAR EXTEND X-REBAR TO TOP OF THE WALL

LEVEL BACKFILL, WITH KEY, FOOTING INTO HILL

ILLUSTRATION ONLY. NOT FOR CONSTRUCTION.

EXECUTIVE ENGINEERING
19510 Ventura Blvd., Suite 110
Tarzana, California 91356-4234
Phone: (818) 708-1962 / Facsimile: (818) 708-1958
E-mail: EE110@aol.com

ILLUSTRATION ONLY. NOT FOR CONSTRUCTION.

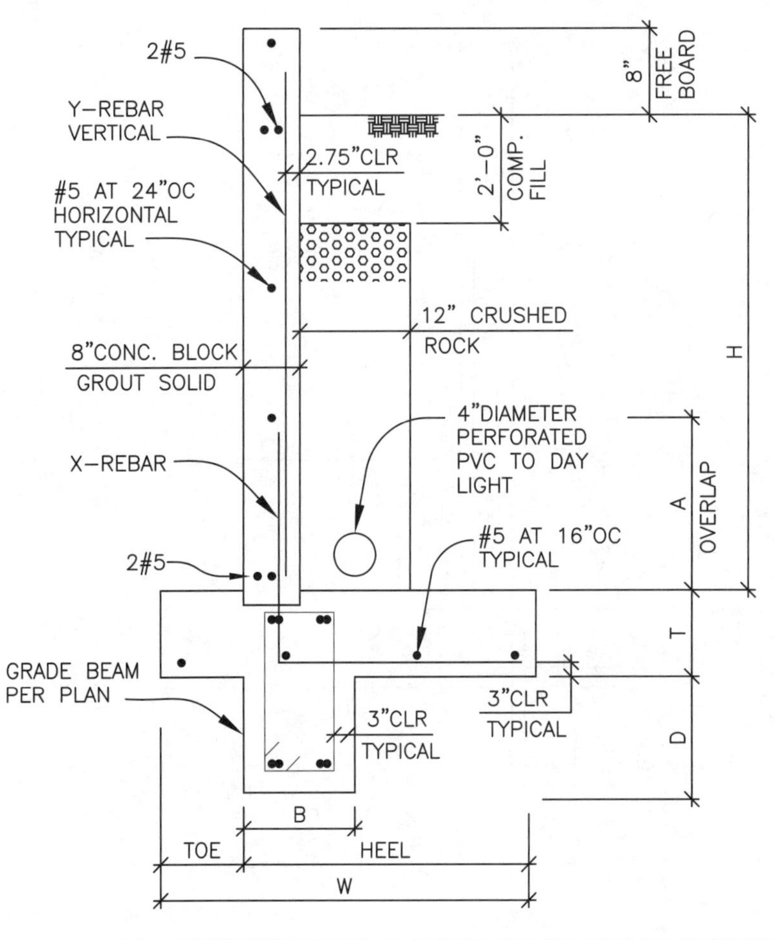

H	A	B	D	T	X	Y	TOE	HEEL	W
6'–0"	2'–6"	2'–0"	2'–0"	1'–0"	#6@8"	#6@8"	6"	2'–6"	3'–0"

LEVEL BACKFILL, WITH KEY, FOOTING INTO HILL

ILLUSTRATION ONLY. NOT FOR CONSTRUCTION.

EXECUTIVE ENGINEERING
19510 Ventura Blvd., Suite 110
Tarzana, California 91356-4234
Phone: (818) 708-1962 / Facsimile: (818) 708-1958
E-mail: EE110@aol.com

ILLUSTRATION ONLY. NOT FOR CONSTRUCTION.

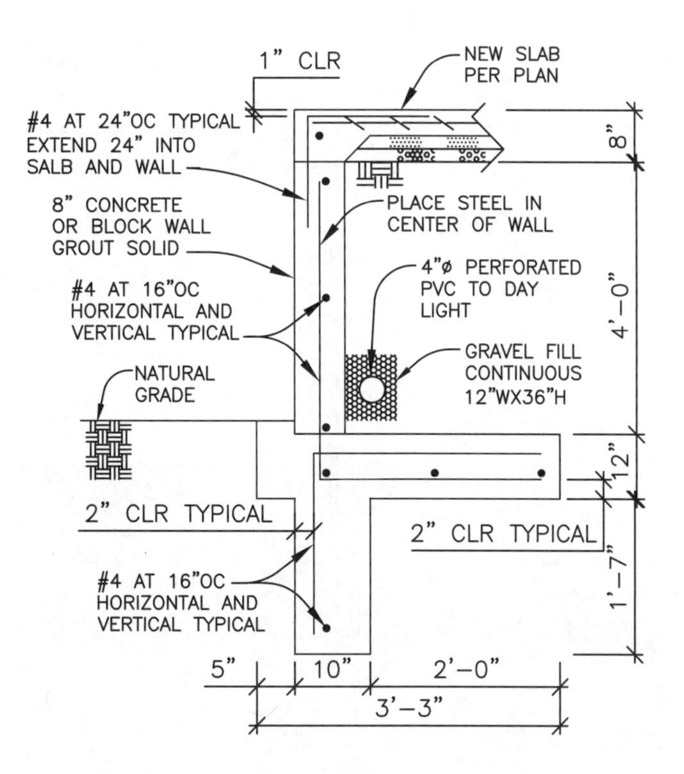

LEVEL BACKFILL, WITH KEY, FOOTING INTO HILL

ILLUSTRATION ONLY. NOT FOR CONSTRUCTION.

EXECUTIVE ENGINEERING
19510 Ventura Blvd., Suite 110
Tarzana, California 91356-4234
Phone: (818) 708-1962 / Facsimile: (818) 708-1958
E-mail: EE110@aol.com

ILLUSTRATION ONLY. NOT FOR CONSTRUCTION.

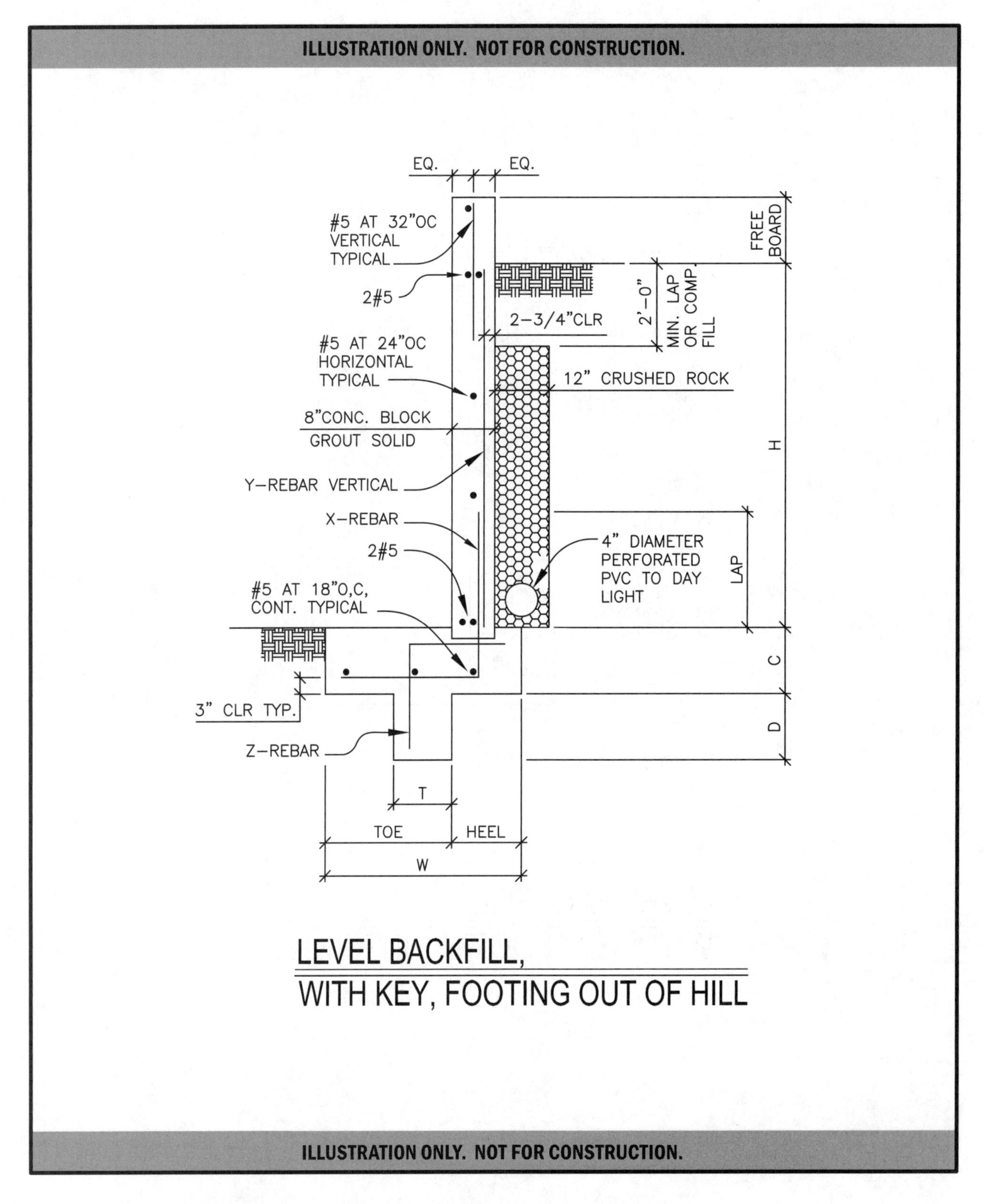

LEVEL BACKFILL,

WITH KEY, FOOTING OUT OF HILL

ILLUSTRATION ONLY. NOT FOR CONSTRUCTION.

EXECUTIVE ENGINEERING
19510 Ventura Blvd., Suite 110
Tarzana, California 91356-4234
Phone: (818) 708-1962 / Facsimile: (818) 708-1958
E-mail: EE110@aol.com

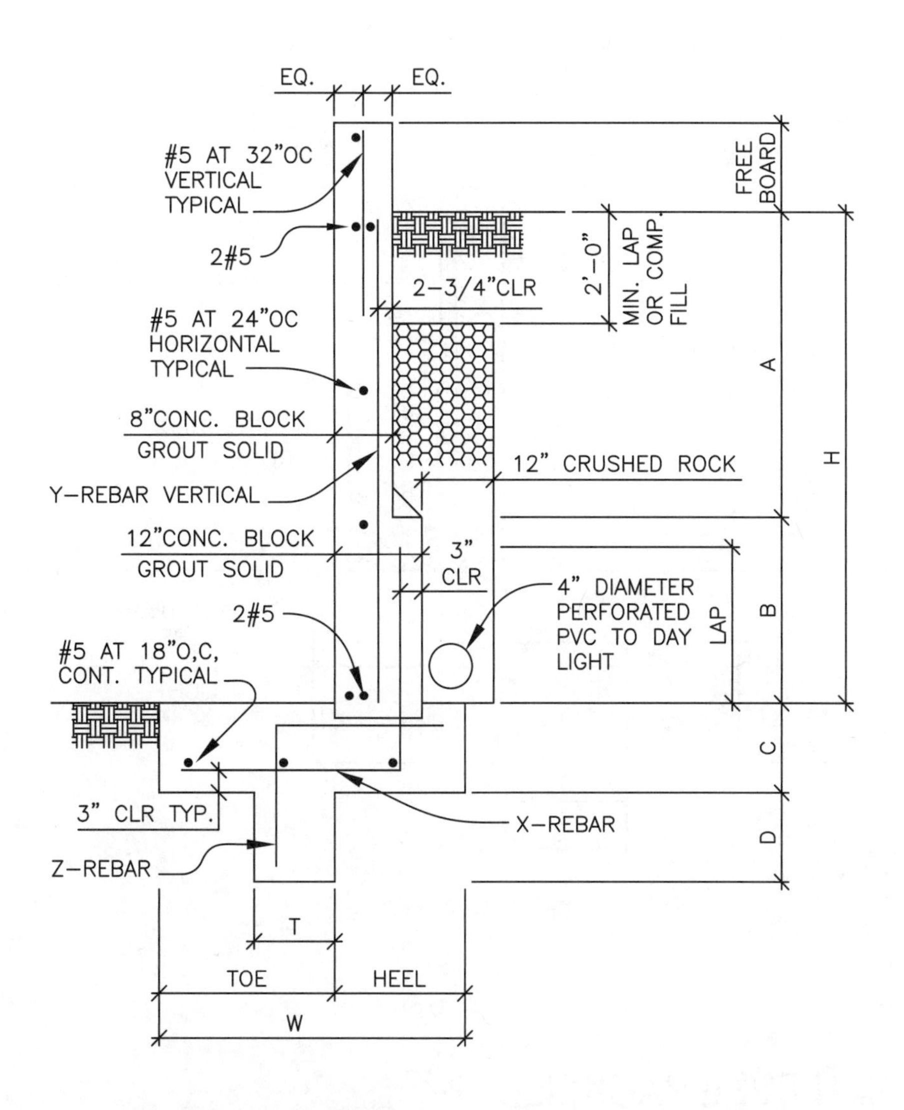

LEVEL BACKFILL, WITH KEY, FOOTING OUT OF HILL

EXECUTIVE ENGINEERING
19510 Ventura Blvd., Suite 110
Tarzana, California 91356-4234
Phone: (818) 708-1962 / Facsimile: (818) 708-1958
E-mail: EE110@aol.com

ILLUSTRATION ONLY. NOT FOR CONSTRUCTION.

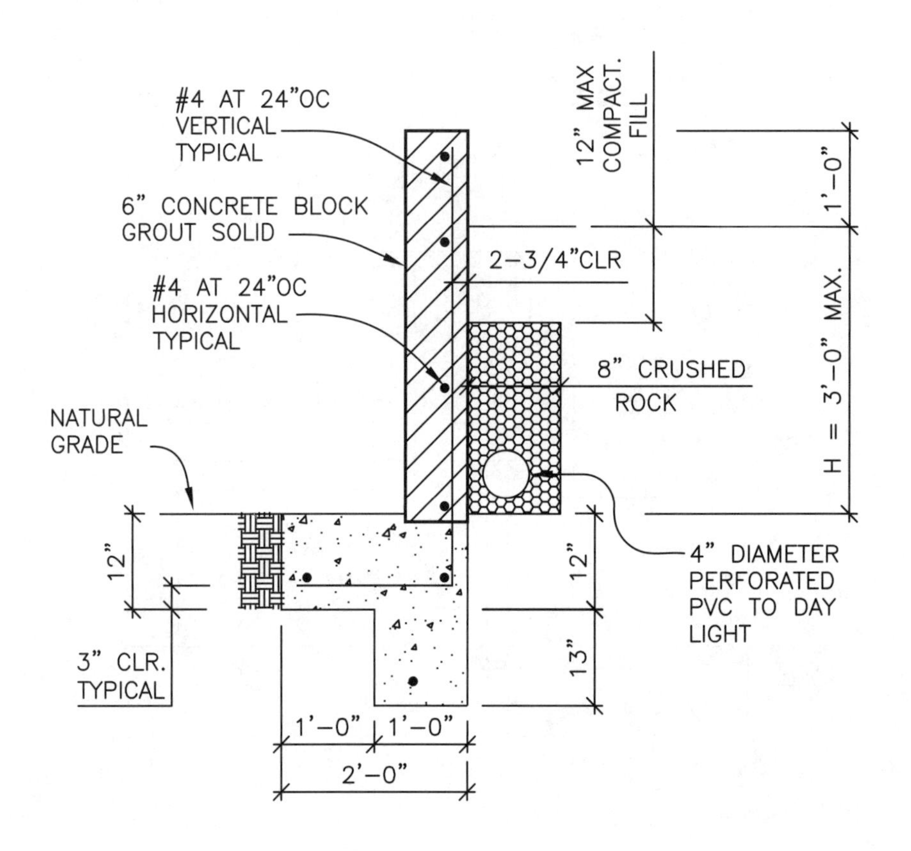

LEVEL BACKFILL,
WITH KEY, FOOTING OUT OF HILL

ILLUSTRATION ONLY. NOT FOR CONSTRUCTION.

EXECUTIVE ENGINEERING
19510 Ventura Blvd., Suite 110
Tarzana, California 91356-4234
Phone: (818) 708-1962 / Facsimile: (818) 708-1958
E-mail: EE110@aol.com

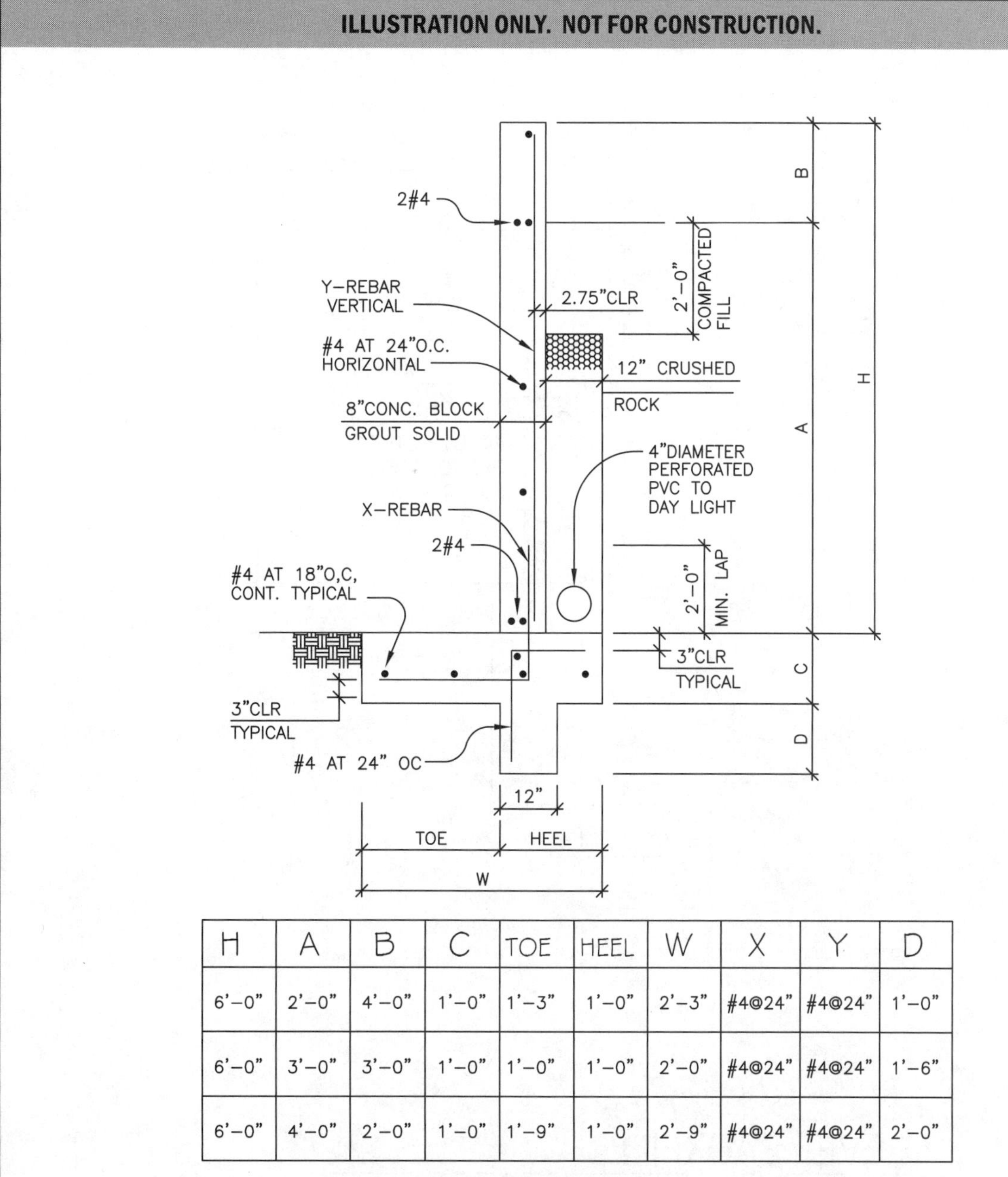

H	A	B	C	TOE	HEEL	W	X	Y	D
6'–0"	2'–0"	4'–0"	1'–0"	1'–3"	1'–0"	2'–3"	#4@24"	#4@24"	1'–0"
6'–0"	3'–0"	3'–0"	1'–0"	1'–0"	1'–0"	2'–0"	#4@24"	#4@24"	1'–6"
6'–0"	4'–0"	2'–0"	1'–0"	1'–9"	1'–0"	2'–9"	#4@24"	#4@24"	2'–0"

LEVEL BACKFILL, WITH KEY, FOOTING OUT OF HILL

ILLUSTRATION ONLY. NOT FOR CONSTRUCTION.

EXECUTIVE ENGINEERING
19510 Ventura Blvd., Suite 110
Tarzana, California 91356-4234
Phone: (818) 708-1962 / Facsimile: (818) 708-1958
E-mail: EE110@aol.com

ILLUSTRATION ONLY. NOT FOR CONSTRUCTION.

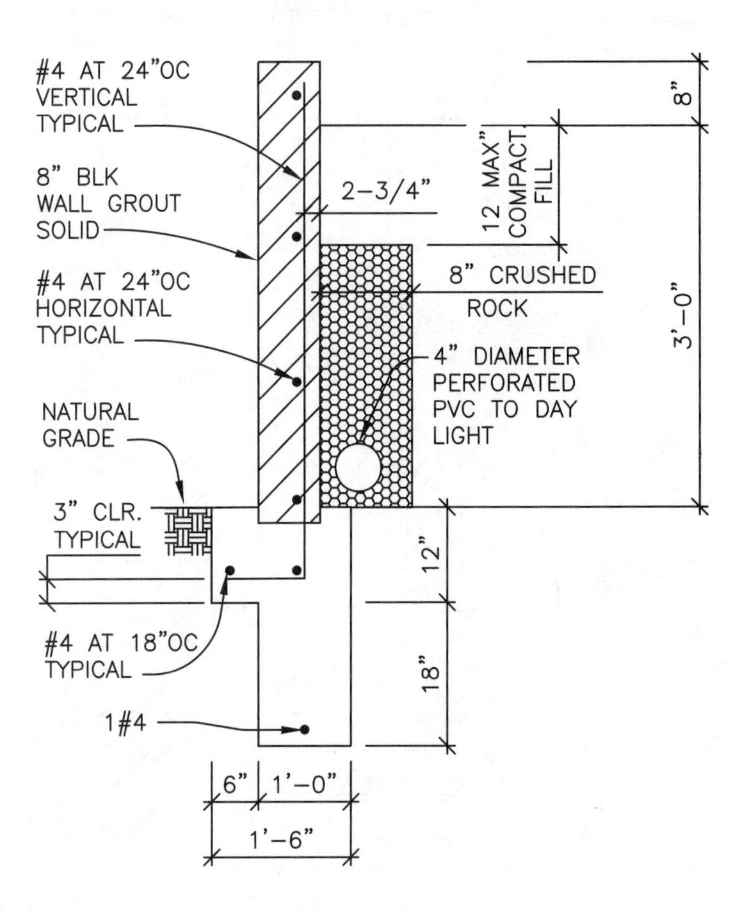

LEVEL BACKFILL,
WITH KEY, FOOTING OUT OF HILL

ILLUSTRATION ONLY. NOT FOR CONSTRUCTION.

EXECUTIVE ENGINEERING
19510 Ventura Blvd., Suite 110
Tarzana, California 91356-4234
Phone: (818) 708-1962 / Facsimile: (818) 708-1958
E-mail: EE110@aol.com

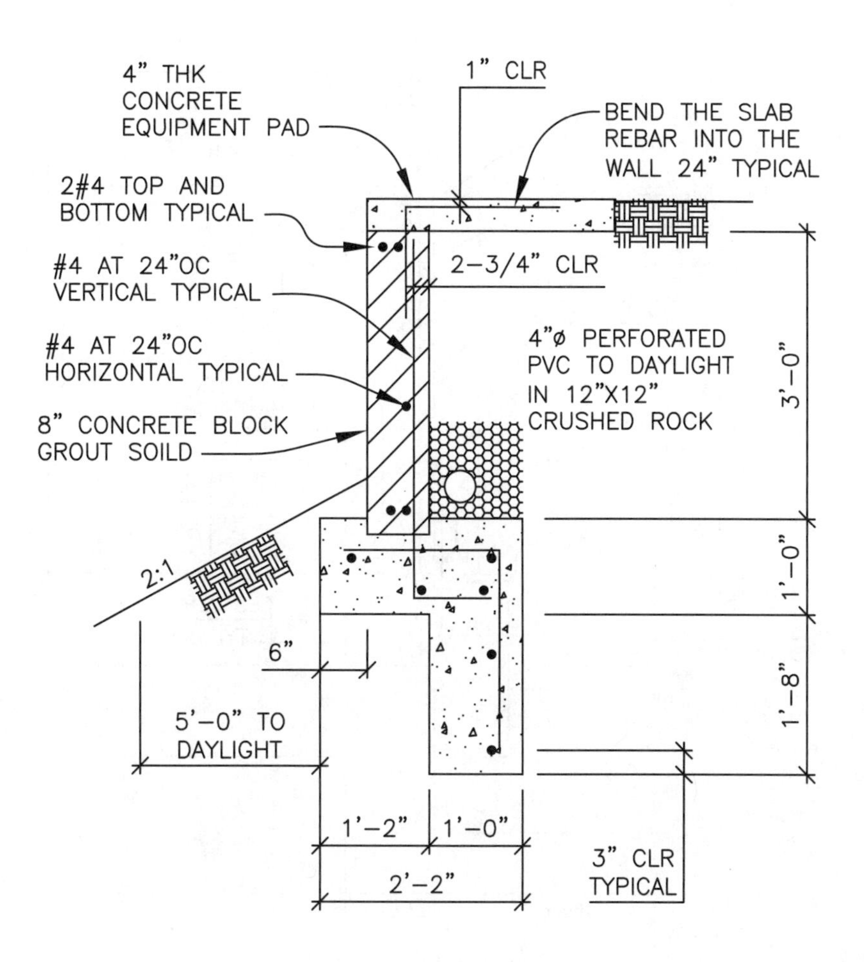

LEVEL BACKFILL,
WITH KEY, FOOTING OUT OF HILL

ILLUSTRATION ONLY. NOT FOR CONSTRUCTION.

EXECUTIVE ENGINEERING
19510 Ventura Blvd., Suite 110
Tarzana, California 91356-4234
Phone: (818) 708-1962 / Facsimile: (818) 708-1958
E-mail: EE110@aol.com

ILLUSTRATION ONLY. NOT FOR CONSTRUCTION.

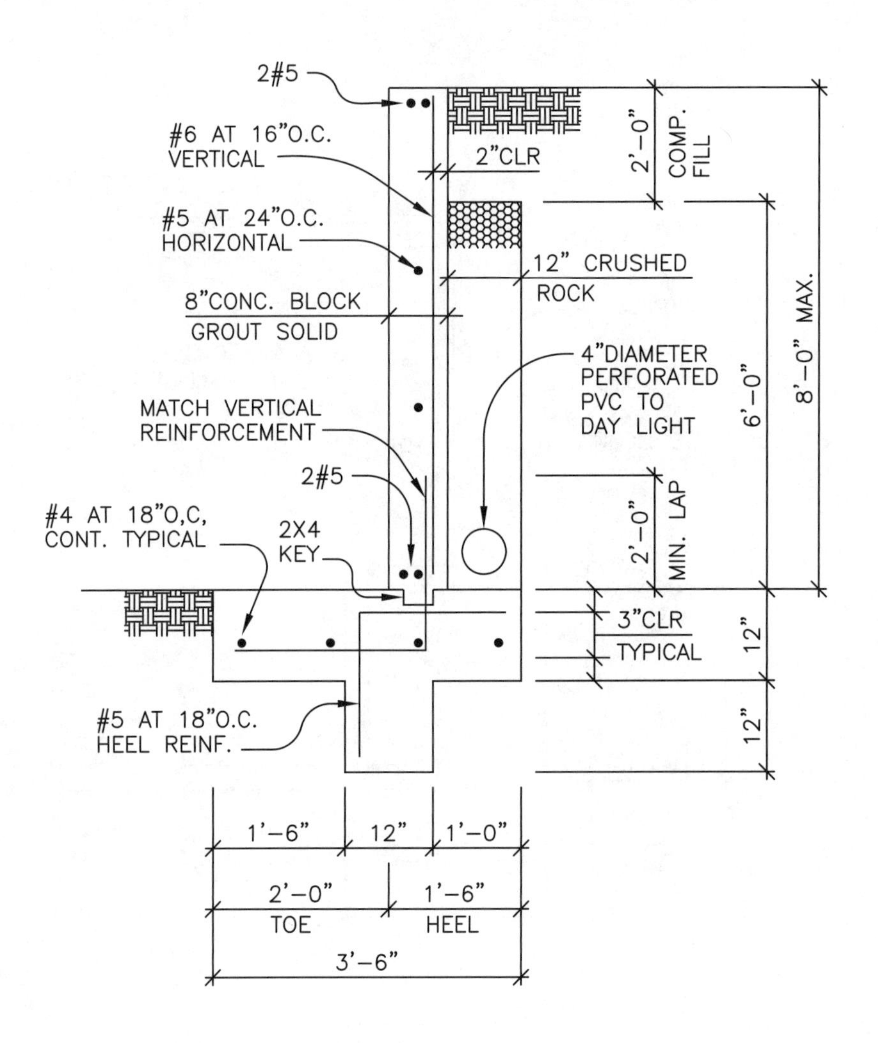

LEVEL BACKFILL,
WITH KEY, FOOTING OUT OF HILL

ILLUSTRATION ONLY. NOT FOR CONSTRUCTION.

EXECUTIVE ENGINEERING
19510 Ventura Blvd., Suite 110
Tarzana, California 91356-4234
Phone: (818) 708-1962 / Facsimile: (818) 708-1958
E-mail: EE110@aol.com

ILLUSTRATION ONLY. NOT FOR CONSTRUCTION.

2#5
#6 AT 16"O.C. VERTICAL
2"CLR
2'–0"
COMP. FILL
#5 AT 24"O.C. HORIZONTAL
6'–0"
8"CONC. BLOCK GROUT SOLID
12" MINIMUM CRUSHED ROCK
8'–0" MAX.
#6 AT 16"O.C.
12" CONC.BLK.GROUT SOLID
2'–0"
12"BLK AND MIN. LAP
2#5
2"CLR
#4 AT 18"O,C, CONT. TYPICAL
2X6 KEY
3"CLR TYPICAL
12"
3"CLR TYP.
3"CLR
4"DIAMETER PERFORATED PVC TO DAY LIGHT
24"
#4 AT 12"O.C. HEEL REINF.
24" 12" 18"
3'–0" TOE
1'–6" HEEL
4'–6"

LEVEL BACKFILL, WITH KEY, FOOTING OUT OF HILL

ILLUSTRATION ONLY. NOT FOR CONSTRUCTION.

EXECUTIVE ENGINEERING
19510 Ventura Blvd., Suite 110
Tarzana, California 91356-4234
Phone: (818) 708-1962 / Facsimile: (818) 708-1958
E-mail: EE110@aol.com

ILLUSTRATION ONLY. NOT FOR CONSTRUCTION.

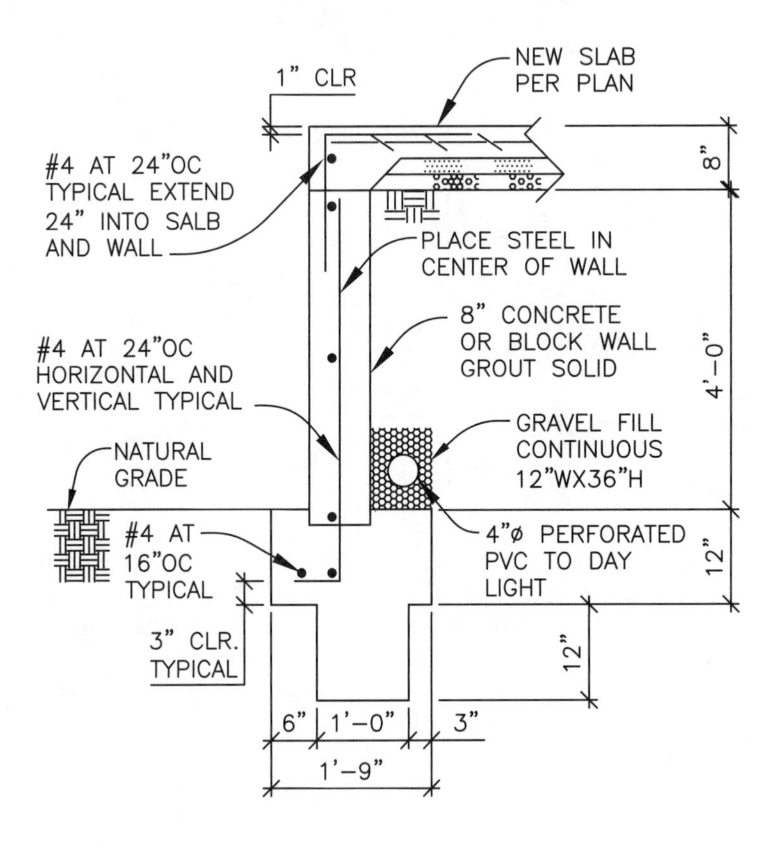

LEVEL BACKFILL,
WITH KEY, FOOTING OUT OF HILL

ILLUSTRATION ONLY. NOT FOR CONSTRUCTION.

EXECUTIVE ENGINEERING
19510 Ventura Blvd., Suite 110
Tarzana, California 91356-4234
Phone: (818) 708-1962 / Facsimile: (818) 708-1958
E-mail: EE110@aol.com

ILLUSTRATION ONLY. NOT FOR CONSTRUCTION.

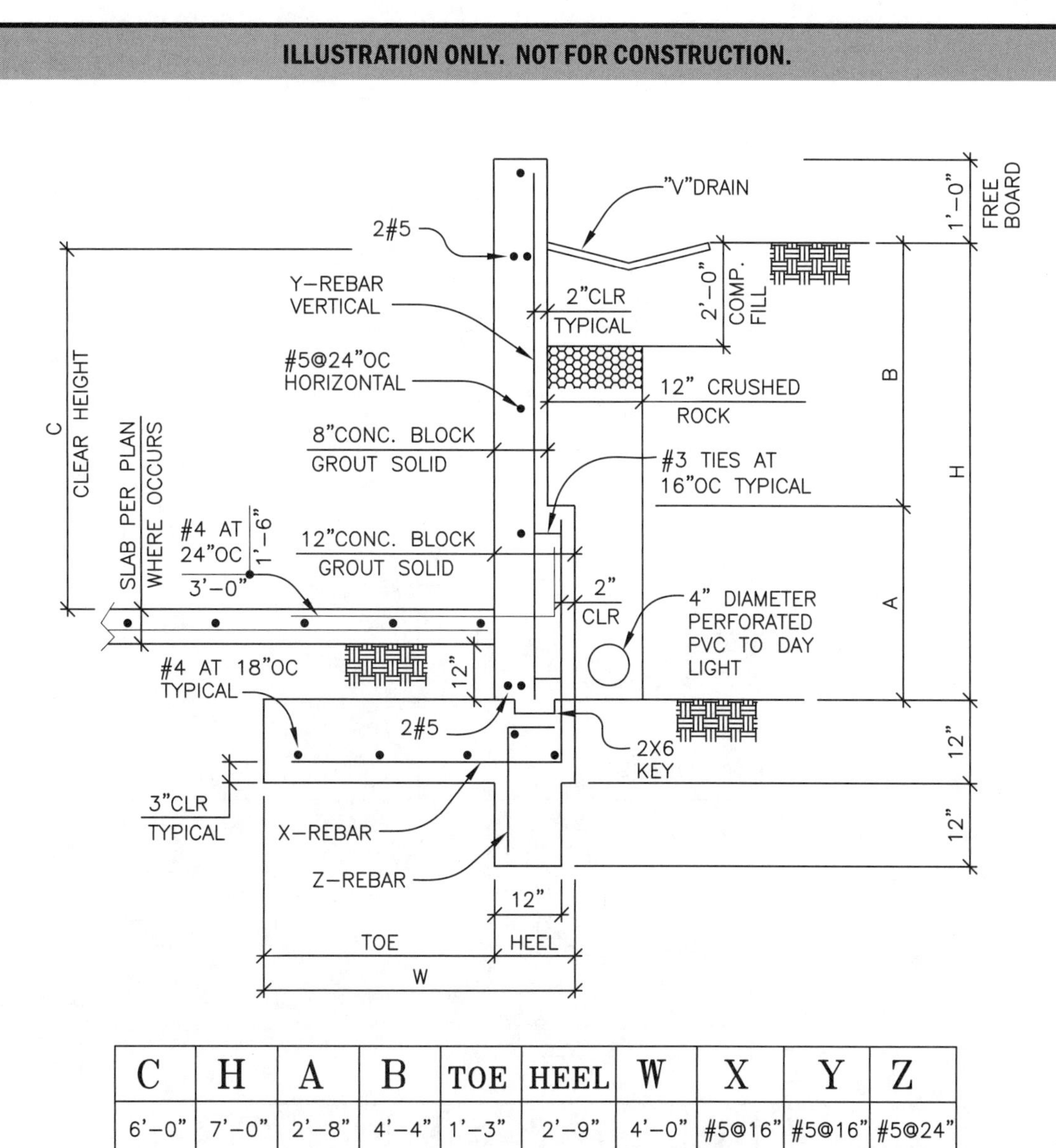

C	H	A	B	TOE	HEEL	W	X	Y	Z
6'–0"	7'–0"	2'–8"	4'–4"	1'–3"	2'–9"	4'–0"	#5@16"	#5@16"	#5@24"
7'–0"	8'–0"	2'–8"	5'–4"	1'–3"	3'–6"	4'–9"	#6@16"	#5@16"	#5@24"

LEVEL BACKFILL,
WITH KEY, FOOTING OUT OF HILL

ILLUSTRATION ONLY. NOT FOR CONSTRUCTION.

EXECUTIVE ENGINEERING
19510 Ventura Blvd., Suite 110
Tarzana, California 91356-4234
Phone: (818) 708-1962 / Facsimile: (818) 708-1958
E-mail: EE110@aol.com

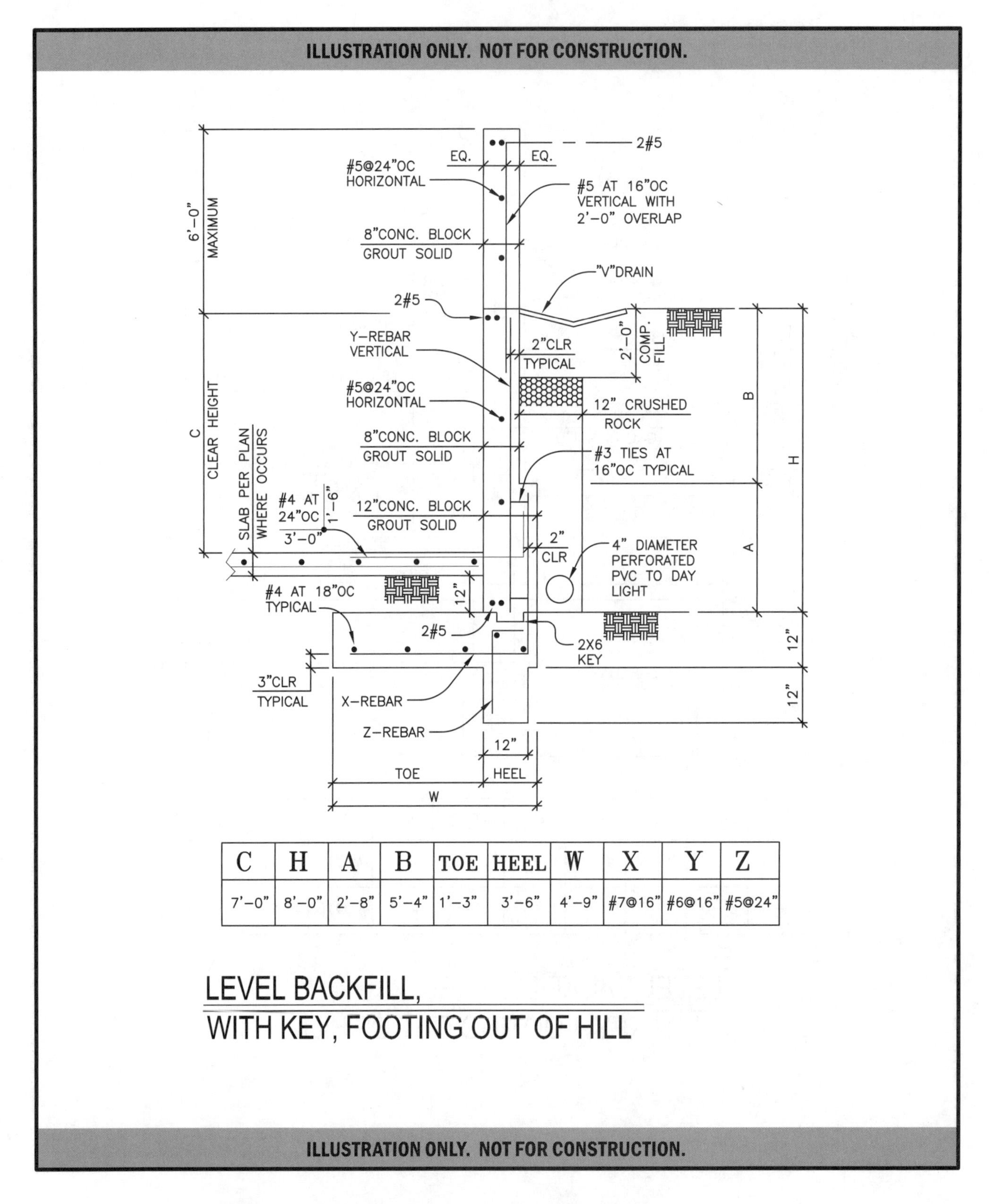

C	H	A	B	TOE	HEEL	W	X	Y	Z
7'-0"	8'-0"	2'-8"	5'-4"	1'-3"	3'-6"	4'-9"	#7@16"	#6@16"	#5@24"

LEVEL BACKFILL, WITH KEY, FOOTING OUT OF HILL

EXECUTIVE ENGINEERING
19510 Ventura Blvd., Suite 110
Tarzana, California 91356-4234
Phone: (818) 708-1962 / Facsimile: (818) 708-1958
E-mail: EE110@aol.com

ILLUSTRATION ONLY. NOT FOR CONSTRUCTION.

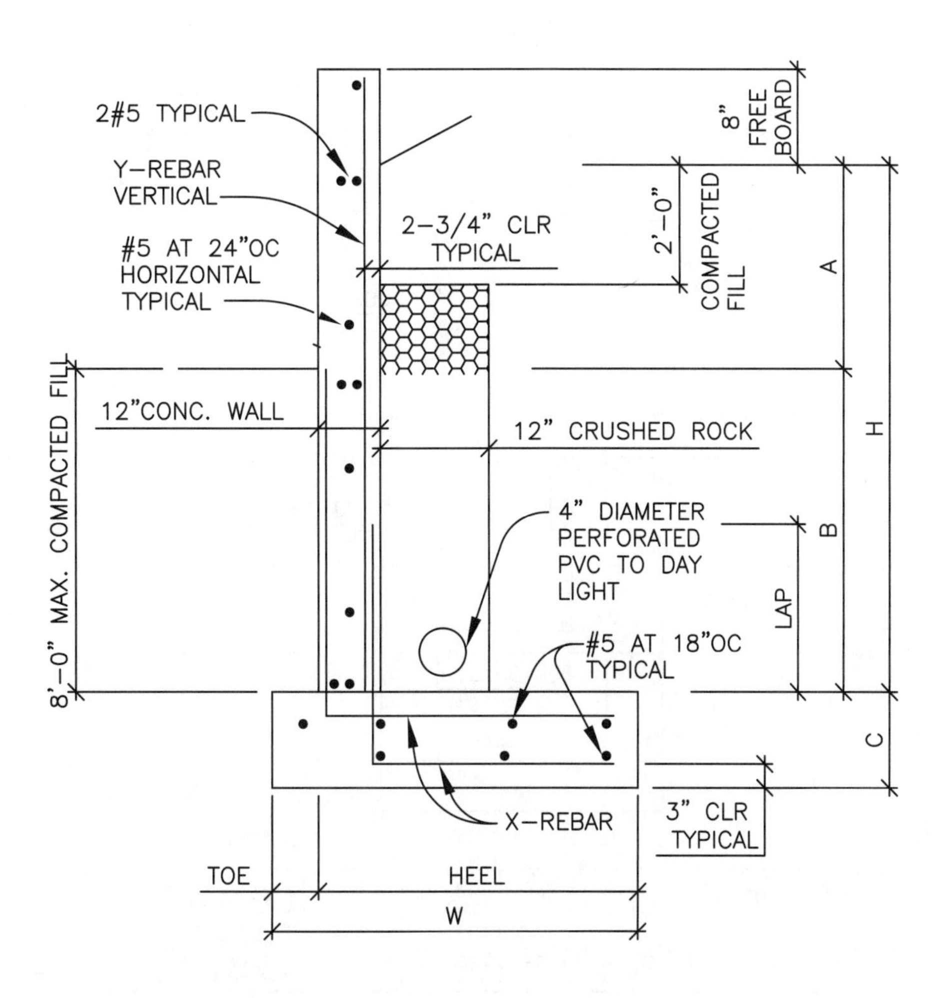

SLOPED BACKFILL,
NO KEY, FOOTING INTO HILL

ILLUSTRATION ONLY. NOT FOR CONSTRUCTION.

EXECUTIVE ENGINEERING
19510 Ventura Blvd., Suite 110
Tarzana, California 91356-4234
Phone: (818) 708-1962 / Facsimile: (818) 708-1958
E-mail: EE110@aol.com

ILLUSTRATION ONLY. NOT FOR CONSTRUCTION.

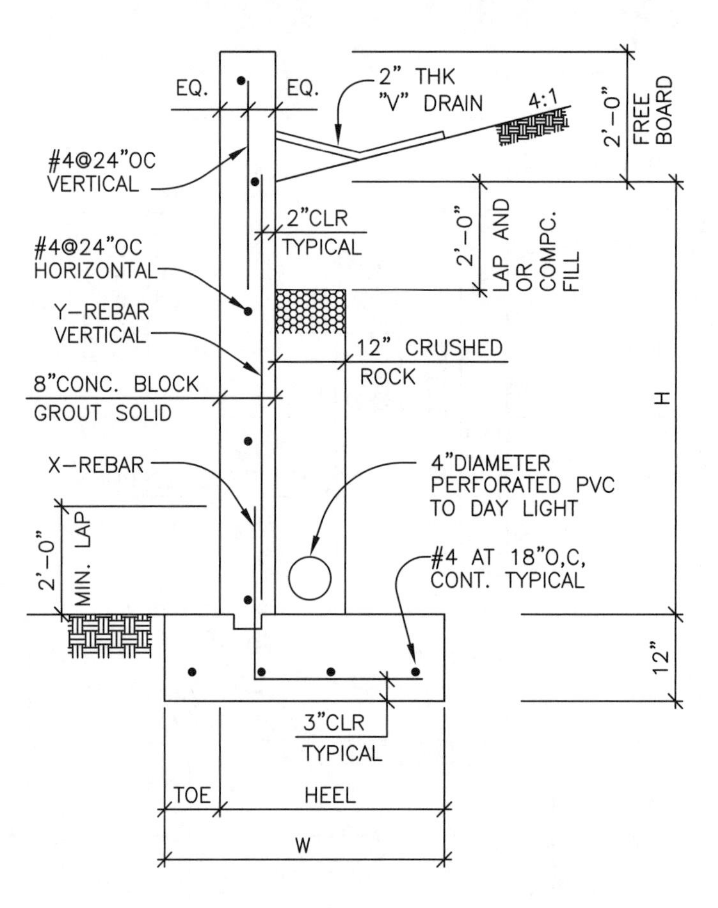

H	TOE	HEEL	W	X	Y	COMMENT
3'-0"	6"	1'-0"	1'-6"	#4@24"	#4@24"	ELIMINATE Y-REBAR EXTEND X-REBAR TO TOP OF THE WALL
4'-0"	6"	1'-6"	2'-0"	#4@24"	#4@24"	ELIMINATE Y-REBAR EXTEND X-REBAR TO TOP OF THE WALL

SLOPED BACKFILL, NO KEY, FOOTING INTO HILL

ILLUSTRATION ONLY. NOT FOR CONSTRUCTION.

EXECUTIVE ENGINEERING
19510 Ventura Blvd., Suite 110
Tarzana, California 91356-4234
Phone: (818) 708-1962 / Facsimile: (818) 708-1958
E-mail: EE110@aol.com

ILLUSTRATION ONLY. NOT FOR CONSTRUCTION.

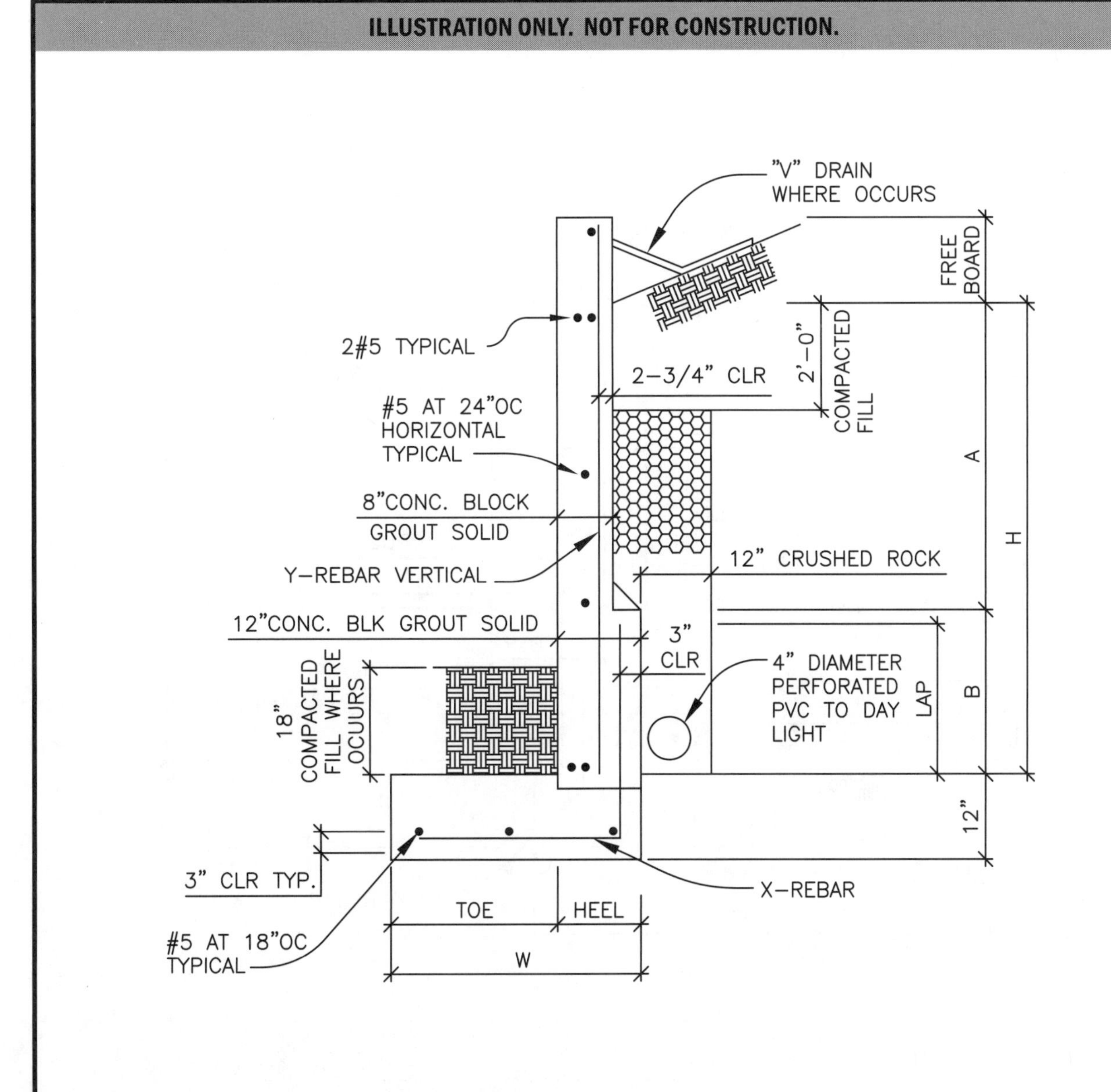

SLOPED BACKFILL,
NO KEY, FOOTING OUT OF HILL

ILLUSTRATION ONLY. NOT FOR CONSTRUCTION.

EXECUTIVE ENGINEERING
19510 Ventura Blvd., Suite 110
Tarzana, California 91356-4234
Phone: (818) 708-1962 / Facsimile: (818) 708-1958
E-mail: EE110@aol.com

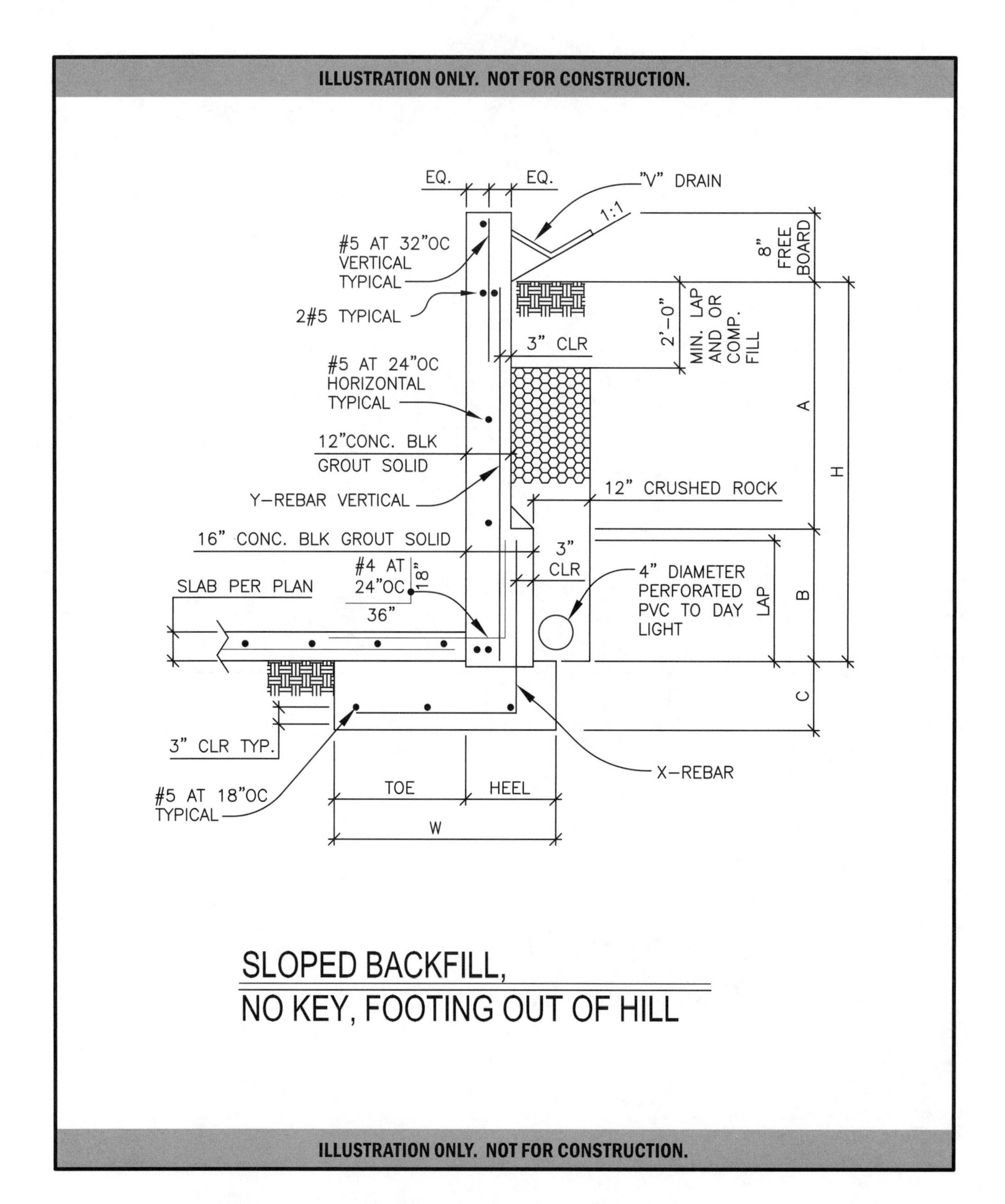

EXECUTIVE ENGINEERING
19510 Ventura Blvd., Suite 110
Tarzana, California 91356-4234
Phone: (818) 708-1962 / Facsimile: (818) 708-1958
E-mail: EE110@aol.com

ILLUSTRATION ONLY. NOT FOR CONSTRUCTION.

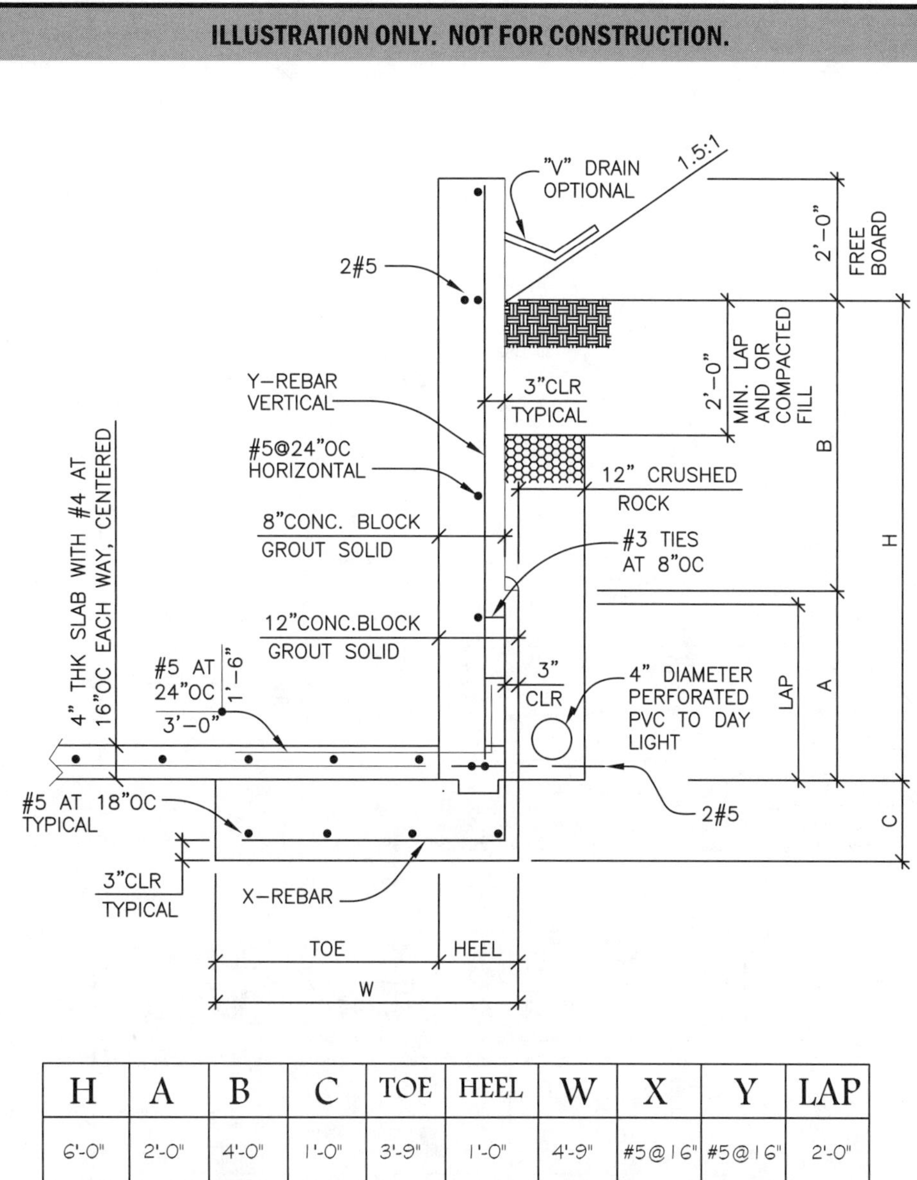

H	A	B	C	TOE	HEEL	W	X	Y	LAP
6'-0"	2'-0"	4'-0"	1'-0"	3'-9"	1'-0"	4'-9"	#5@16"	#5@16"	2'-0"

SLOPED BACKFILL, NO KEY, FOOTING OUT OF HILL

ILLUSTRATION ONLY. NOT FOR CONSTRUCTION.

EXECUTIVE ENGINEERING
19510 Ventura Blvd., Suite 110
Tarzana, California 91356-4234
Phone: (818) 708-1962 / Facsimile: (818) 708-1958
E-mail: EE110@aol.com

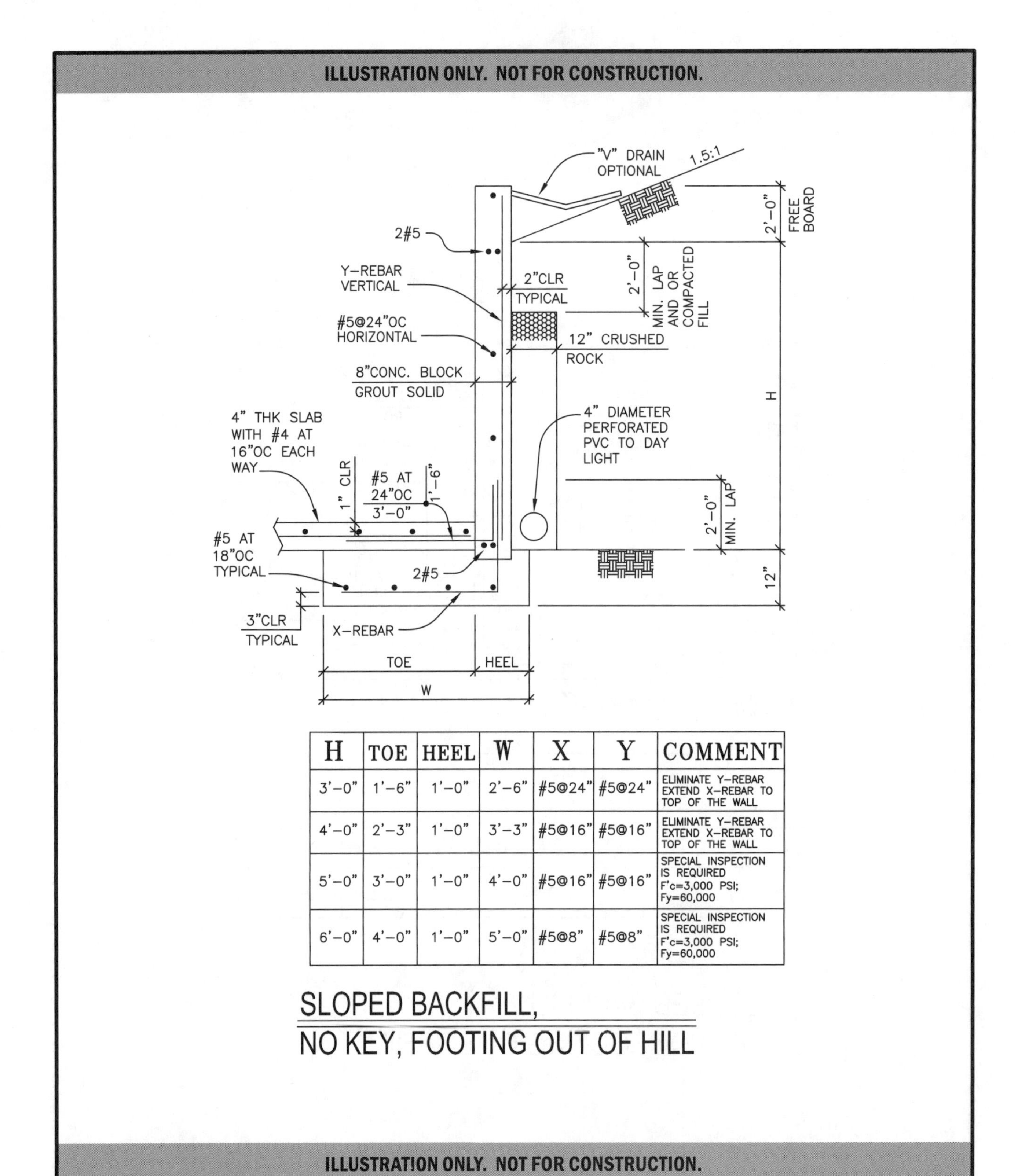

H	TOE	HEEL	W	X	Y	COMMENT
3'-0"	1'-6"	1'-0"	2'-6"	#5@24"	#5@24"	ELIMINATE Y-REBAR EXTEND X-REBAR TO TOP OF THE WALL
4'-0"	2'-3"	1'-0"	3'-3"	#5@16"	#5@16"	ELIMINATE Y-REBAR EXTEND X-REBAR TO TOP OF THE WALL
5'-0"	3'-0"	1'-0"	4'-0"	#5@16"	#5@16"	SPECIAL INSPECTION IS REQUIRED F'c=3,000 PSI; Fy=60,000
6'-0"	4'-0"	1'-0"	5'-0"	#5@8"	#5@8"	SPECIAL INSPECTION IS REQUIRED F'c=3,000 PSI; Fy=60,000

SLOPED BACKFILL,
NO KEY, FOOTING OUT OF HILL

ILLUSTRATION ONLY. NOT FOR CONSTRUCTION.

EXECUTIVE ENGINEERING
19510 Ventura Blvd., Suite 110
Tarzana, California 91356-4234
Phone: (818) 708-1962 / Facsimile: (818) 708-1958
E-mail: EE110@aol.com

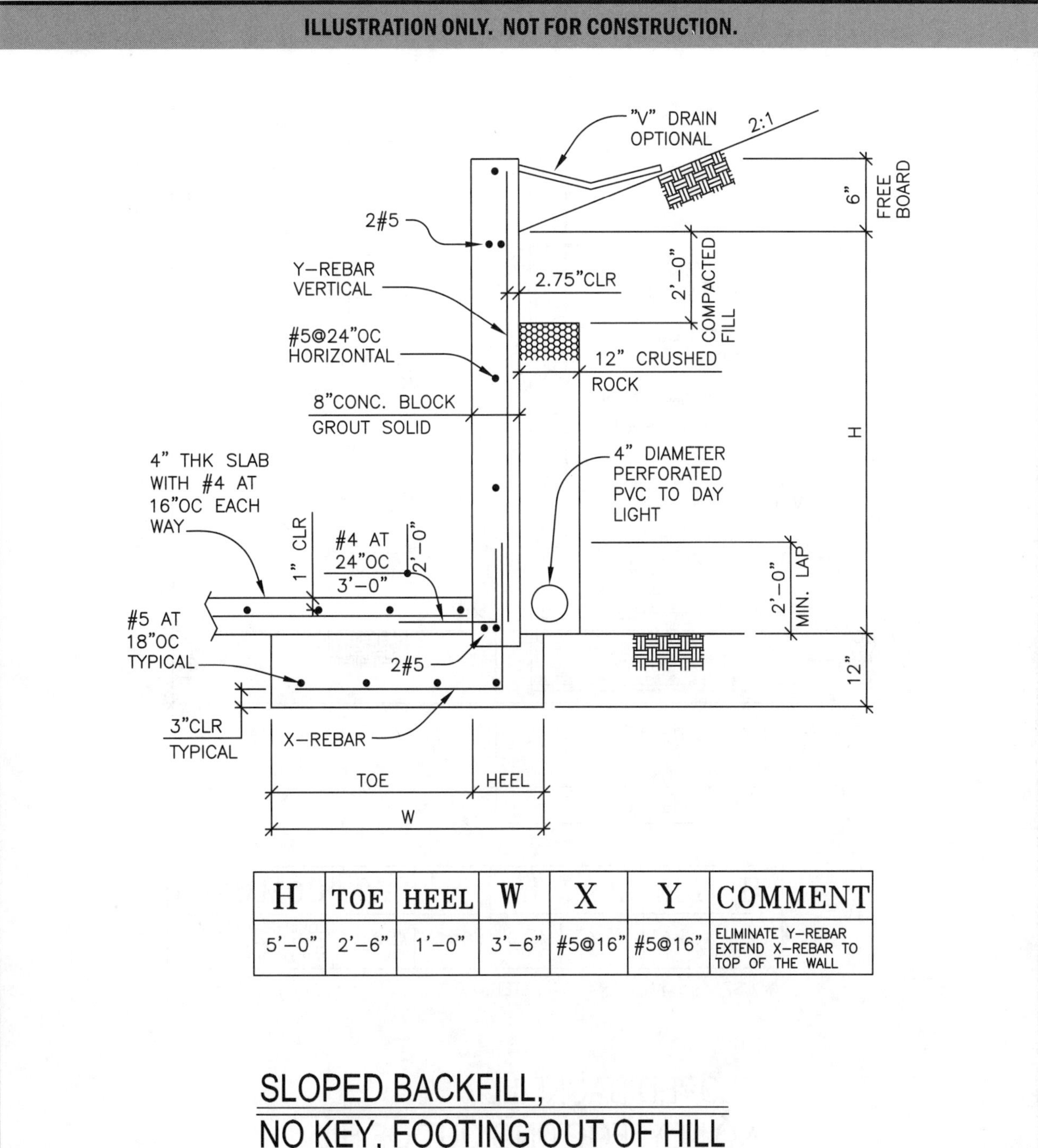

H	TOE	HEEL	W	X	Y	COMMENT
5'-0"	2'-6"	1'-0"	3'-6"	#5@16"	#5@16"	ELIMINATE Y-REBAR EXTEND X-REBAR TO TOP OF THE WALL

SLOPED BACKFILL, NO KEY, FOOTING OUT OF HILL

ILLUSTRATION ONLY. NOT FOR CONSTRUCTION.

EXECUTIVE ENGINEERING
19510 Ventura Blvd., Suite 110
Tarzana, California 91356-4234
Phone: (818) 708-1962 / Facsimile: (818) 708-1958
E-mail: EE110@aol.com

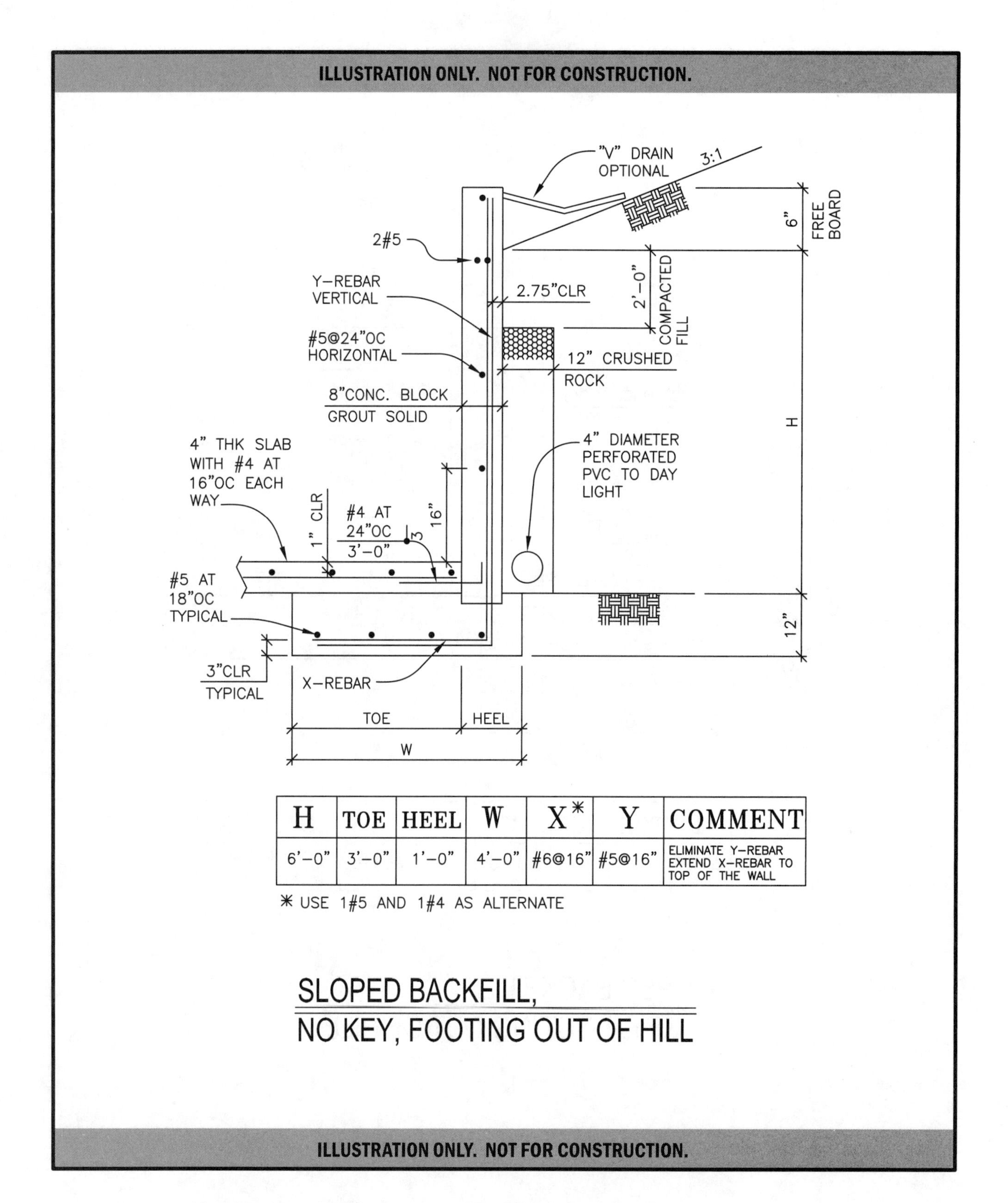

H	TOE	HEEL	W	X*	Y	COMMENT
6'-0"	3'-0"	1'-0"	4'-0"	#6@16"	#5@16"	ELIMINATE Y-REBAR EXTEND X-REBAR TO TOP OF THE WALL

* USE 1#5 AND 1#4 AS ALTERNATE

SLOPED BACKFILL, NO KEY, FOOTING OUT OF HILL

EXECUTIVE ENGINEERING
19510 Ventura Blvd., Suite 110
Tarzana, California 91356-4234
Phone: (818) 708-1962 / Facsimile: (818) 708-1958
E-mail: EE110@aol.com

ILLUSTRATION ONLY. NOT FOR CONSTRUCTION.

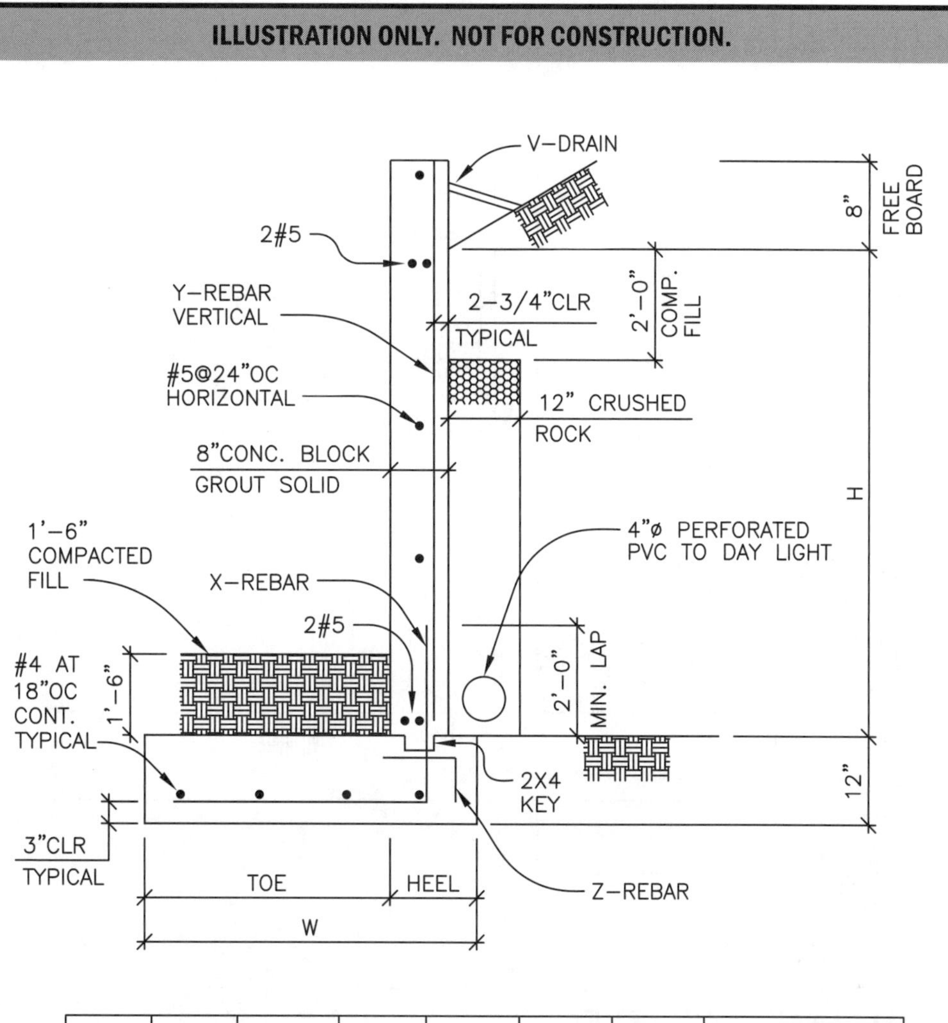

H	TOE	HEEL	W	X	Y	Z	COMMENT
5'-0"	2'-6"	1'-3"	3'-9"	#5@8"	#5@8"	#5@16"	ELIMINATE Y-REBAR EXTEND X-REBAR TO TOP OF THE WALL

SLOPED BACKFILL, NO KEY, FOOTING OUT OF HILL

ILLUSTRATION ONLY. NOT FOR CONSTRUCTION.

EXECUTIVE ENGINEERING
19510 Ventura Blvd., Suite 110
Tarzana, California 91356-4234
Phone: (818) 708-1962 / Facsimile: (818) 708-1958
E-mail: EE110@aol.com

ILLUSTRATION ONLY. NOT FOR CONSTRUCTION.

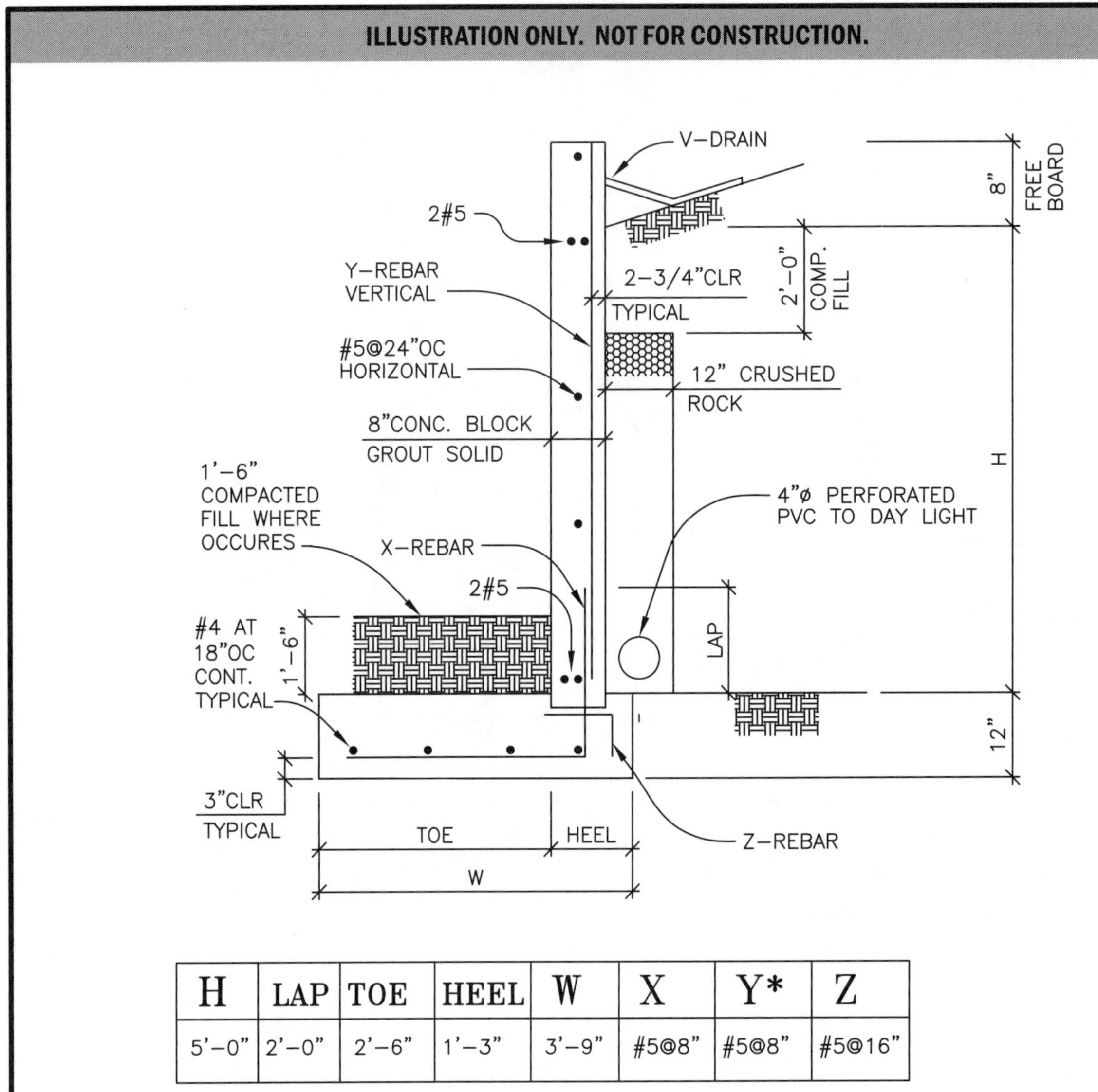

H	LAP	TOE	HEEL	W	X	Y*	Z
5'-0"	2'-0"	2'-6"	1'-3"	3'-9"	#5@8"	#5@8"	#5@16"

*COMMENT: ELIMINATE Y-REBAR. EXTEND X-REBAR TO THE TOP OF THE WALL

SLOPED BACKFILL, NO KEY, FOOTING OUT OF HILL

ILLUSTRATION ONLY. NOT FOR CONSTRUCTION.

EXECUTIVE ENGINEERING
19510 Ventura Blvd., Suite 110
Tarzana, California 91356-4234
Phone: (818) 708-1962 / Facsimile: (818) 708-1958
E-mail: EE110@aol.com

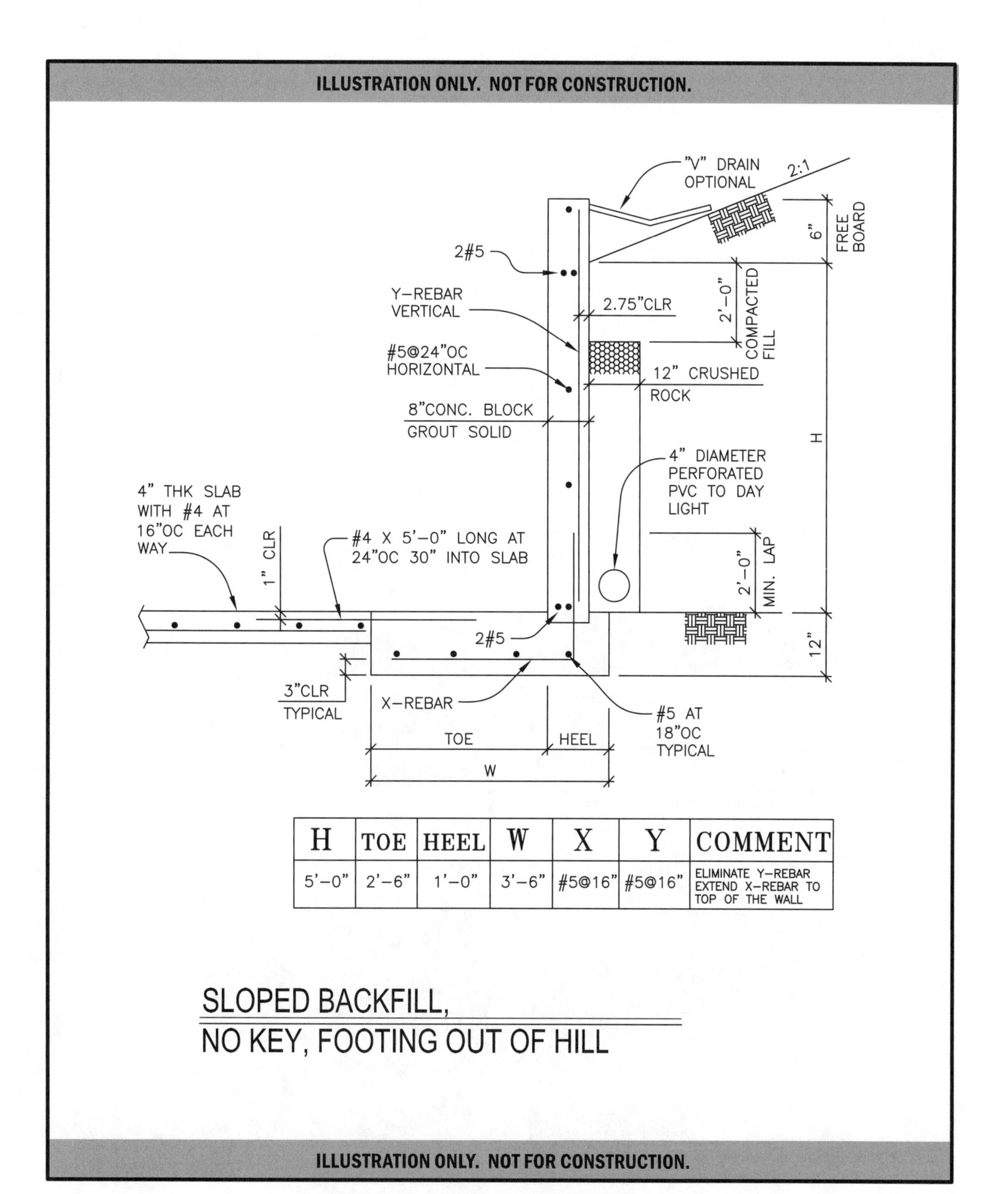

H	TOE	HEEL	W	X	Y	COMMENT
5'-0"	2'-6"	1'-0"	3'-6"	#5@16"	#5@16"	ELIMINATE Y-REBAR EXTEND X-REBAR TO TOP OF THE WALL

SLOPED BACKFILL,
NO KEY, FOOTING OUT OF HILL

EXECUTIVE ENGINEERING
19510 Ventura Blvd., Suite 110
Tarzana, California 91356-4234
Phone: (818) 708-1962 / Facsimile: (818) 708-1958
E-mail: EE110@aol.com

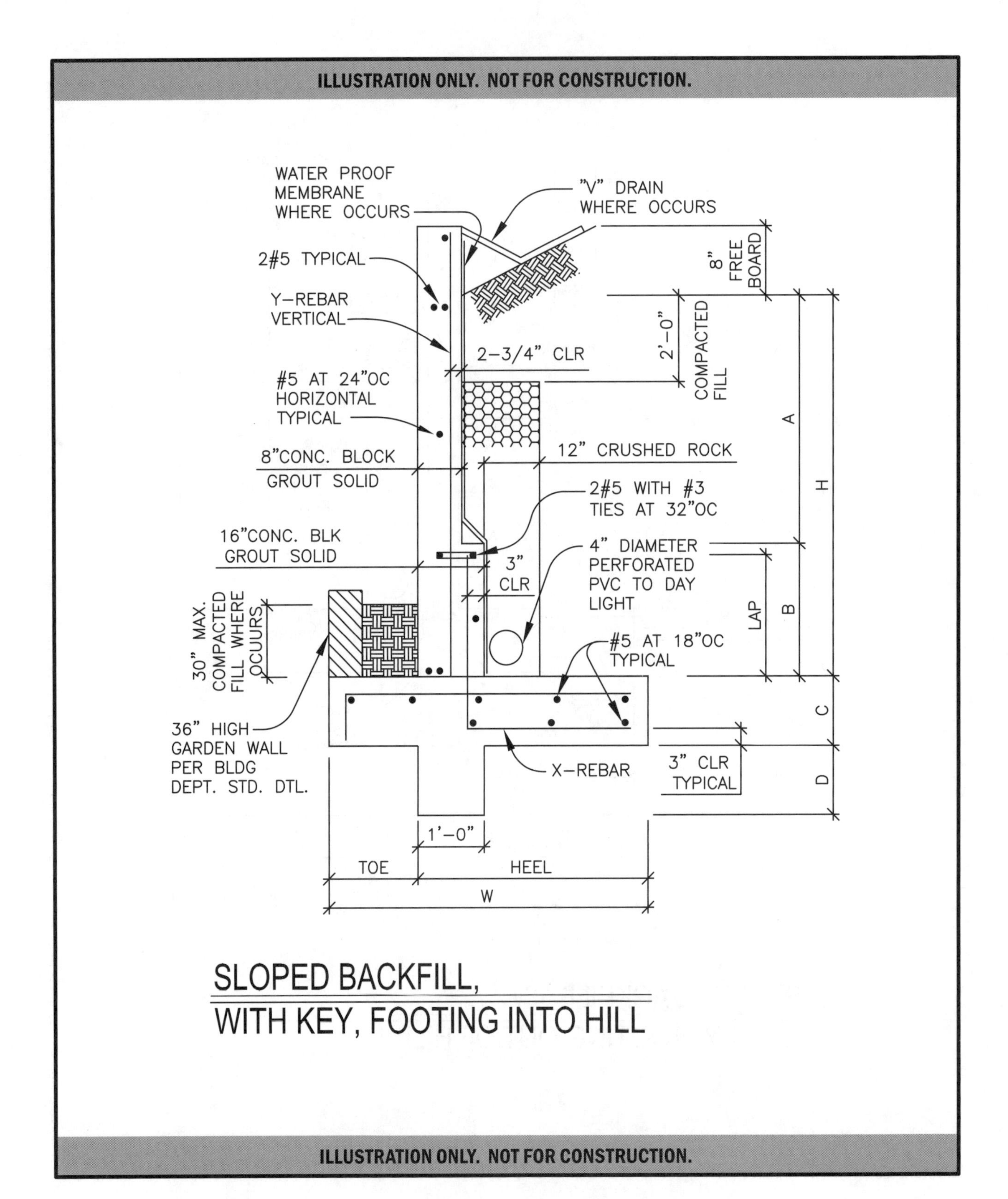

SLOPED BACKFILL,
WITH KEY, FOOTING INTO HILL

EXECUTIVE ENGINEERING
19510 Ventura Blvd., Suite 110
Tarzana, California 91356-4234
Phone: (818) 708-1962 / Facsimile: (818) 708-1958
E-mail: EE110@aol.com

ILLUSTRATION ONLY. NOT FOR CONSTRUCTION.

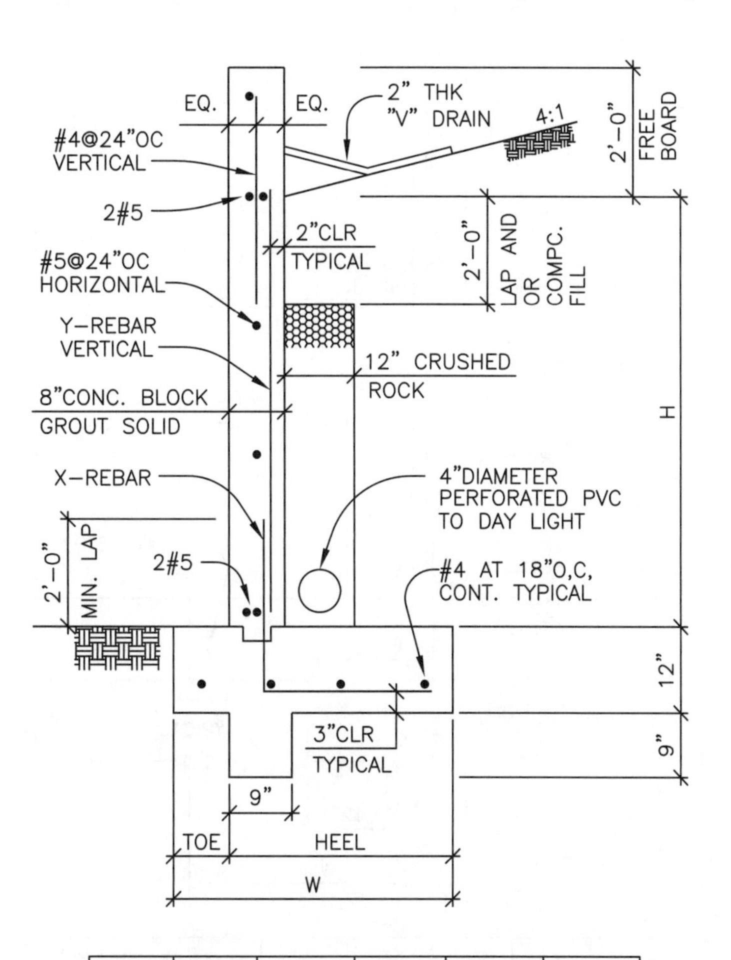

H	TOE	HEEL	W	X	Y
5'–0"	6"	2'–6"	3'–0"	#5@16"	#4@16"
6'–0"	6"	3'–0"	3'–6"	#5@16"	#5@8"

SLOPED BACKFILL, WITH KEY, FOOTING INTO HILL

ILLUSTRATION ONLY. NOT FOR CONSTRUCTION.

EXECUTIVE ENGINEERING
19510 Ventura Blvd., Suite 110
Tarzana, California 91356-4234
Phone: (818) 708-1962 / Facsimile: (818) 708-1958
E-mail: EE110@aol.com

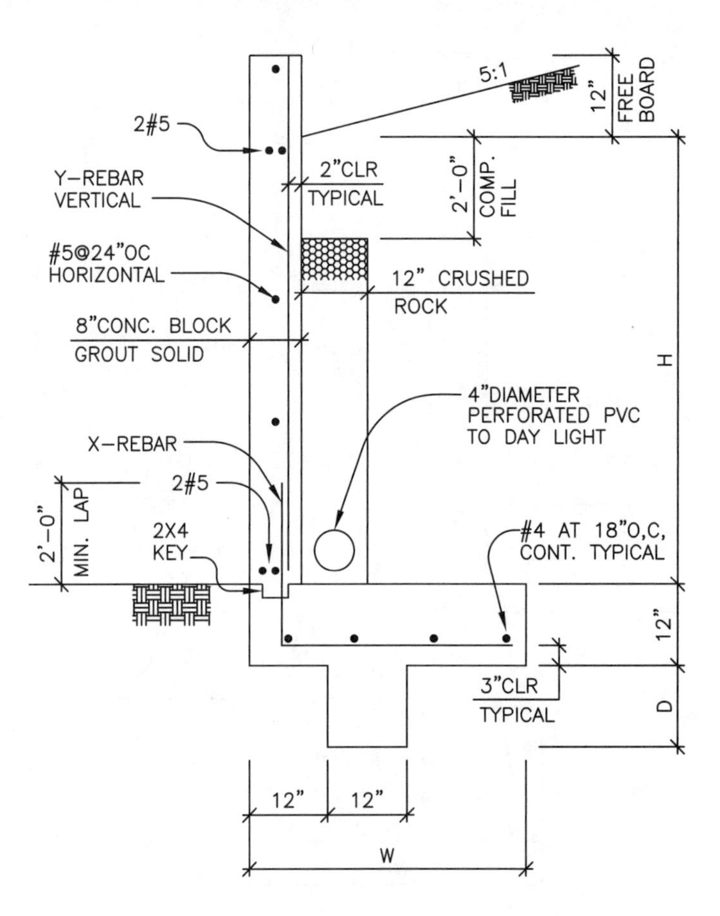

H	D	W	X	Y	COMMENT
4'–0"	12"	2'–8"	#5@24"	#5@24"	ELIMINATE Y–REBAR EXTEND X–REBAR TO TOP OF THE WALL
5'–0"	12"	3'–9"	#5@16"	#5@16"	ELIMINATE Y–REBAR EXTEND X–REBAR TO TOP OF THE WALL
6'–0"	12"	5'–0"	#5@8"	#5@8"	ELIMINATE Y–REBAR EXTEND X–REBAR TO TOP OF THE WALL

SLOPED BACKFILL, WITH KEY, FOOTING INTO HILL

ILLUSTRATION ONLY. NOT FOR CONSTRUCTION.

EXECUTIVE ENGINEERING
19510 Ventura Blvd., Suite 110
Tarzana, California 91356-4234
Phone: (818) 708-1962 / Facsimile: (818) 708-1958
E-mail: EE110@aol.com

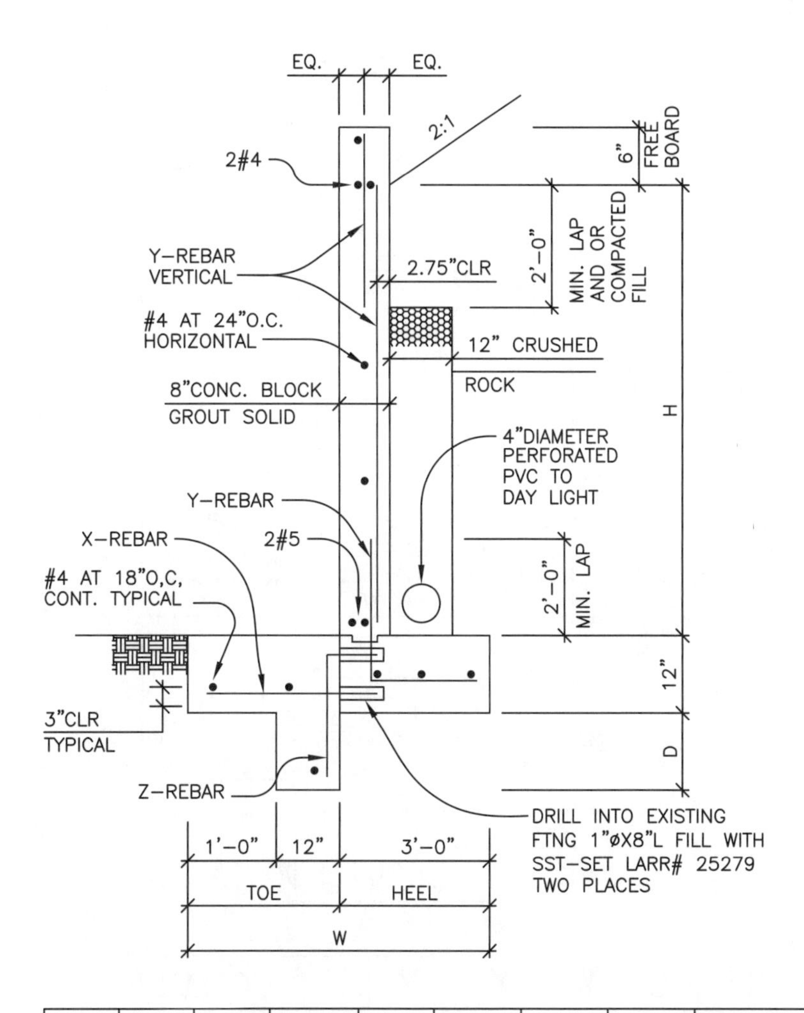

H	D	TOE	HEEL	W	X	Y	Z	COMMENT
5'-0"	2'-0"	2'-0"	3'-0"	5'-0"	#6@18"	#4@8"	#6@18"	ELIMINATE Y-REBAR EXTEND X-REBAR TO TOP OF THE WALL

SLOPED BACKFILL, WITH KEY, FOOTING INTO HILL

ILLUSTRATION ONLY. NOT FOR CONSTRUCTION.

EXECUTIVE ENGINEERING
19510 Ventura Blvd., Suite 110
Tarzana, California 91356-4234
Phone: (818) 708-1962 / Facsimile: (818) 708-1958
E-mail: EE110@aol.com

ILLUSTRATION ONLY. NOT FOR CONSTRUCTION.

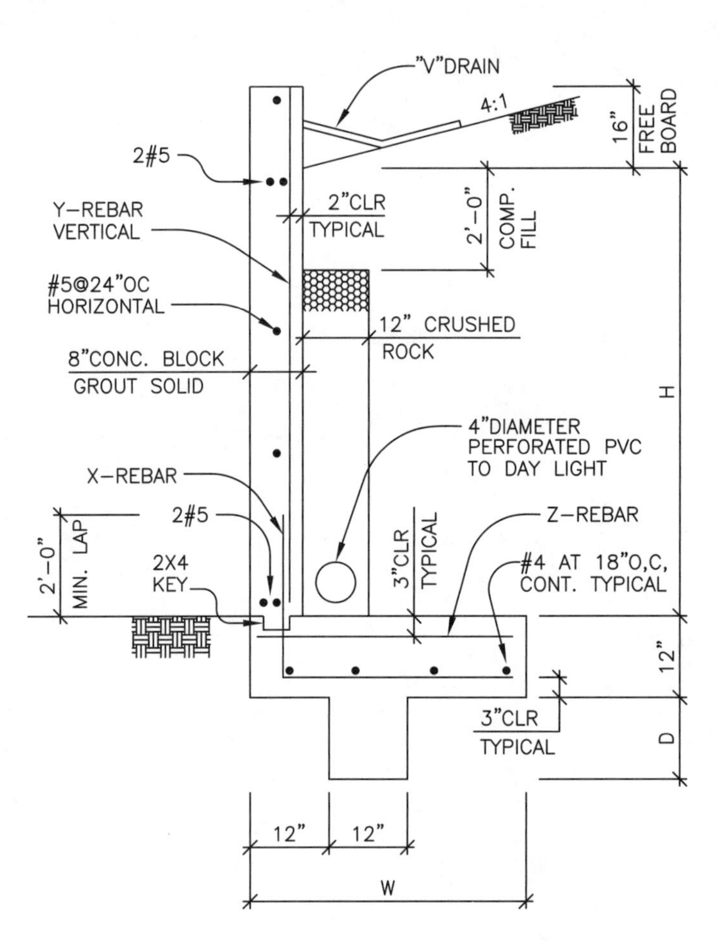

H	D	W	X	Y	Z	COMMENT
4'–0"	14"	2'–8"	#5@16"	#5@16"	#5@16"	ELIMINATE Y–REBAR EXTEND X–REBAR TO TOP OF THE WALL
5'–0"	18"	3'–8"	#5@16"	#5@16"	#5@16"	ELIMINATE Y–REBAR EXTEND X–REBAR TO TOP OF THE WALL

SLOPED BACKFILL, WITH KEY, FOOTING INTO HILL

ILLUSTRATION ONLY. NOT FOR CONSTRUCTION.

EXECUTIVE ENGINEERING
19510 Ventura Blvd., Suite 110
Tarzana, California 91356-4234
Phone: (818) 708-1962 / Facsimile: (818) 708-1958
E-mail: EE110@aol.com

ILLUSTRATION ONLY. NOT FOR CONSTRUCTION.

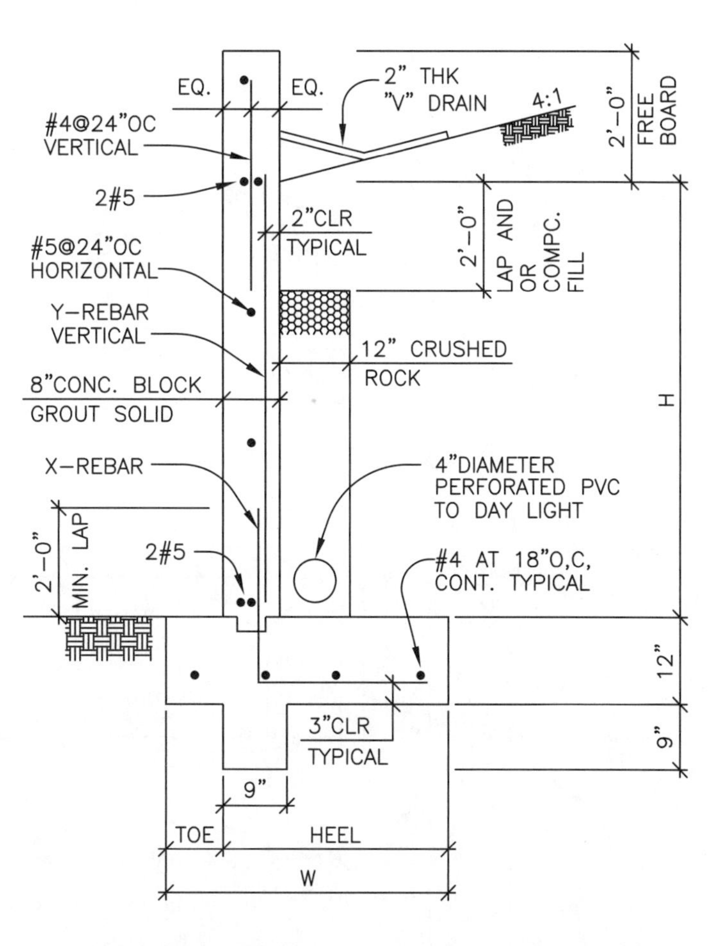

H	TOE	HEEL	W	X	Y
5'-0"	6"	2'-6"	3'-0"	#5@16"	#4@16"
6'-0"	6"	3'-0"	3'-6"	#5@16"	#5@8"

SLOPED BACKFILL, WITH KEY, FOOTING INTO HILL

ILLUSTRATION ONLY. NOT FOR CONSTRUCTION.

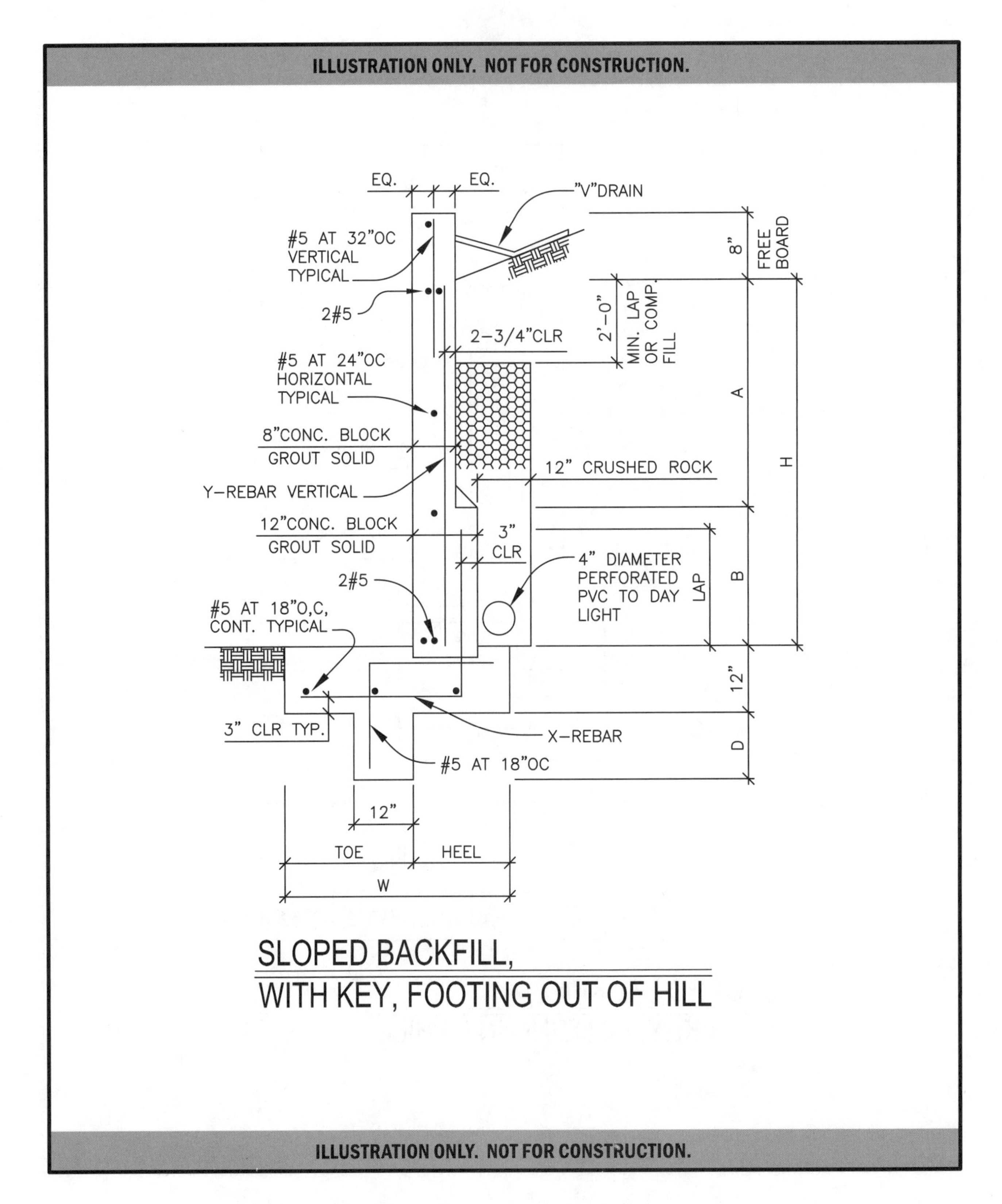

EXECUTIVE ENGINEERING
19510 Ventura Blvd., Suite 110
Tarzana, California 91356-4234
Phone: (818) 708-1962 / Facsimile: (818) 708-1958
E-mail: EE110@aol.com

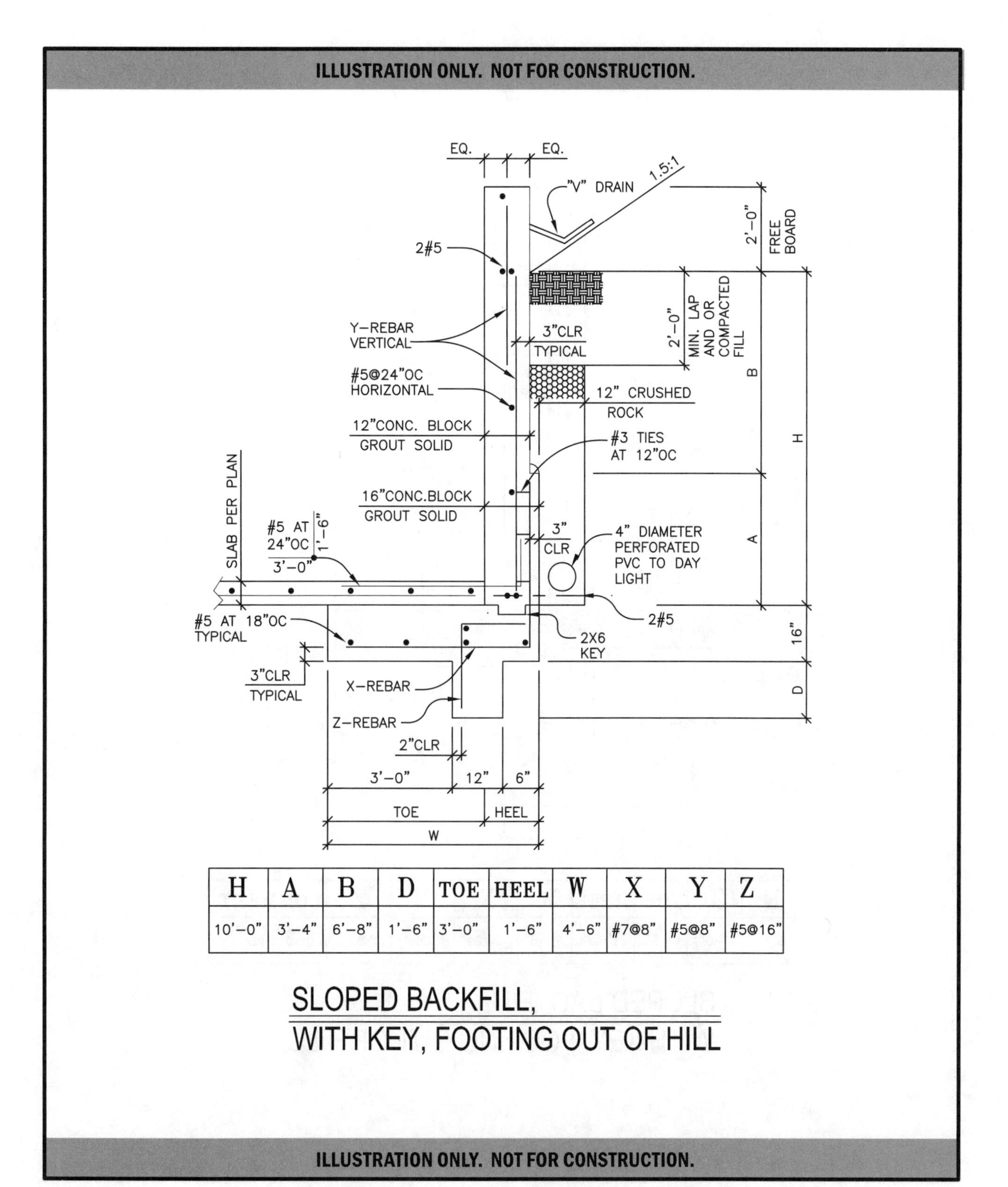

H	A	B	D	TOE	HEEL	W	X	Y	Z
10'–0"	3'–4"	6'–8"	1'–6"	3'–0"	1'–6"	4'–6"	#7@8"	#5@8"	#5@16"

SLOPED BACKFILL, WITH KEY, FOOTING OUT OF HILL

EXECUTIVE ENGINEERING
19510 Ventura Blvd., Suite 110
Tarzana, California 91356-4234
Phone: (818) 708-1962 / Facsimile: (818) 708-1958
E-mail: EE110@aol.com

ILLUSTRATION ONLY. NOT FOR CONSTRUCTION.

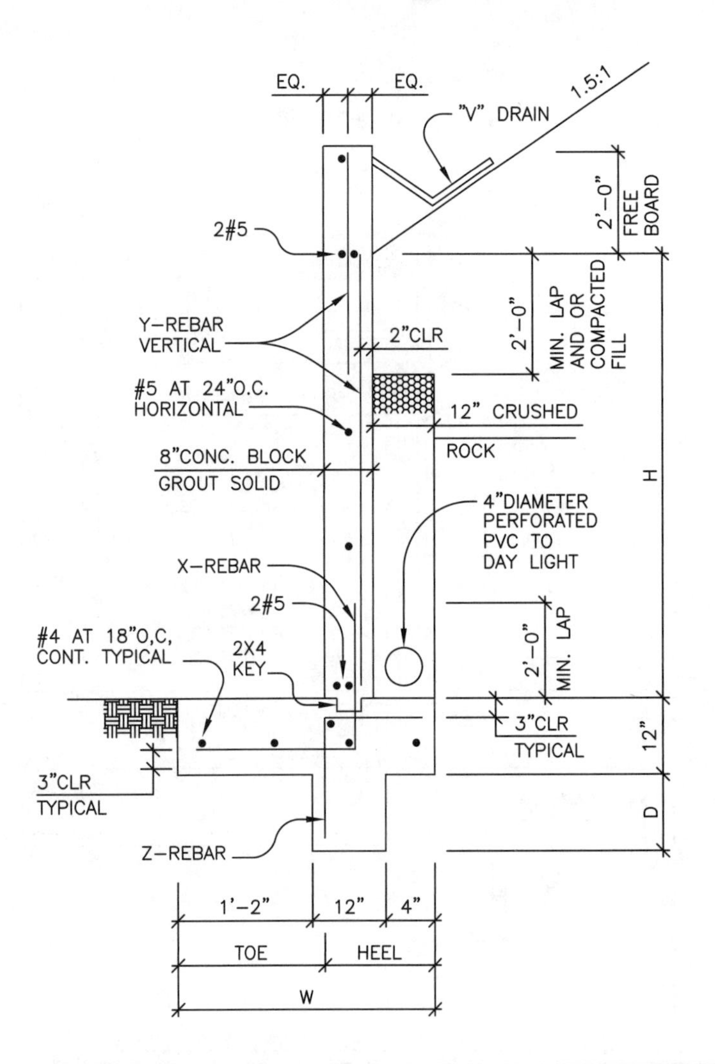

H	D	TOE	HEEL	W	X	Y	Z	COMMENT
5'-0"	1'-6"	1'-6"	1'-0"	2'-6"	#5@24"	#5@24"	#5@24"	ELIMINATE Y-REBAR EXTEND X-REP

SLOPED BACKFILL, WITH KEY, FOOTING OUT OF HILL

ILLUSTRATION ONLY. NOT FOR CONSTRUCTION.

EXECUTIVE ENGINEERING
19510 Ventura Blvd., Suite 110
Tarzana, California 91356-4234
Phone: (818) 708-1962 / Facsimile: (818) 708-1958
E-mail: EE110@aol.com

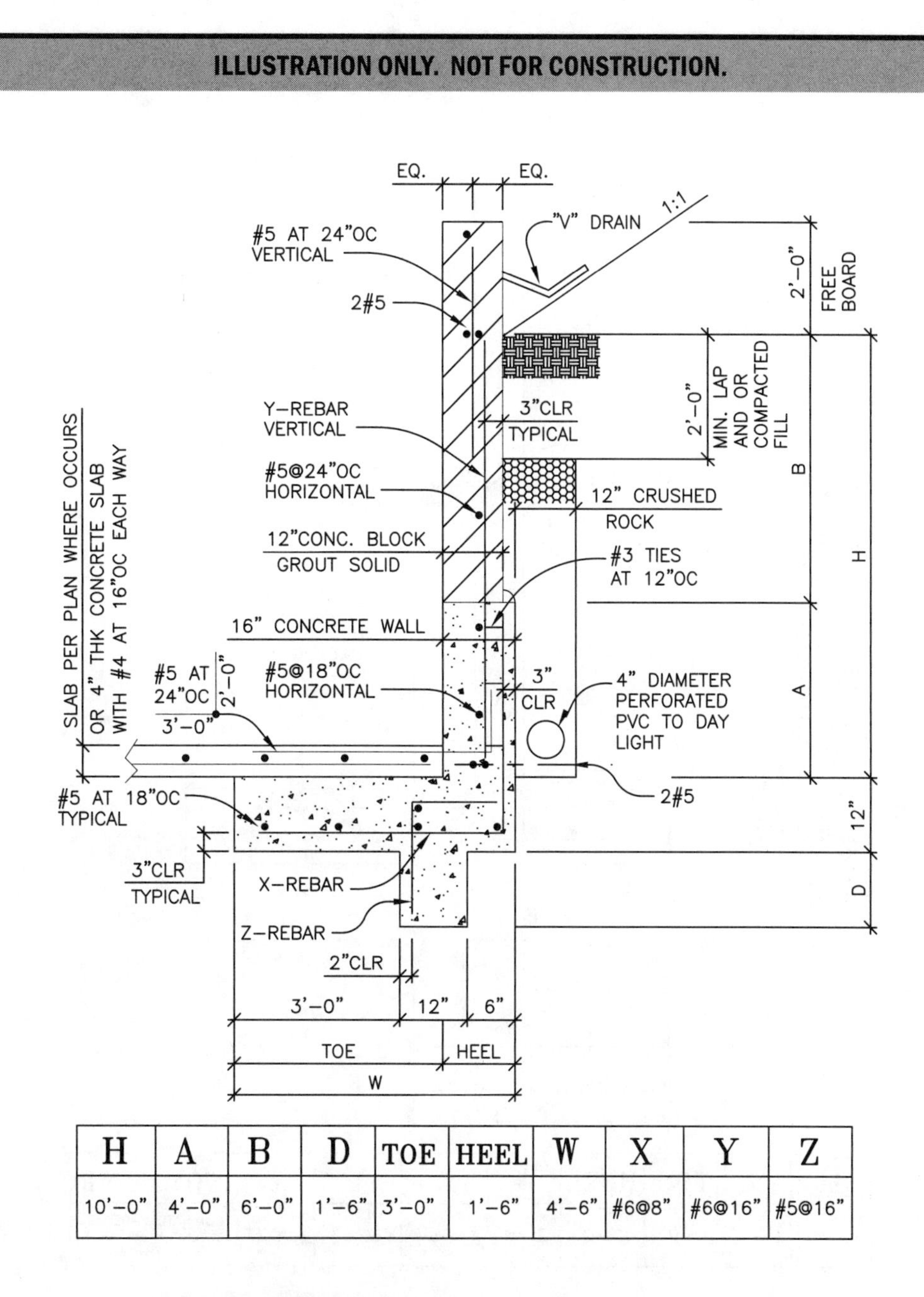

H	A	B	D	TOE	HEEL	W	X	Y	Z
10'-0"	4'-0"	6'-0"	1'-6"	3'-0"	1'-6"	4'-6"	#6@8"	#6@16"	#5@16"

SLOPED BACKFILL,
WITH KEY, FOOTING OUT OF HILL

ILLUSTRATION ONLY. NOT FOR CONSTRUCTION.

EXECUTIVE ENGINEERING
19510 Ventura Blvd., Suite 110
Tarzana, California 91356-4234
Phone: (818) 708-1962 / Facsimile: (818) 708-1958
E-mail: EE110@aol.com

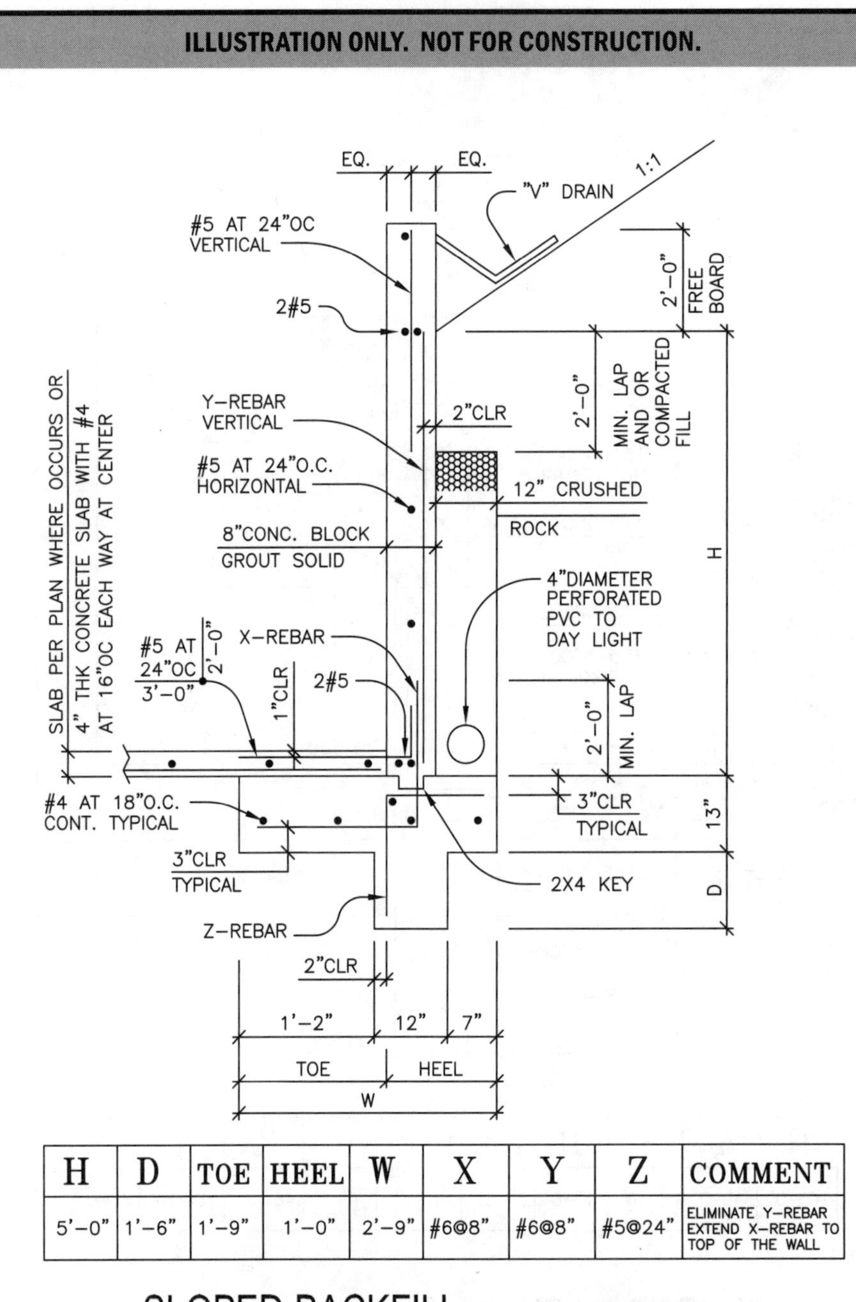

H	D	TOE	HEEL	W	X	Y	Z	COMMENT
5'-0"	1'-6"	1'-9"	1'-0"	2'-9"	#6@8"	#6@8"	#5@24"	ELIMINATE Y-REBAR EXTEND X-REBAR TO TOP OF THE WALL

SLOPED BACKFILL, WITH KEY, FOOTING OUT OF HILL

ILLUSTRATION ONLY. NOT FOR CONSTRUCTION.

EXECUTIVE ENGINEERING
19510 Ventura Blvd., Suite 110
Tarzana, California 91356-4234
Phone: (818) 708-1962 / Facsimile: (818) 708-1958
E-mail: EE110@aol.com

EQ. EQ.
"V" DRAIN
1:1
#5 AT 24"OC VERTICAL
2#5
2'-0" FREE BOARD
Y-REBAR VERTICAL
2"CLR
2'-0" MIN. LAP AND OR COMPACTED FILL
#5 AT 24"O.C. HORIZONTAL
12" CRUSHED ROCK
8"CONC. BLOCK GROUT SOLID
H
8" COMPACTED FILL
4"DIAMETER PERFORATED PVC TO DAY LIGHT
X-REBAR
2#5
3'-0" MIN. LAP
#4 AT 18"O.C. CONT. TYPICAL
3"CLR TYPICAL
18"
3"CLR TYPICAL
2X4 KEY
D
Z-REBAR
2"CLR
1'-11" 12" 7"
TOE HEEL
W

H	D	TOE	HEEL	W	X	Y	Z	COMMENT
5'-0"	1'-6"	2'-6"	1'-0"	3'-6"	#6@8"	#6@8"	#5@24"	ELIMINATE Y-REBAR EXTEND X-REBAR TO TOP OF THE WALL

SLOPED BACKFILL, WITH KEY, FOOTING OUT OF HILL

ILLUSTRATION ONLY. NOT FOR CONSTRUCTION.

EXECUTIVE ENGINEERING
19510 Ventura Blvd., Suite 110
Tarzana, California 91356-4234
Phone: (818) 708-1962 / Facsimile: (818) 708-1958
E-mail: EE110@aol.com

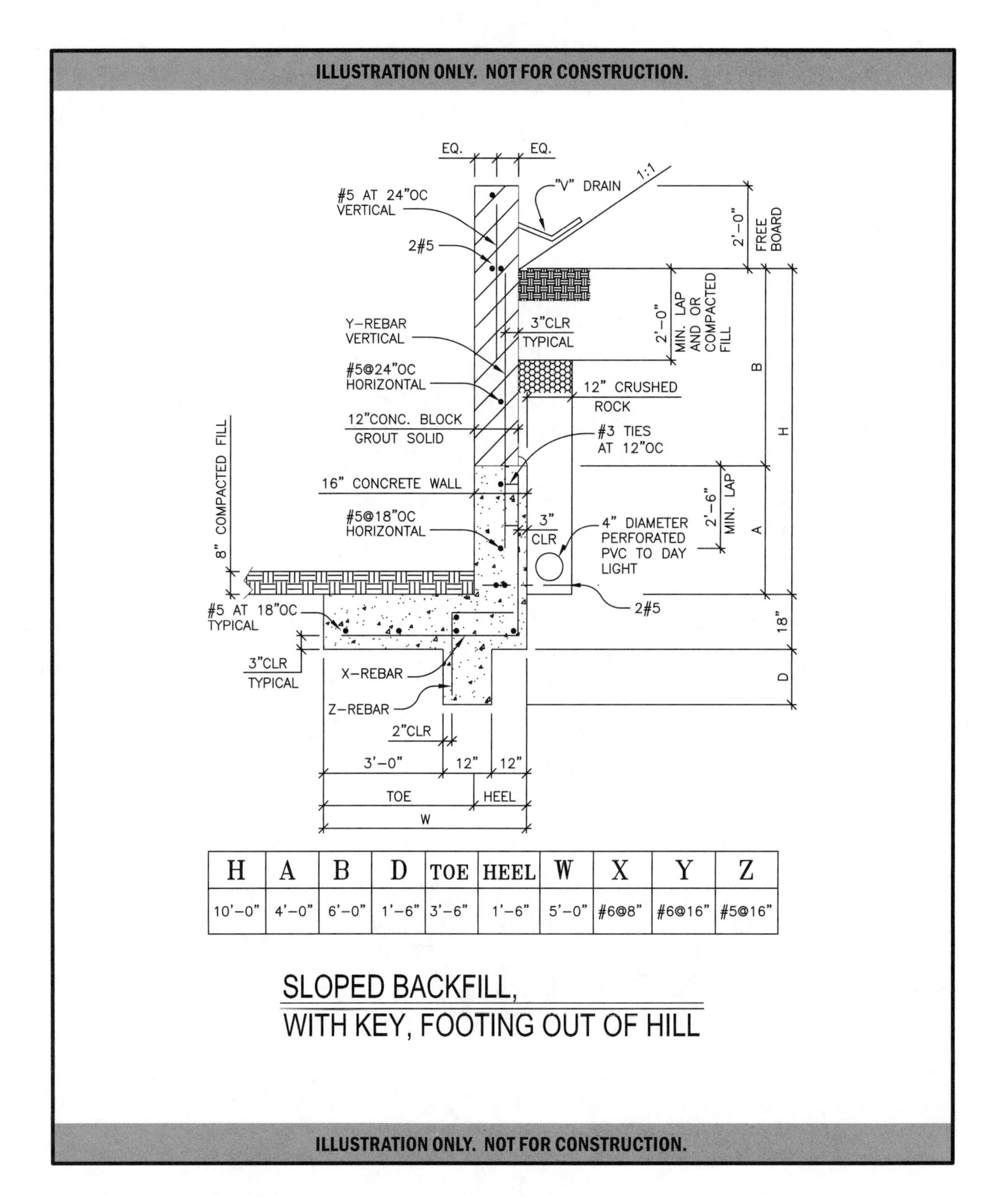

H	A	B	D	TOE	HEEL	W	X	Y	Z
10'–0"	4'–0"	6'–0"	1'–6"	3'–6"	1'–6"	5'–0"	#6@8"	#6@16"	#5@16"

SLOPED BACKFILL, WITH KEY, FOOTING OUT OF HILL

EXECUTIVE ENGINEERING
19510 Ventura Blvd., Suite 110
Tarzana, California 91356-4234
Phone: (818) 708-1962 / Facsimile: (818) 708-1958
E-mail: EE110@aol.com

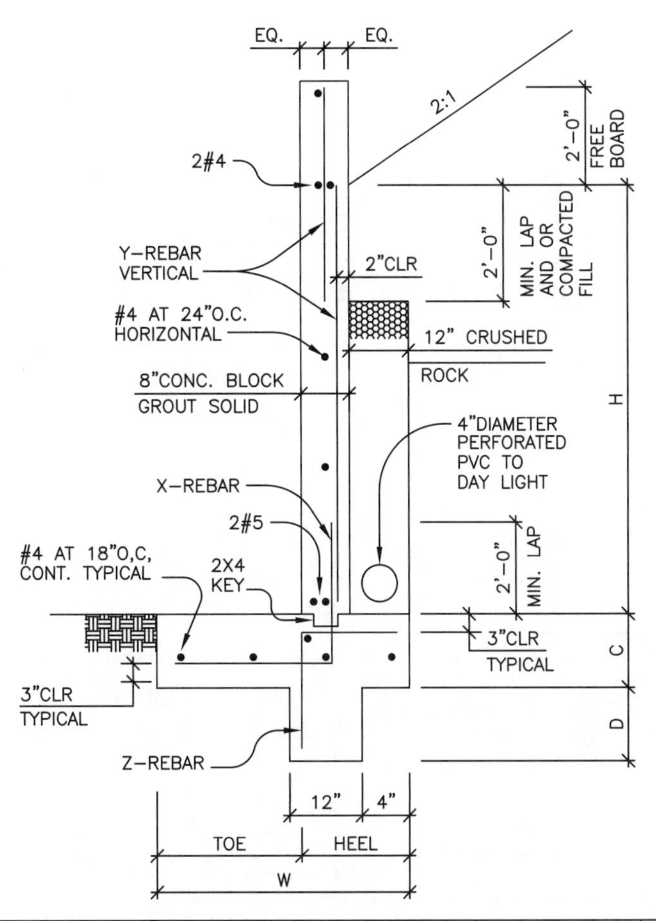

H	D	C	TOE	HEEL	W	X	Y	Z	COMMENT
2'–0"	1'–0"	1'–0"	1'–6"	8"	2'–2"	#4@24"	#4@24"	#4@48"	ELIMINATE Y–REBAR EXTEND X–REBAR TO TOP OF THE WALL
3'–0"	1'–6"	1'–6"	2'–0"	8"	2'–8"	#4@24"	#4@24"	#4@48"	ELIMINATE Y–REBAR EXTEND X–REBAR TO TOP OF THE WALL

SLOPED BACKFILL, WITH KEY, FOOTING OUT OF HILL

ILLUSTRATION ONLY. NOT FOR CONSTRUCTION.

EXECUTIVE ENGINEERING
19510 Ventura Blvd., Suite 110
Tarzana, California 91356-4234
Phone: (818) 708-1962 / Facsimile: (818) 708-1958
E-mail: EE110@aol.com

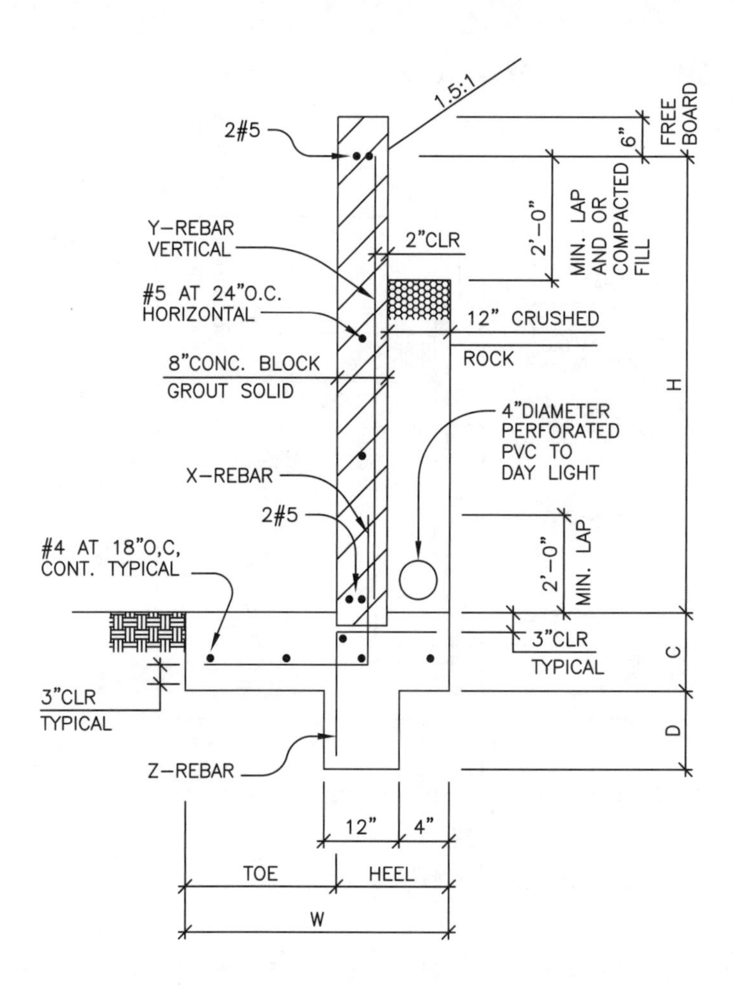

H	D	C	TOE	HEEL	W	X	Y	Z	COMMENT
4'–0"	2'–0"	1'–0"	2'–0"	1'–0"	3'–0"	#5@16"	#5@16"	#4@24"	ELIMINATE Y–REBAR EXTEND X–REBAR TO TOP OF THE WALL

SLOPED BACKFILL,
WITH KEY, FOOTING OUT OF HILL

ILLUSTRATION ONLY. NOT FOR CONSTRUCTION.

EXECUTIVE ENGINEERING
19510 Ventura Blvd., Suite 110
Tarzana, California 91356-4234
Phone: (818) 708-1962 / Facsimile: (818) 708-1958
E-mail: EE110@aol.com

ILLUSTRATION ONLY. NOT FOR CONSTRUCTION.

'V" DRAIN OPTIONAL

2:1

8" FREE BOARD

2#4

2'–0" COMPACTED FILL

#4 AT 16"O.C. VERTICAL

2.75"CLR

#4 AT 24"O.C. HORIZONTAL

12" CRUSHED ROCK

8"CONC. BLOCK GROUT SOLID

6'–0"

5'–4"

4"DIAMETER PERFORATED PVC TO DAY LIGHT

#4 AT 16"OC

2#4

2'–0" MIN. LAP

#4 AT 18"O,C, CONT. TYPICAL

3"CLR TYPICAL

12"

3"CLR TYPICAL

24"

#4 AT 24" OC

12"

1'–9"

1'–0"

2'–9"

SLOPED BACKFILL, WITH KEY, FOOTING OUT OF HILL

ILLUSTRATION ONLY. NOT FOR CONSTRUCTION.

EXECUTIVE ENGINEERING
19510 Ventura Blvd., Suite 110
Tarzana, California 91356-4234
Phone: (818) 708-1962 / Facsimile: (818) 708-1958
E-mail: EE110@aol.com

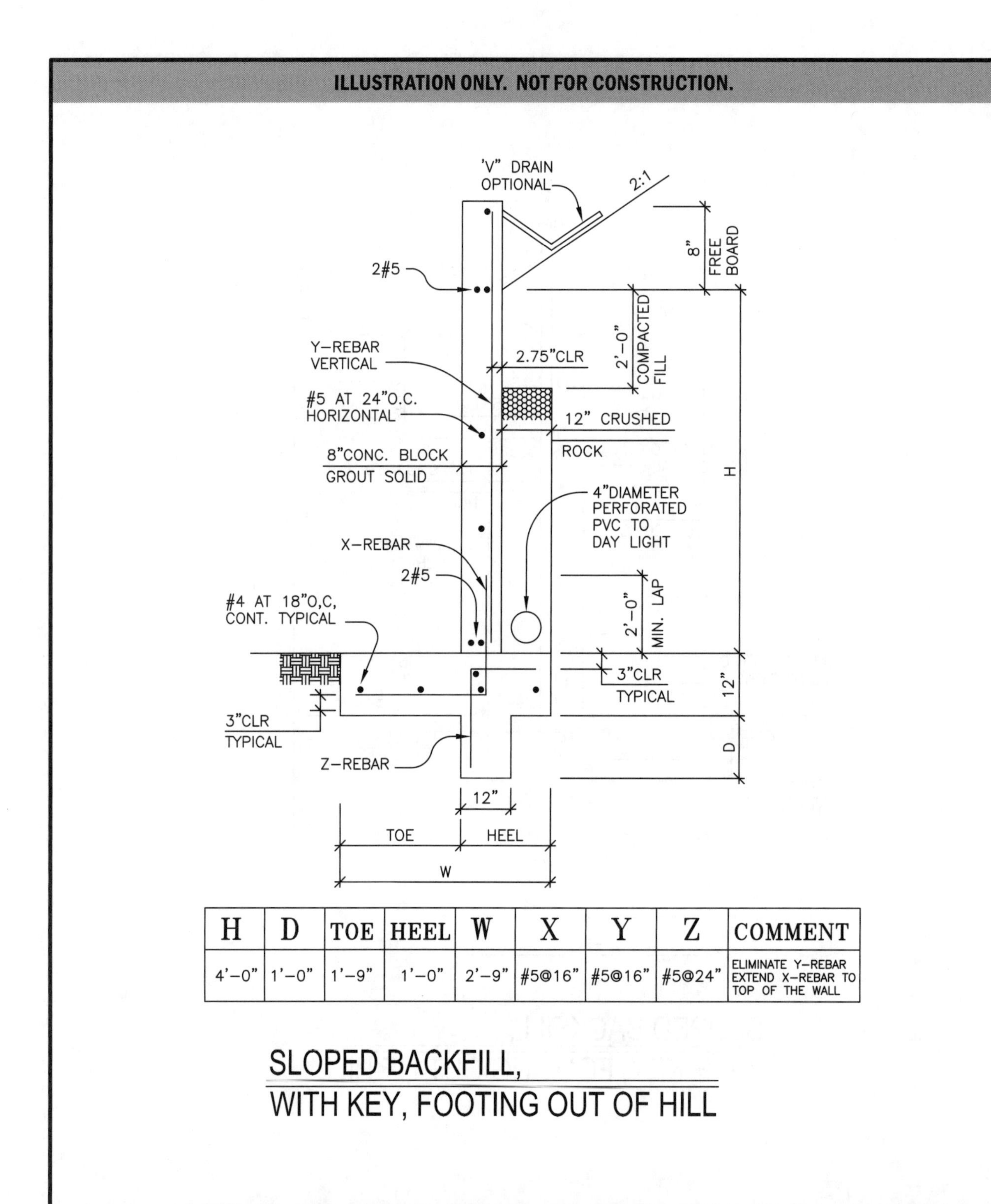

H	D	TOE	HEEL	W	X	Y	Z	COMMENT
4'–0"	1'–0"	1'–9"	1'–0"	2'–9"	#5@16"	#5@16"	#5@24"	ELIMINATE Y–REBAR EXTEND X–REBAR TO TOP OF THE WALL

SLOPED BACKFILL,
WITH KEY, FOOTING OUT OF HILL

ILLUSTRATION ONLY. NOT FOR CONSTRUCTION.

EXECUTIVE ENGINEERING
19510 Ventura Blvd., Suite 110
Tarzana, California 91356-4234
Phone: (818) 708-1962 / Facsimile: (818) 708-1958
E-mail: EE110@aol.com

ILLUSTRATION ONLY. NOT FOR CONSTRUCTION.

3:1 MAX. SLOPE

8"

#5 AT 24"OC
VERTICAL

24 MAX" COMPACT. FILL

2-3/4"

8" CONCRETE BLOCK
GROUT SOLID

#5 AT 24"OC
HORIZONTAL

8" CRUSHED
ROCK

H = 4'-0" MAX.

4" DIAMETER
PERFORATED
PVC TO DAY
LIGHT

NATURAL
GRADE

12"

12"

12"

3" CLR.
TYPICAL

1'-3"

1'-0"

2'-3"

SLOPED BACKFILL, WITH KEY, FOOTING OUT OF HILL

ILLUSTRATION ONLY. NOT FOR CONSTRUCTION.

EXECUTIVE ENGINEERING
19510 Ventura Blvd., Suite 110
Tarzana, California 91356-4234
Phone: (818) 708-1962 / Facsimile: (818) 708-1958
E-mail: EE110@aol.com

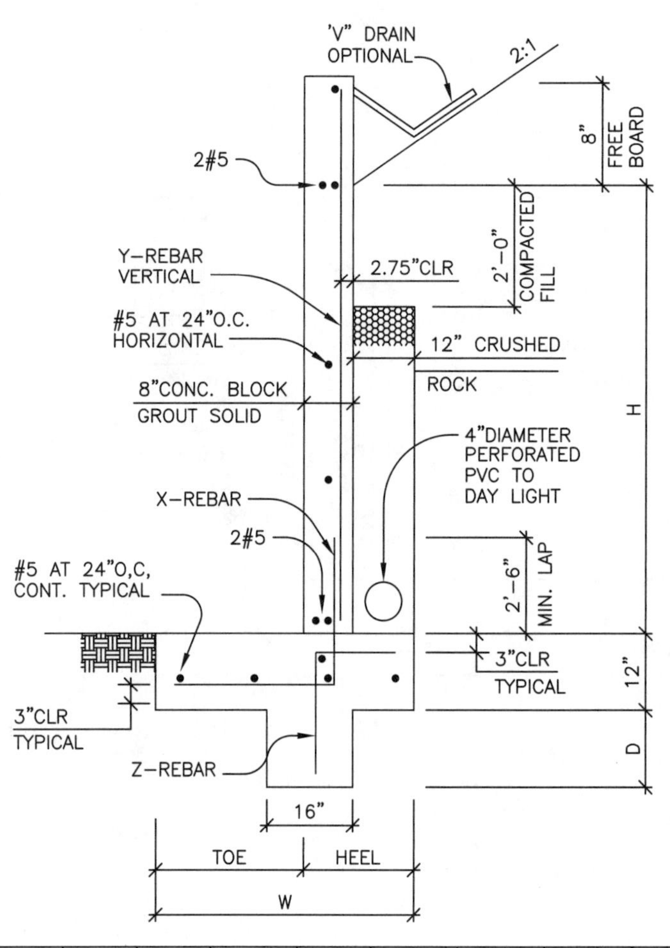

H	D	TOE	HEEL	W	X	Y	Z	COMMENT
4'-0"	2'-0"	2'-6"	1'-0"	3'-6"	#5@16"	#5@16"	#5@24"	ELIMINATE Y-REBAR EXTEND X-REBAR TO TOP OF THE WALL

SLOPED BACKFILL,
WITH KEY, FOOTING OUT OF HILL

EXECUTIVE ENGINEERING
19510 Ventura Blvd., Suite 110
Tarzana, California 91356-4234
Phone: (818) 708-1962 / Facsimile: (818) 708-1958
E-mail: EE110@aol.com

Index

I

K

L

M

N

O

P

R

S

T

U

V

W

Y

BUILDER'S GUIDE TO

Drainage & Retaining Walls

Using the PDF Version

BASIC USAGE

To use the PDF files on the enclosed CD-ROM, you need the Adobe Reader (available for free download from http://www.adobe.com/products/acrobat/readstep2.html). As the size of the PDF file for the book is quite large, you may achieve better performance (and find it more convenient) to copy the PDF file to your hard drive and access it from there. (Under the Limited License below, you are allowed to copy this PDF file to a maximum of two (2) personal computer hard drives which you use for your personal or business research.)

Once you open the PDF file, you can navigate the book by using the links in the left-hand list of "bookmarks." There are also links in the book's Table of Contents and Index. You can also use Adobe Reader's Search capabilities, to find pages containing words or phrases that interest you.

IMPORTANT NOTES ABOUT PRINTING

- As the CD-ROM is provided as a companion to the printed book, **printing from the PDF Version has been disabled for the main book** – *Builder's Guide to Drainage & Retaining Walls.*
- A second PDF file ("Retaining Wall Details") has been provided, reproducing the contents of the Appendix. Printing has not been disabled in this PDF file, so that you may – subject to the Limited License and Terms and Conditions of Use below – print pages from this PDF file.

Builder's Book, Inc.

BOOKSTORE • PUBLISHER

8001 Canoga Avenue / Canoga Park, CA 91304

1-800-273-7375 / www.buildersbook.com